HEALTH AND INDUSTRY

HEALTH AND INDUSTRY
A Behavioral Medicine
Perspective

edited by

MICHAEL F. CATALDO
The Johns Hopkins University School of Medicine
Baltimore, Maryland

THOMAS J. COATES
University of California School of Medicine
San Francisco, California

A Wiley-Interscience Publication

JOHN WILEY & SONS

New York / Chichester / Brisbane / Toronto / Singapore

Library of Congress Cataloging-in-Publication Data:

Main entry under title:

Health in industry.

  (Wiley series on health psychology/behavioral
medicine)
  "A Wiley-Interscience publication."
  Includes indexes.
  1. Occupational health services. 2. Health
promotion. 3. Medicine and psychology. I. Cataldo,
Michael F. II. Coates, Thomas J. III. Series.
  [DNLM: 1. Behavioral Medicine--methods. 2. Health
Promotion--methods. 3. Occupational Health Services.
WA 412 H4339]

RC968.H43 1986   362.1'0425   85-17948
ISBN 0-471-80921-7

Printed in the United States of America

10 9 8 7 6 5 4 3 2 1

To my father
—M.F.C.

To my parents
—T.J.C.

Contributors

DAVID B. ABRAMS, Ph.D.
Assistant Professor of Psychiatry and Human Behavior
The Center for Health Promotion and Maintenance
The Miriam Hospital and Brown University Program in Medicine
Providence, Rhode Island

W. STEWART AGRAS, M.D.
Professor of Psychiatry
Stanford University Medical School
Stanford, California

LYNN M. ARTZ, M.D.
The Memorial Hospital and Pawtucket Heart Health Program
Pawtucket, Rhode Island

WILLIAM A. AYER, D.D.S., Ph.D.
Professor, Behavioral Science & Community Dentistry
Northwestern University Dental School
Chicago, Illinois

JUDY BEREK
Vice President and Director
Legislative and Professional Programs, District 1199
National Union of Hospital and Health Care Employees
New York, New York

JOSEPH V. BRADY, Ph.D.
Professor, Behavioral Biology
The Johns Hopkins University School of Medicine
Baltimore, Maryland

KELLY D. BROWNELL, Ph.D.
Associate Professor, Department of Psychiatry
University of Pennsylvania School of Medicine
Philadelphia, Pennsylvania

PATRICIA M. BROWNSTEIN, R.N., M.S.N.
Coordinator, Employee Health Service
Health Insurance Plan of Greater New York
New York, New York

RICHARD A. CARLETON, M.D.
Professor of Medicine
Brown University—The Memorial Hospital
Providence, Rhode Island

MICHAEL F. CATALDO, Ph.D.
Associate Professor
The Johns Hopkins University School of Medicine
Director of Psychology
The John F. Kennedy Institute
Baltimore, Maryland

MARGARET CHESNEY, Ph.D.
Director of Behavioral Medicine
Stanford Research Institute International
Stanford University
Menlo Park, California

THOMAS J. COATES, Ph.D.
Director, Behavioral Medicine Unit
University of California School of Medicine
San Francisco, California

DANA DAVIS, M.A.
Executive Director
American School Health Association
Kent, Ohio

DEBORAH DEATRICK, M.P.H.
Director, Office of Dental Health
Maine Department of Health and Human Services
Augusta, Maine

NANCY MARWICK DeMUTH, Ph.D., M.B.A.
Research Associate
The Johns Hopkins University School of Hygiene and Public Health
Baltimore, Maryland

JOHN P. ELDER, Ph.D.
Assistant Professor of Community Health
Brown University—The Memorial Hospital
Providence, Rhode Island

JONATHAN E. FIELDING, M.D., M.P.H.
Professor, Schools of Public Health & Medicine
University of California at Los Angeles
Los Angeles, California

MARIAN W. FISCHMAN, Ph.D.
Associate Professor
The Johns Hopkins University School of Medicine
Baltimore, Maryland

RUSSELL E. GLASGOW, Ph.D.
Research Scientist
Oregon Research Institute
Eugene, Oregon

WILLIS B. GOLDBECK
President
Washington Business Group on Health
Washington, DC

LAWRENCE W. GREEN, D.P.H.
Center for Health Promotion Research and Development
University of Texas Health Science Center
Houston, Texas

TYLER D. HARTWELL, Ph.D.
Director
Center for Medical Environment and Energy Statistics
Research Triangle Institute
Research Triangle Park, North Carolina

J. ALAN HERD, M.D.
Medical Director
Sid W. Richardson Institute of Preventive Medicine
The Methodist Hospital
Professor Baylor College of Medicine
Houston, Texas

ROBERTA B. HOLLANDER, Ph.D.
Assistant Professor
University of Maryland Department of Health Education
College Park, Maryland

BILL HOPKINS, Ph.D.
Professor
Department of Human Development
University of Kansas
Lawrence, Kansas

MARJANE JENSEN, M.S.
Ph.D. Candidate in Industrial Organizational Psychology
Purdue University
West Lafayette, Indiana

STANISLAV V. KASL, Ph.D.
Professor of Epidemiology
Yale University School of Medicine
New Haven, Connecticut

ROBERT C. KLESGES, Ph.D.
Center for Applied Psychological Research
Department of Psychology, Memphis State University
Memphis, Tennessee

JUDITH L. KOMAKI, Ph.D.
Associate Professor of Psychological Sciences
Purdue University
West Lafayette, Indiana

THOMAS M. LASATER, Ph.D.
Assistant Professor of Community Health
Brown University—The Memorial Hospital
Providence, Rhode Island

DAVID LeGRANDE
Director, Occupational Safety & Health
Communications Workers of America (CWA)
Washington, DC

SANDRA M. LEVY, Ph.D.
Associate Professor of Psychiatry and Medicine
Program Director, Behavioral Medicine in Oncology
Western Psychiatric Institute and Clinic
University of Pittsburgh School of Medicine
Pittsburgh, Pennsylvania

VERNON MacDOUGAL, M.D.
Executive Director
Workers Institute for Safety and Health
Washington, DC

JUNE McMAHON
Director of Research
Service Employees International Union (SEIU)
Washington, DC

MURRAY P. NADITCH, Ph.D.
Director, Healthcare Industry
Control Data Corporation
Minneapolis, Minnesota

PETER NATHAN, Ph.D.
Chairman, Department of Clinical Psychology
Graduate School of Applied and Professional Psychology
Rutgers University
Piscataway, New Jersey

MICHAEL S. NEALE
Department of Psychology
Yale University
New Haven, Connecticut

REBECCA S. PARKINSON, M.S.P.H.
Staff Manager, Employee Health Education
AT&T
New York

KENNETH R. PELLETIER, Ph.D.
Department of General Internal Medicine
University of California School of Medicine
San Francisco, California

PHILIP V. PISERCHIA, M.S.
Senior Research Statistician
Research Triangle Institute
Research Triangle Park, North Carolina

KNUT RINGEN, D.P.H.
Program Director, Field Programs
National Cancer Institute
Bethesda, Maryland

GARY E. SCHWARTZ, Ph.D.
Professor of Psychology and Psychiatry
Yale University
New Haven, Connecticut

JEANNE SCHWARTZ
Clinical Coordinator
Yale Behavioral Medicine Clinic
Yale University
New Haven, Connecticut

SUSAN SEFFRIN, M.S.
Assistant Director
Bureau of Health Education and Audiovisual Services
American Dental Association
Chicago, Illinois

GUNNAR SEVELIUS, M.D.
Medical Director
Lockheed Missiles and Space Company, Inc.
Sunnyvale, California

JEFFERSON A. SINGER
Department of Psychology
Yale University
New Haven, Connecticut

BARRY R. SNOW, Ph.D.
Clinical Coordinator
Research & Training Services
Department of Behavioral Medicine
Hospital for Joint Diseases/Orthopaedic Institute
New York, New York

JEANNE M. STELLMAN, Ph.D.
Associate Professor
Columbia University School of Public Health
Executive Director
Woman's Occupational Health Resource Center
New York, New York

ALBERT J. STUNKARD, M.D.
Professor of Psychiatry
University of Pennsylvania School of Medicine
Philadelphia, Pennsylvania

C. BARR TAYLOR, M.D.
Associate Professor (Clinical) of Psychiatry
Stanford University Medical School
Stanford, California

CURTIS S. WILBUR, Ph.D.
Director, LIVE FOR LIFE™
Johnson & Johnson
New Brunswick, New Jersey

ROD WILFORD
Health & Safety Director
International Brotherhood of Painters and Allied Trades (IBPAT)
Washington, DC

GERALDINE WIRTHMAN, M.P.H.
Assistant Director, Office of Dental Health
Maine Department of Health & Human Services
Augusta, Maine

Series Preface

This series is addressed to clinicians and scientists who are interested in human behavior relevant to the promotion and maintenance of health and the prevention and treatment of illness. *Health psychology* and *behavioral medicine* are terms that refer to both the scientific investigation and inter-disciplinary integration of behavioral and biomedical knowledge and technology to prevention, diagnosis, treatment and rehabilitation.

The major and purposely somewhat general areas of both health psychology and behavioral medicine which will receive greatest emphasis in this series are: theoretical issues of bio-psycho-social function, diagnosis, treatment, and maintenance; issues of organizational impact on human performance and an individual's impact on organizational functioning; development and implementation of technology for understanding, enhancing, or remediating human behavior and its impact on health and function; and clinical considerations with children and adults, alone, in groups, or in families that contribute to the scientific and practical/clinical knowledge of those charged with the care of patients.

The series encompasses considerations as intellectually broad as psychology and as numerous as the multitude of areas of evaluation treatment and prevention and maintenance that make up the field of medicine. It is the aim of the series to provide a vehicle which will focus attention on both the breadth and the interrelated nature of the sciences and practices making up health psychology and behavioral medicine.

THOMAS J. BOLL

The University of Alabama in Birmingham
Birmingham, Alabama

Preface

In his article on corporate health benefits for *Fortune*, Louis S. Richman said, "True control over medical costs will require more than mere tightening up," and he noted that some companies have already taken the initiative to avoid a cumulative health care crisis. The envisioned crisis about which Richman speaks—the baby boom, a demographic bulge now entering middle-age and experiencing age-related chronic health ailments—and its resolution are the focus of *Health Promotion in Industry*. The "penicillin" in the year 2000, suggest the authors, will probably be preventive medicine; that is, behavioral or lifestyle programs implanted in the workplace where the health care problems of a large and captive adult population can be addressed effectively. To spread the word about behavioral medicine and its application in the workplace is the reason this book was written.

"Wellness" programs already initiated by future-oriented companies, such as Control Data's STAYWELL and Johnson & Johnson's LIVE FOR LIFE™, are discussed at length in this book. In addition, a cross section of health care experts—psychologists, physicians, biomedical scientists, and business and labor leaders—discusses scientific principles, research data, practical applications, policy issues, as well as the actual programs in progress. As this book emphasizes, the adaptation of preventive or behavioral medicine programs at the worksite is a relevant and timely concept.

The book takes a multidisciplinary approach, having evolved from scientific research and professional symposia, and should be of value to a wide-ranging audience: corporate managers interested in establishing programs, college professors interested in training future leaders in health care, their students, and, of course, health care practitioners and biomedical scientists, interested in up-to-date findings pertaining to critical analyses of research, program efficacy, and policy issues.

The authors of this collection of articles are to be congratulated for their formative work in the field and for the clear and concise manner in which it is described in *Health Promotion in Industry*.

MICHAEL F. CATALDO
THOMAS J. COATES

Baltimore, Maryland
San Francisco, California
January 1986

Acknowledgements

Assistance and support from several people must be acknowledged, beginning with the Society of Behavioral Medicine, which designated one annual scientific meeting to highlight "Health and Industry." We are indebted to the Society's Board for encouraging us to develop a book based on the meeting's activities; to those members who helped plan the meeting: Barry Snow, Francis Keefe, Craig Ewart, Cheryl Perry, and John Parrish; and especially to the Society's administrative director, Jude Woodward. We are particularly appreciative to our Wiley editor, Herb Reich, for his patience and wisdom. We thank Margaret Grulich and Deborah Baniadam for their assistance at the University of California office. Finally, we owe a deep debt of gratitude to Joyce Meals at the Hopkins office for her dedication to the seemingly never-ending task of editing and organization, without which this book would not have been possible.

M.F.C.
T.J.C.

Contents

FOUR   POLICY ISSUES

HEALTH AND INDUSTRY

Introduction

MICHAEL F. CATALDO and THOMAS J. COATES

Beginning with the earliest written records in which the ancient Babylonians and Greeks describe how people attempt to understand and influence biological functioning, great interest has been shown in the effect of the mind on the body. This concern about the mind/body relationship has included not only attitudes and thoughts, but also the environmental events and contingencies that affect behavior.

Throughout the centuries this interest in mind/body relationships has taken many forms and has been subjected, more or less, to scientific inquiry. In the last 50 years, different disciplines have explored mind/body relationships. Psychosomatic medicine, for example, has been primarily the domain of psychiatry; health psychology is obviously an area of psychology and has been formalized as the 38th Division of the American Psychological Association; behavioral medicine, more interdisciplinary in nature, has been the subject of various organizational efforts and attempts to define its activity. An early definition of behavioral medicine was offered by the participants of a three-day conference held at Yale University in 1977:

> Behavioral medicine is the field concerned with the *development* of *behavioral science* knowledge and techniques relevant to the understanding of *physical health* and *illness* and the *application* of this knowledge and these techniques to prevention, diagnosis, treatment and rehabilitation. Psychosis, neurosis, and substance abuse are included only insofar as they contribute to physical disorders as an end point. (Schwartz & Weiss, 1978)

1

Shortly thereafter, a second meeting was held at the National Academy of Sciences in Washington DC, and an amended definition resulted:

> Behavioral medicine is the *interdisciplinary* field concerned with the development and *integration* of behavioral and biomedical science knowledge and techniques relevant to health and illness and the application of this knowledge and these techniques to prevention, diagnosis, treatment and rehabilitation.

The meeting also resulted in the formation of the first professional organization devoted to the area of behavioral medicine: the Academy of Behavioral Medicine Research. Partly because the Academy was a closed, by-invitation-only group, another organization was spawned under the administrative and financial umbrella of the Association for the Advancement of Behavior Therapy (AABT). Soon to break off from the AABT, this new open-membership organization, the Society of Behavioral Medicine, adopted the following as its purposes:

> The Society of Behavioral Medicine, Inc. (the "Society") is intended to serve the needs of all health professionals interested in the integration of the behavioral and biomedical sciences in clinical areas. The functions of the Society shall include, but shall not be limited to, the following:
>
> (1)   to encourage and coordinate communication and fraternization among various health professionals without regard to specific discipline loyalties;
>
> (2)   to maintain liaison with related professional organizations;
>
> (3)   to develop guidelines for implementation of behavioral medicine activities in various health settings;
>
> (4)   to stimulate clinical and research activities through formal meetings, collaborative undertakings, and awards for meritorious effort; and
>
> (5)   to serve as a resource for any public or private agencies seeking information about behavioral medicine.

During this time, several books on behavioral medicine began to appear, some of which also offered definitions of behavioral medicine. Somewhat unique was the definition offered by Ovide Pomerleau and John Paul Brady (1979) in their edited text:

> *Behavioral medicine* can be defined as (a) the clinical use of techniques derived from the experimental analysis of behavior—behavior therapy and behavior modification—for the evaluation, prevention, management, or treatment of physical disease or physiological dysfunction; and (b) the conduct of research contributing to the functional analysis and understanding of behavior associated with medical disorders and problems in health care. (p. XII)

Definitions are important; they serve an organizational function. However, a field or area of endeavor is more likely to be defined by the activity of its participants. In terms of both health-related interventions and scientific research, contemporary activities of a health-related nature are conceptually depicted as a 2 X 2 table presented in Figure I.1. The table is constructed by considering variables commonly used in any clinical or scientific endeavor: in essence, variables that are either manipulated or changed (independent variables) and variables that are altered as a function of such manipulation (dependent variables). The table is further constructed by the conceptually distinct features that characterize behavioral medicine; that is, biological aspects on one hand, and behavioral aspects (the environmental contingencies, attitudes, lifestyle variables, etc.) on the other. Thus the upper left-hand box depicts those activities in which biological independent variables are manipulated and the primary subject matter of interest is the resulting biological dependent variables; thus this box describes most of the work that has been conducted in biomedical science and the practice of medicine. The lower right-hand part of the table describes activities in which the independent and dependent variables were both behavioral in nature; thus this box characterizes in general most of psychology and, in particular, those subareas of the experimental analysis of behavior and applied behavior analysis. Behavior analysis approaches have proven to be relevant to medical problems, such as adherence to medical regimens and modification of lifestyle variables.

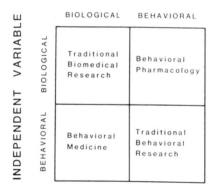

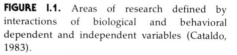

FIGURE I.1. Areas of research defined by interactions of biological and behavioral dependent and independent variables (Cataldo, 1983).

The upper right-hand part of the table describes work in which biological factors are manipulated and resulting changes in behavior are noted. One area described by this box that has proven to be beneficial to medicine and health care is that of behavioral pharmacology. The lower left-hand part of the table is the area of activity that, in our opinion, has the greatest potential for advancing knowledge about health and making significant changes in medicine. The box describes activities that seek to make changes in biological function or dysfunction by manipulating behavioral factors.

The set of activities in Figure I.1 involves health aspects related to industry.

Industry has been a focus of attention because, to study the health problems of American adults and how such problems could be related to environment and behavior, it has also been important to consider where people spend most of their time. The workplace is where both men and women spend most of their waking hours. At a more important level, industry has been the focus of behavioral medicine study about the characteristics of the environment that adversely affect biological functioning and therefore the worker's health. Such focus on industry has been primarily from science. Both science and industry have considered how the workplace can be restructured to impact positively on the health problems of workers, regardless of whether those problems are related to characteristics of the work environment.

To complete a behavioral medicine analysis of the workplace, it has been necessary not only to consider how to analyze the effects of the workplace environment on health and behavior, but also to identify how programs of research and intervention can be integrated successfully in the industrial environment. Thus we had to consider not only how the workplace affects the worker's health and behavior, but also how to analyze the functioning of industry with regard to effecting changes in the workplace.

To these considerations must be added the financial problems of providing high-technology health care to the American people. With the current financial climate concerned about weathering ever-increasing federal deficits, the exponentially increasing costs of health care will be difficult to maintain. This significant financial problem can provide an advantage and opportunity for behavioral medicine approaches to health in industry because it offers the promise of producing healthier workers and thereby reducing costs associated with ill health and disability. The costs related to promoting a healthy work force provide an important incentive for industry in terms of insurance premiums, workers' benefits, productivity, disability, compensation, and so forth.

We, therefore, find ourselves at a unique point in history—a long history concerned with mind/body relationships—where the science of human behavior has matured sufficiently to be integrated with biomedical science so that we can begin to understand and favorably alter health. Fortunately, industry is becoming both an enthusiastic partner in this knowledge-generating activity and a part of the solution.

This book has developed from these recent activities. The 1983 Annual Meeting of the Society of Behavioral Medicine focused on the topic of "Behavioral Medicine in Industry." In fact, some of the chapters in this book resulted from a symposia at that meeting. The meeting included both behavioral and biomedical scientists, clinicians, directors of health programs in industry, and representatives of management and labor. The timeliness of this issue was also attested to by the presence of that bastion representing the general population—the media. The conference was covered by UPI, AP, *Time Magazine, Fortune Magazine,* and many newspapers.

Despite the interest in and increasing number of behavioral medicine programs in industry, no comprehensive work describing this effort has occurred—until now. This book is comprised of four sections. We designed this book both to document current practices in the workplace and to present and discuss enduring principles and dilemmas that guide and develop from health promotion programs in the workplace. The first section describes the basic premises, scientific principles and methodologies, and the underlying uniqueness of a behavioral medicine approach to health promotion in industry. In Chapter 1, Joseph Brady describes the behavioral science principles and procedures underlying intervention strategies. Continued reflection on these principles in relation to the workplace environment should provide additional insight into novel and more effective methods. Chapter 2 extends these principles and demonstrates their application in a framework for conceptualizing and planning interventions for specific settings. In Chapter 3, Stanislav Kasl presents a scholarly critique from an occupational epidemiological perspective on the evidence related to stress and disease in the workplace. Finally, in Chapter 4, Judith Komaki presents an important alternative to traditional experimental designs. Adherence to the principle of randomization requires sufficient units to randomize, possibly encouraging only large and expensive studies. The time-series design is an important but often ignored alternative that generates data sufficiently sensitive to elucidate process and valid to draw conclusions about program efficacy.

The second and third sections present up-to-date summaries of research in specific problem areas. The second section describes the health problem areas of greatest concern and impact on the American people and how their health can be affected by the workplace. The major leading causes of death and disease among adults in the United States today are cardiovascular disease, cerebrovascular disease, and cancer. The chapters in Section II discuss interventions for the major risk factors for these diseases—obesity, stress, smoking, lack of physical activity, hypertension, and exposure to carcinogenic toxins. In addition, we have included chapters on psychosocial well-being, since recent epidemiological surveys are relating such factors as stress, depression, and social support to morbidity and mortality.

Job safety is also an important consideration because of its enormous impact on disability and death. Finally, dental health promotion is an area usually relegated to the dentist's office, but which might profitably be addressed in the workplace as well.

The third section describes and critically analyzes exemplary behavioral medicine programs in industry. Kenneth Pelletier presents the results of his survey of health promotion programs in 150 companies. In addition, two exemplary programs, STAYWELL from Control Data Corporation and Live for Life™ from Johnson & Johnson, with controlled outcomes are presented. In addition, Jonathan Fielding presents six questions related to workplace health promotion and provides examples of programs meeting these criteria.

Finally, the last section is concerned with the policy issues and future problems that must be addressed. Michael Cataldo and colleagues address the interdisciplinary dilemmas in industrial health promotion programs. Perspectives are provided from health education, medicine, management, and labor on basic assumptions, priorities, and barriers. Labor's issues and concerns are addressed by Barry Snow in collaboration with four people directly involved in the labor movement. Nancy DeMuth and colleagues address the concepts, methods, and policies underlying workplace programs and highlight possibilities for innovation and progress.

Finally, it should be noted that this is a book of research results. We are well aware that the practice in this area is expanding at a geometric rate. It is essential that programs be efficacious and that the efficacy be documented. We hope that this volume will enhance efficacy by disseminating research, encouraging more research, and building a bridge between research and practice.

REFERENCES

Cataldo, M. F. (1983, August). *The health Armageddon of the twenty-first century: Priorities for the experimental analysis of biobehavioral interaction.* Invited address to the 91st Annual Convention of the American Psychological Association, Anaheim, CA.

Pomerleau, O. F., & Brady, J. P. (Eds.). (1979). *Behavioral medicine: Theory and practice.* Baltimore: Williams & Wilkins.

Schwartz, G. E., & Weiss, S. M. (1978). Yale Conference on Behavioral Medicine: A proposed definition and statement of goals. *Journal of Behavioral Medicine, 1*(1), 3–12.

ONE

The Scientific Foundation

1

Biobehavioral Principles, Behavioral Medicine, and the Workplace

JOSEPH V. BRADY and MARIAN W. FISCHMAN

The development of industrial organizations in the United States has been characterized by a continuing focus on issues of productivity, and more recently, on the related concerns of both labor and management with the improvement and maintenance of health and well-being in the work force. Defined broadly, the health concept embraces more than simply freedom from physical illness and pain. Indeed, substantial investments have been made in health management organizations to provide not only for medical intervention as needed, but also for behaviorally oriented preventive education. These initiatives can be expected to have an impact not only on health and productivity in the workplace, but also on all aspects of an individual's lifestyle. Wide gaps still exist, however, that separate the available data base from theoretical formulations providing the framework for such programmatic efforts. An understanding of the relationship between the development of work-related prevention and treatment procedures and the principles on which an integrated biobehavioral approach to general health problems must be based can help to bridge these gaps. This essay, therefore, provides an overview of these biobehavioral principles derived over the past several decades from scientific observation and experiments.

The perspective from which this treatment of basic principles proceeds is the now commonplace observation that health-related biological processes interact, often in profound and enduring ways, with environmental circumstances and behavioral activities. Analyzing these fundamental relationships

9

and their application to behavioral medicine in the workplace has benefited from the strong empirical influence of the experimental laboratory. The conceptual foundation of this biobehavioral perspective on health in the workplace is based on two fundamental premises: (1) knowledge comes from experience and (2) behavior is governed by its consequences. These two constructs about human nature define a philosophy of social optimism: that is, if you want people to be a certain way or to do certain things, circumstances can be arranged. These two ideas coalesced in nineteenth-century England and foreshadowed the emergence of modern behaviorism. Their influence on medicine in general and industrial health problems in particular developed slowly, but their impact is now beginning to find expression in the emergence of a behavioral perspective that encompasses virtually all aspects of health and disease (Pomerleau & Brady, 1979).

BACKGROUND AND BASIC OBSERVATIONS

The important influence of the experimental laboratory in establishing the data base, which now provides the foundations for these assertions, can be traced to the contributions of I. P. Pavlov, who focused on the role of environmental circumstances and behavioral activities in the biochemical and physiological adaptations and adjustments of the milieu interieur. But equally important was the foundation that Pavlov's work provided for conceptualizing behavioral interactions within the framework of an orderly and systematic body of scientific knowledge based on observation and experiment. The contrast between this objective approach to the analysis of behavior and the more traditional emphasis on unobserved and unobservable mental processes is worth emphasizing.

But what, in fact, have we learned in the laboratory about the nature of those behavioral activities at the interface between individuals and their environments, and to what extent are these findings relevant to medicine and health problems in the workplace? Fundamentally, *two basic modes*, reactive and active, characterize this biobehavioral process. In the first instance, a *reactive mode* is clearly rooted in the biochemical and physiological adaptations of the organism to the influences of a changing environment (i.e., the environment acts upon the organism and the organism reacts). Since the time of Pavlov, this respondent paradigm has provided the basis for describing and experimentally analyzing increasingly more complex interactions of direct relevance to clinical medicine in general, and to industrial health in particular. Early respondent conditioning studies (Deane & Zeaman, 1958; Dykman & Gantt, 1958), provided systematic accounts of how neutral environmental stimuli (e.g., tones and lights), which initially produced only minimal changes in somatic activity, could elicit conditional physiological responses (e.g., heart rate increases) of substantial magnitude and duration when paired repeatedly with unconditional environmental stimulus events (e.g., food or electric shock) that normally elicited such changes. If such conditional tone or light stimuli (CS) are subsequently presented a number of times without the unconditional food or shock stimuli (UCS), the magnitude and frequency of the conditional heart rate increase response (CR) elicited by the CS diminish, and respondent extinction occurs. When a period of time intervenes between such extinction and subsequent presentations of the CS, however, spontaneous recovery of the CR is observed in the form of temporary reappearance of the response elicited by the CS.

The power to elicit a CR, which is developed in one CS by conditioning, extends to other stimuli, with the degree of this *stimulus generalization* determined by the similarities and differences between the other stimuli and the CS. Because stimuli other than the CS differ with respect to the magnitude and frequency with which they elicit the CR, *stimulus discrimination* also occurs. Indeed, discrimination can be made increasingly more pronounced by repeated pairings of the UCS only with a specific CS (i.e., respondent conditioning), while ensuring that the occurrence of other stimuli is not paired with the UCS.

These basic observations with regard to the reactive or respondent conditioning mode have been elaborated in numerous laboratory and clinical-experimental studies since Russian researchers first introduced this systematic approach to behavior analysis. It has been convincingly demonstrated, for example, that second-or higher-order conditioning can occur when a well-established CS is paired with a neutral stimulus. The neutral stimulus acquires the power to elicit the reactive CR. Although it has not been empirically determined just how far this process can be carried, the development of eliciting properties by the CS two or three steps removed from the original UCS is not uncommon. And the intensive investigative effort, primarily Russian in origin, to extend the conceptual framework of such classical or Pavlovian (i.e., respondent) conditioning to encompass verbal stimuli and somatic responses (Razran, 1961) suggests potentially important directions for development of theory and practice in medicine and its application in industrial settings.

Elicited responses of the type that have provided the primary focus for basic respondent conditioning analyses constitute only a relatively small proportion of the behavioral repertoire of higher organisms. The most prominent aspects of such advanced performances are represented by the *second* basic, and generally more *active* than reactive, *mode*. These behavioral processes focus on the operations performed by organisms on their internal and external environments rather than on their "reflex" reactions to such environmental influences. Technically, this active mode has been explicated within the framework of a three-term contingency analysis, which emphasizes the temporal relationships between *organismic performances* (R), *reinforcing consequences* (S^R), and the *environmental context* (S^D) in which the $R \longrightarrow S^R$ relationship occurs. The major contributions to the experimental analysis of such operant (i.e., the organism operating on the environment) behavior interactions have been identified with the work of B. F. Skinner, as well as his students and colleagues. The dominant relationship between these component terms emphasizes the governance of action (the likelihood of a response) by the contingently occurring effects of that action (its reinforcing consequences). Emergent relations between behavior and the environmental context in which it occurs are specified to the extent that *response-consequence* contingency relations depend on environmental context events. More complex interrelationships between these terms have, of course, been elaborated, and along with historical variables, have enhanced a precise definitional account of such behavioral contingencies. *Within the framework of these empirical referents, the likelihood, strength, and persistence of behavior can be accounted for more readily than by any other means.*

FUNDAMENTAL PRINCIPLES

During the past three decades, a broad range of animal laboratory and human experimental studies has provided important insights into the principles that determine the acquisition, maintenance, and modification of operant behavior (Honig, 1966; Honig & Staddon, 1976). The basic observation is that the rate of an operant response already in the organism's repertoire can be readily increased by reinforcement (operant conditioning). Beyond this, it has been possible to make explicit the process called *shaping*, whereby operant conditioning can extend existing simple responses into new and more complex performances. A reinforcer not only strengthens the particular response that precedes it, but also produces an increase in the frequency of many other similar bits of behavior, and, in effect, raises the individual's general activity level. This is critical for the shaping process.

The shaping of behavior proceeds as reinforcers are presented after a response similar to or approximating the desired one. Since this technique tends to increase the strength of various other similar behaviors, a response still closer to the desired one can be selected from this new array and can be reinforced. Continued narrowing and refinement of the response criteria required for reinforcement leads progressively to new arrays of available behavior. In this way, by successive and progressive approximation, a new and desired performance can be shaped. The importance of this simple, but fundamental and powerful, shaping process for the development and modification of behavior can not be overstated. The weight of available evidence suggests that a careful and systematic application of such procedures with effective reinforcers is sufficient to establish or alter any operant performance of which the organism is physically capable. This shaping process has enormous clinical importance in behavioral medicine. Its applications to industrial health are substantial since many performances, both work and nonwork, can effectively be changed only in this way. Without shaping, one might wait for inordinately long periods before performance of some critical work–health-related behavior could be reinforced.

The fact that changes in behavior are not always due to deliberate and systematic manipulation of the environment, however, has led to an analysis of *superstitious behavior*. A potentially reinforcing environmental event may, by chance, follow a response, resulting in the adventitious strengthening of that response. If this sequence of events recurs even infrequently, the individual may learn quite elaborate sequences of superstitious behavior that have nothing to do with production of the event influencing the frequency of the behavior. The elaborate rituals of the gambler do not produce winning dice combinations any more than native dances produce rain; they persist because they are occasionally followed by "7 or 11," in the first instance, and precipitation in the second.

The powerful effects of reinforcement in establishing and maintaining operant behavior suggest that witholding such reinforcing consequences (*extinction*) will have comparably powerful effects on the strength of previously reinforced responses. Indeed, such extinction procedures do reduce the frequency of response, although the reduction is not usually immediate. Rather, after the onset of extinction, the initial effect is often a brief increase in the frequency as well as in the force and variability of the response previously followed by reinforcement. The extent to which operant responding persists in the absence of reinforcing environmental consequences (resistance to extinction) depends, of course, on the interaction of many complex influences, including motivational factors (level of deprivation). But both laboratory and clinical experimental evidence now confirm that the single most important variable affecting the course of operant extinction is the schedule of reinforcement on which the performance was previously acquired and maintained.

SOME IMPORTANT FUNCTIONAL RELATIONS

Whenever a reinforcing environmental stimulus follows some but not all occurrences of an operant response, a *schedule of intermittent reinforcement* is operating. Intermittent reinforcement, therefore, is defined when only selected occurrences of an operant are followed by a reinforcer. Every reinforcer occurs according to some schedule or rule, although some schedules are so complicated that detailed analysis is required to formulate them precisely. Simple schedules of intermittent reinforcement can be classified into two broad categories: those based on number (ratio) and those based on time (interval). Ratio schedules prescribe that a certain number of responses be emitted before one response is reinforced; the term *ratio* refers to the relationship between the required response total (e.g., 50) and the one response followed by the reinforcing event (e.g., piece-work schedule requiring 49 discrete labor units before the single 50th performance is followed by payoff). Interval schedules on the other hand prescribe that a given interval of time elapse before an emitted response can be followed by a reinforcing stimulus. The relevant interval can be measured from any event, but the occurrence of a previous reinforcer is usually used (e.g., salaried pay schedules). Interval schedules, under which the mere passage of even long time intervals provides an opportunity for a single response to be followed by a reinforcer, have recuperative properties. This contrasts with the strain potential of high ratio schedule requirements under which the performance may extinguish before a sufficient number of responses are emitted and one is followed by reinforcement (Rachlin, 1970).

Even simple ratio and interval schedules can, in turn, be classified into two general categories based on whether the required number of responses or lapse of time are fixed or variable, and all known schedules of reinforcement can be reduced to variations of these basic ratio and interval parameters. Each schedule variant, simple or complex, generates and maintains its own characteristic performance, and when reinforcement is discontinued, the course and character of extinction are prominently influenced by the preceding schedule of reinforcement. Furthermore, as evident in the laboratory, the clinic, and the workplace, the *frequency or rate of a given operant performance can be more effectively controlled by reinforcement schedule manipulation than by any other means.*

The detailed experimental analysis of reinforcement schedules has also served to emphasize another important set of relationships between operant performances and environmental events encompassed within the general conceptual framework of *stimulus control.* The occurrence of a reinforcer following an operant not only increases the likelihood that the response will recur, but it also contributes to bringing that performance under the control of other environmental stimuli present when the operant is reinforced. After the responses comprising the operant have been reinforced in the presence of a particular stimulus, that stimulus begins to control the operant. The frequency of those responses is higher in the presence of that stimulus complex and lower in its absence. A discriminative stimulus is thus defined as one in whose presence a particular operant performance is highly probable because the behavior has previously been reinforced in its presence. Nevertheless, discriminative stimuli do not elicit performances as in the respondent or reflex case, but rather set the occasion for operant responses. They provide the circumstances under which the performance has previously been reinforced. The control over driving behaviors by traffic signals occasioning vehicle braking and accelerating occurs because of systematic relationships between such performances and their consequences (e.g., fines and accidents) not because of any eliciting properties of red, green, and yellow lights. *This controlling power of a discriminative stimulus* develops gradually, and at least several occurrences of the reinforcer following the response in the presence of the stimulus are required before the stimulus effectively controls the performance.

Such discriminative stimulus control is not an entirely selective process, however, since reinforcement of a performance in the presence of one stimulus increases the tendency to respond not only to that stimulus but also in the presence of other stimuli with similar properties. This is *stimulus generalization*. It is not always clear from simple observation, of course, which stimulus or which property of a stimulus is controlling an operant performance and both laboratory and clinical experiences have documented the hazard of assuming that the similarity casually observed between stimuli provides an adequate explanation of such generalization. There is unfortunately no substitute for experiment in differentiating the many detailed aspects of a stimulus complex that may exercise critical control. Furthermore, related *response generalization* effects have also been observed to occur, when following an operant with a reinforcer, which results not only in an increase in the frequency of the responses composing that operant but also in an increase in the frequency of similar responses.

This sensitivity to the differential aspects of stimulus and response complexes provides the basis for the other major aspect of the stimulus control process identified as *discrimination*. A discrimination between two stimuli has occurred when an organism behaves differently in the presence of each. Such stimulus discrimination is pronounced under conditions that provide differential reinforcement. A discrimination is formed when there is a high probability that a reinforcer will follow a particular response in the presence of one stimulus, and a low or zero probability that reinforcement will follow the response in the presence of another stimulus. The extent of the generalization between two stimuli will influence the rapidity and stability with which a discrimination can be formed. The careful application of differential reinforcement procedures can produce remarkably precise control of an operant performance by highly selective aspects of a stimulus complex. This attention to specific properties of a stimulus can be facilitated and enhanced by the use of *instructional stimuli*, which reveal features of the environment that are currently relevant to the occasioning of reinforcement. (An example is a treasure map.) *Imitation* and *modeling*, considered analytically, appear to represent special instances of such instructional control. Furthermore, this precise stimulus control attention can be transferred from one group of stimuli or stimulus properties to another by simultaneous presentation of the two followed by the gradual withdrawal or *fading* of the original stimulus.

The intimate and continuing association between discriminative environmental stimulus events and the occurrence of reinforcement endows at least some originally nonreinforcing stimuli with acquired reinforcing properties. These stimuli, money and stock market quotations, for example, are called *secondary or conditioned reinforcers* to distinguish them from innate, primary, or unconditioned reinforcers, which require no experience to be effective. Such conditioned reinforcers can be either appetitive, strengthening responses by their appearance, or aversive, in which case their removal or postponement is reinforcing. The development or acquisition of conditioned reinforcing properties by a stimulus is usually a gradual process, as is the case with discriminative stimuli in general.

Response chaining refers to the observationally and experimentally verified occurrence of a series of performances joined together by environmental stimuli that act both as conditioned reinforcers and as discriminative stimuli. A chain (e.g., party-going) usually begins with the occurrence of a discriminative stimulus (e.g., phone invitation) in the presence of which an appropriate response (e.g., acceptance) is followed by a conditioned reinforcer (e.g., "Glad you can make it."). This conditioned reinforcer is also the discriminative stimulus occasion for succeeding responses (e.g., bathing and dressing), which in turn is followed by another conditioned reinforcer (e.g., leaving the house, catching a cab) that is also a discriminative stimulus for the next response (e.g., joining the party), and so on. Although the entirety of such chains is probably maintained by the terminal occurrence of potent environmental consequences (e.g., social interactions, food, sex), laboratory experiments have clearly demonstrated that the overlapping links in the chain (i.e., discriminative stimulus ⟶ operant response ⟶ conditioned reinforcer) are held together primarily by the dual (and demonstrably separable) discriminative and conditioned reinforcing functions of environmental stimuli. The significance of this general chaining principle is that virtually all behavioral interactions occur as chains of greater or lesser length, and that even performances usually treated as unitary phenomena can be analyzed usefully at various component levels (e.g., machine operators, product assembly, quality control checks) for purposes of modification or proficiency enhancement.

Perhaps the most important aspect of this complex analysis of environment-behavior interactions is the clear implication that the factors limiting conditioned-reinforcer potency can be overcome by the formation of conditioned reinforcers based on two or more primary reinforcers. Such conditioned stimulus events (*generalized reinforcers*) gain potency from all the reinforcers on which they are based. The most prominent operant performances in the human repertoire, as well as the most valued stimulus consequence in the social environment (money), can be seen to share these broadly based discriminative and generalized conditioned-reinforcing properties.

This abbreviated overview of experimentally derived behavioral concepts and principles relevant to industrial health has thus far maintained the traditionally accepted differentiation between active (operant) and reactive (respondent) behavioral interaction modes based primarily on procedural distinctions identified in the laboratory. The independent and distinctive features of these two coextensive processes are seldom apparent, however, in the course of even detailed natural observation. The complex interactions between these active and reactive modes are most obvious in the experimental analysis of *aversive control* procedures represented by the technical terms punishment, escape, avoidance, and their emotional and motivational corollaries.

BASIC ASPECTS OF AVERSIVE CONTROL

Empirical and theoretical accounts of those aspects of behavioral medicine and industrial health concerned with disordered perfomances frequently assign a central role to historical and contemporary environmental interactions involving aversive circumstances and conditions. Aversive stimuli are defined as environmental events that decrease the subsequent frequency of the operant responses they follow, on the one hand, and/or increase the subsequent frequency of operant responses that remove or postpone them. When an aversive stimulus follows an operant, and decreases the likelihood that such performances will recur, a *punishment* condition is defined. Punishment may be made contingent on the occurrence of an operant that has never before been followed by a reinforcer, an operant currently being maintained by appetitive or aversive reinforcement, or an operant that is undergoing extinction. Under each condition, the short- and long-term effects of punishment will vary as a function of complex operant-respondent interactions, and both discriminative stimulus control and reinforcement schedule factors may operate to further influence the subsequent form and frequency of the performance.

An *escape* condition is defined when a response terminates an aversive stimulus after the stimulus has appeared. The interaction between operants and respondents is especially prominent in escape situations, since the aversive stimulus usually elicits reflexive responses that eventually result in or accompany an operant performance followed by withdrawal of the aversive stimulus. Strong generalization effects appear during initial exposures to escape situations, but the gradual development of discriminative properties by the aversive stimulus narrow the performance, and low intensities of the aversive stimulus may eventually maintain an operant escape performance that requires a more intense aversive stimulus to establish. Reinforcement schedule effects similar in all essential respects to the appetitive conditions described above are observed when withdrawal of an aversive stimulus is the reinforcer. Extinction of an operant escape response occurs rapidly when presentation of the aversive stimulus is discontinued, or more slowly and erratically if the occurrence of the operant is no longer reinforced by withdrawal of the recurring aversive stimulus.

An *avoidance* condition is defined by the occurrence of an operant response that postpones an aversive stimulus. Avoidance performances may be established and maintained either in the presence or absence of a warning stimulus preceding the aversive stimulus. When an exteroceptive warning stimulus precedes the aversive stimulus, respondent conditioning effects operate to endow the warning stimulus with aversive properties, the termination of which following the operant avoidance response probably combines with the continued absence of the aversive stimulus to act as a reinforcer. The complexity of the avoidance process is suggested by the functionally simultaneous properties acquired by the conditioned aversive warning stimulus as (1) an eliciting environmental event for respondent behaviors, (2) a conditioned aversive punisher, withdrawal of which strengthens the operant avoidance performance effective in removing it, and (3) a discriminative stimulus, which provides the occasion for the operant avoidance response to be followed by a reinforcer. In the absence of an exteroceptive warning stimulus, a temporal respondent conditioning process provides disciminative cues, and the temporal stimulus correlated with the aversive environmental event acquires the same three simultaneous functions as an exteroceptive stimulus.

Such an analysis of aversive control emphasizes the simultaneous operation of active and reactive processes in ongoing behavior segments. Whenever the conditioned stimulus in a respondent conditioning procedure is an appetitive or aversive consequence, operant conditioning occurs at the same time as respondent conditioning. Similarly, whenever the reinforcer in an operant procedure is an unconditioned stimulus, respondent conditioning proceeds at the same time as operant conditioning. Thus, if the eliciting and reinforcing stimulus classes are composed of the same environmental events, operant and respondent processes are coextensive.

SOME BIOBEHAVIORAL APPLICATIONS

Relevant applications of these basic behavior analysis principles to medicine in general and to workplace health problems in particular have emerged in three major forms defining, in part, pertinent biobehavioral inter-relationships as represented graphically in the Introduction to this book. *In the first instance*, both the independent and dependent variables are behavioral, and the effects on productivity are studied: For example, the way in which behavioral shaping, maintenance, and modification procedures affect health-related performances, such as food intake, exercise, and compliance with medically prescribed regimens. In the compliance case, for example, several of the contributions to this book document both experimentally and clinically the potent effects of scheduling conditions, stimulus control, and chaining, which determine under what circumstances and in accordance with what behavioral requirements a valued commodity or substance, such as food, drugs, money, or social interactions, can be obtained. All such consequences are subject to this type of rule governance, which can be complex. But they all appear to be variations and/or combinations of a few basic types, and a great deal has been learned about their properties and effects, both in the experimental laboratory and in the natural ecology.

The two major classes into which such effects can be categorized appear to be either schedule-maintained, on the one hand, or schedule-induced on the other. The curious side effects of reward-enhancing intermittent schedules and complex historical circumstances (e.g., maintenance of shock-producing performances and great strengthening of adjunctive or ancillary behaviors) certainly provide examples of direct relevance to medicine and industrial health (Falk, 1971; Kelleher & Morse, 1968). It is most appropriate, in the present context, to study the ways in which environmental constraints imposed by such scheduling maintain remarkably persistent performances. These properties of a behavioral interaction frequently appear to be the most baffling and recalcitrant aspects of the health-related repertoire (e.g., smoking, overeating, etc.).

Figure 1.1 illustrates a typical segment of a cumulative record from an experiment in whcih a chimpanzee sustained performance on a ratio-schedule that required 120,000 responses on a heavy push-button manipulandum for access to food (Findley & Brady, 1965). After every 4,000 responses toward the total requirement, a brief flash of light was presented—the same light that was illuminated continuously during food access once the total ratio was completed. Of particular interest is the pause that follows each

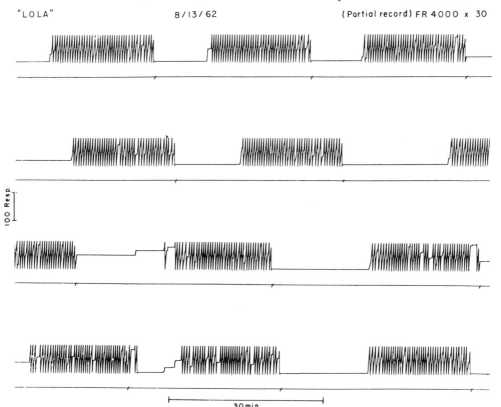

FIGURE 1.1. Cumulative record of responses (vertical excursion of stepping pen—reset after each 100 responses) over time (horizontal baseline-paper speed) for chimpanzee "Lola" showing pause following each flash of light (conditioned reinforcer) after a block of 4,000 responses toward the total requirement of 120,000 for access to food. (From Findley & Brady, 1965. Copyright 1965 by the Society for the Experimental Analysis of Behavior, Inc. Reprinted by permission.)

flash of light after a block of 4,000 responses, illustrating the control acquired by this conditioned-reinforcing stimulus event. Subsequent extension to a 250,000 response ratio and manipulations involving removal and reintroduction of the light flash after each 10,000 responses documented the critical interactions between rule-governance and stimulus control in the establishment and maintenance of such remarkably persistent performance repertoires.

It is important to recognize that while such unusual and extreme examples of schedule and stimulus control conditions may appear to push the limits of adaptive functions, they are not tricks or circus acts. They represent the orderly and lawful operation of general relationships common to all behavioral interactions, including those related to health in the workplace, and they appear to be of particular relevance to the excessive aspects of such performances.

The relevance of such behavioral research to industrial health issues can be demonstrated by studies carried out with human research volunteers living in a programmed environment laboratory for periods of two-to-three weeks. This facility is a five-room residential laboratory divided into individual living areas plus work and social areas, which are shared by all subjects. Extended experimental control, objective recording, and the maintenance of realistic and naturalistic conditions are conducive to the assessment of a broad range of behavioral processes. Subjects can either engage in activities designed for experimental laboratory purposes or can live under conditions closely approximating those found outside of the laboratory.

Four three-person groups participated in a study (Brady, Emurian, & Fischman, 1984) in which subjects could select a one-to-two-hour "work trip" to be performed on an individual basis. This work consisted of five performance tasks, which included signal detection, monitoring, matching and computation tasks (for a complete description, see: Emurian, Emurian, & Brady, 1978) in which accurate responses produced points.

The consequences of completing a work trip were varied to assess the effects of aversive and appetitive consequences on behavior of the subjects. Under appetitive conditions, each work trip completed by an individual subject produced an increment in a group "bank account" that was divided evenly among the subjects at the conclusion of the experiments. Under the aversive condition, completion of work trips did not produce increments in the group account. In fact, each group was assigned a criterion number of work trips to accomplish during a 24-hour period, and uncompleted trips below the criterion produced a decrement in the group account identical in magnitude to the increments produced during the appetitive condition. The criterion during avoidance days was based on group productivity observed during the immediately preceding appetitive days. Thus actual work requirements were programmed to be comparable under both appetitive and aversive conditions. There was not a larger work requirement during the aversive experimental conditions. Appetitive and aversive conditions alternated in three-day blocks during this study.

During the aversive condition, subjects behaved in an emotionally stress-ful manner, providing reports of extreme disturbance with the program, great annoyance with the experimenters, and substantial hostility to other subjects. In addition, self-reported mood scales showed high Depression scores during the aversive condition, and work behavior was seriously disrupted for some subjects. In one group there was a severe emotional outburst by one subject, followed by an unwillingness to work. These effects were observed to be reversible when the appetitive condition was reintroduced, indicating clearly that the consequences of the work task, rather than the performance require-ments or the residential laboratory, were the most significant factor in generating and maintaining the stressful environment. The implications for programming of work tasks in the industrial environment are obvious.

The second major form, which relevantly applies basic behavior analysis principles to clinical medicine and workplace health, emphasizes the effects of behavioral independent variables on biological dependent variables. Procedures have been developed for active rather than reactive behavioral control of visceral, somatomotor, and central nervous system processes based on the arrangement of explicit contingency relationships between specific antecedent physiological events and programmed environmental consequences. It has been convincingly demonstrated that such behavioral "biofeedback" intervention can produce reliable bidirectional control over both increases and decreases in cardiac rate (DiCara, & Miller, 1969; Engel & Gottlieb, 1970) and blood pressure (Benson, Herd, Morse, & Kelleher, 1969; Pappas, DiCara, & Miller, 1970). Large magnitude and enduring elevations in heart rate (Harris, Gilliam, & Brady, 1976) and blood pressure (Harris, Findley & Brady, 1971) have also been described in more chronic operant conditioning studies.

The functional relations that characterize such behavioral control of biological processes can be explicated within the same three-term contingency framework that has provided the basis for operant performance analyses of the more traditional variety. In this systematic application of behavioral procedures to effect biological changes, the response is identified by the measurement of events that are generally localized within the organism. A good demonstration of such procedures documents the effectiveness of operant shaping techniques in systematically altering both the amplitude and duration of blood pressure elevations in small progressive steps to diastolic pressures up to 40 mm Hg above preexperimental resting levels.

Figure 1.2, for example, shows the relative frequency distributions of diastolic blood pressure from an experiment in which a baboon learned to increase and maintain blood pressure elevations in order to obtain food and avoid shock (Turkkan & Harris, 1981). The shaping procedure involved delivery of food pellets for accumulation of 600 seconds of time above the diastolic pressure criterion level, and delivery of a single electric shock to the

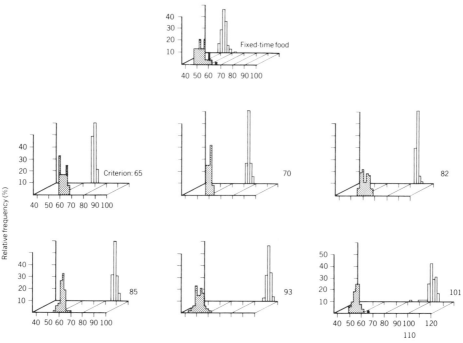

FIGURE 1.2.   Relative frequency distributions of 40-minute average pressures for baboon "#82" during a baseline condition (fixed-time food), and at successively higher diastolic criteria (columns, from left to right). Open bars represent diastolic pressure levels from six experimental sessions, while fixed bars represent data from six associated postsession periods. (From Turkkan & Harris, 1981. Copyright 1981 by Elsevier Scientific Publishing Co. Reprinted by permission.)

tail for accumulation of 240 seconds of time below that criterion level. When the pressure level was above criterion, white light appeared on the animal's work panel, and when pressure was below criterion, a red light accompanied by 1,000 Hz tone was presented. Experimental sessions began at noon each day, and ended at midnight. Criterion levels beginning at 65 mm Hg (i.e., pre-experimental baseline average diastolic pressure level) were progressively elevated at a rate approximating 2–3 mm Hg per week. The systematic shaping of diastolic pressure elevations over a 10–12 week conditioning period, illustrated in Figure 1.2, compares the diastolic pressure levels recorded *during* sessions (open bars) with the levels recorded during the 12-hour intervals *between* sessions (filled bars) under baseline conditions (top segment) and during successive stages of conditioning. At the highest criterion (lower right segment), diastolic pressures were elevated above 100 mm Hg in order to maintain a food-abundant environment throughout the 12-hour experimental session, during which less than one shock per hour was delivered. Remarkably, there was no overlap between the distributions of pressure levels recorded at this highest criterion and those recorded during the baseline period.

These observations clearly reflect the participation of an active behavioral process in the development and maintenance of biological functions traditionally considered under more reactive control. However, these are clearly not the only operative processes. Multiple mechanisms, both behavioral and physiological, must be presumed operative in the mediation of such complex psychophysiological interactions. Both operant and respondent conditioning can be considered cooperative, if the internal and external environmental events involved in these processes have common functional properties.

The third and perhaps most familiar form, indicating the relevance of behavior analysis principles to workplace health, involves the effects of biological independent variables (such as the biochemical and physiological changes associated with drug administration) on dependent performance measures at the interface between organism and environment. The programmed residential environment laboratory (described above) has been useful in demonstrating the ways in which drugs can affect different behaviors. In addition to the traditional performance and learning effects generally seen to occur after drug administration (e.g., Blaine, Meacham, Janowsky, Schoor, & Bozzetti, 1976; Darley & Tinklenberg, 1974), *Figure 1.3* illustrates effects on food intake averaged over five-day periods during which either two active

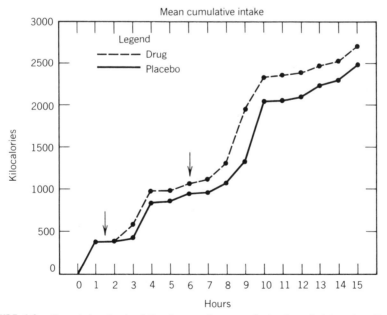

FIGURE 1.3. Cumulative intake following marijuana and placebo administration. Data are averaged over five days when the subject received two doses (indicated by the arrows) of marijuana or placebo.

(1.89 percent) or two placebo marijuana cigarettes were smoked. Arrows indicate smoking of a cigarette. There was a consistent 300 kcal increase in intake on days when active marijuana cigarettes were smoked indicating that drugs can consistently affect different aspects of behavior.

Developments in the area of behavioral pharmacology during the past decade have revealed, for example, that pharmacologic agents can participate in behavioral interactions under the same conditions that govern the relationship between organisms and other stimulus events. As environmental stimulu, drugs can be seen to have *eliciting* properties in reactive behavioral modes, and both *reinforcing* and *discriminative* properties in more active behavioral modes. The *eliciting properties* of drugs as stimuli that *precede* biochemical, physiological, and behavioral responses in a reactive mode have traditionally provided the basis for most pharmacological and toxicological evaluations, including the development of increasingly more sensitive and reliable laboratory behavior models (Barrett, 1980; Geller, Stebbins, & Wayner, 1979; Iverson, Iverson, & Snyder, 1977). The *reinforcing properties* of drugs as stimulus events, which increase the frequency of the responses they *follow*, have been convincingly demonstrated with a range of pharmacologic agents that function to maintain strongly active drug-seeking and drug-taking behaviors (Brady & Griffiths, 1977; Fischman & Schuster, 1982; Griffiths, Bigelow, & Henningfield, 1980; Yanagita, 1973). And the *discriminative properties* of drugs as stimuli, which increase the probability of those responses that have previously been reinforced in their *presence*, have now been extensively analyzed within the methodological framework of drug discrimination studies (Colpaert & Rosecrans, 1978; Schuster & Balster, 1977; Winter, 1978). The results of these latter experiments have shown, for example, that the occurrence of a response can be controlled (i.e., occasioned) by administration of a pharmacologic agent if the response in question has previously been reinforced when (and only when) the administered drug has been "on board."

Of course, some biological changes may have no effect at all on behavior, whether they precede, accompany, or follow responses. Nevertheless, the eliciting, reinforcing, and discriminative properties of a biological event may all participate simultaneously in complex behavioral interactions. The reinforcing and eliciting functions of alcohol, for example, have been well documented by observations and experiments involving excessive self-administration and impaired performance in both episodic and chronic abusers. But the discriminative properties of this CNS depressant in occasioning characteristically drunken behavior patterns in social situations have been vastly underrated. One can get away with a lot more stereotypically intoxicated behavior after drinking alcohol than when one has not been drinking at all. In fact, the discriminative functions of the drug in someone else are frequently sufficient to occasion strange behaviors in what might be characterized as merely "participant observers" of the cocktail party scene.

The idea that the presence of a drug may occasion or legitimize behavior that might better be characterized as a concomitant of drug administration rather than as a direct result cannot be regarded as novel. In medicine, the term *sick role* is widely used to denote a pattern of responses occasioned by prior exposure to pathogen, trauma, or disorder. Indeed, these behavioral concomitants of illness are tolerated if, and only if, associated with illness. They are concomitants because they are not induced by pathogens. Viruses do not induce a demand to be waited on or give license to express one's views on hospital food by throwing it at nurses. Just as prior exposure to a pathogen may license certain behaviors in the afflicted individual, those with whom the patient interacts also adjust their behavior. Illness in one individual may indirectly occasion changes in the behavior of others sometimes characterized as solicitous or care-giving. Drugs, like illness, can occasion behaviors, which then serve to occasion still other changes in other individuals. This discriminative functional property is of particular import- ance when considering drug effects on child behavior occurring pre- dominantly in a social setting, where the most critical relationships involve responses of other people.

These basic relationships reveal the *functional* interactions between bio- logical processes and behavior, versus the *topographical* emphasis, which has frequently characterized traditional models for the analysis of biobehavioral effects. The distinction between topography and function is the distinction between the formal characteristics of a performance (e.g., muscle contractions and extensions involved in an arm movement) and the temporal relationships between the performance, on the one hand, and its environmental antecedents and consequences, on the other (e.g., the sight- ing of friend or foe, the ministering of greeting or blow). For example, the scream for help, the scream for demonstration purposes, and the scream for attention may, when viewed topographically, be identical. However, an analysis of their functional relations to the environment (e.g., their effects on an audience) reveals important differences. Conversely, two topographically different performances (e.g., passing one's fingers over raised letters on a cardboard and moving one's eyes over print on a page) may be functionally identical to the extent that they are both maintained by the consequences of reading.

The differences highlighted by the topographic-functional distinction are directly relevant to critical issues involving the *generality* of a behavioral perspective on health in the workplace. Expressed concerns about the limita- tions on generalizing from an experimental animal's lever pressing to the verbal performances of a human, for example, may be quite valid when the analysis is restricted to the *topography* of responses, which are so obviously different. But when events are considered *functionally* in terms of their relations to antecedents and consequences, it is quite possible to cut across performances and environmental stimuli with entirely different topographies to generalize between species and apply data from laboratory behavior models to human health research and practice. The meaning of a behavioral assessment following a biological intervention is greatly enhanced by an analysis that transcends the topographic features of a performance to reveal its functional significance.

From a broader perspective on behavior and health in the workplace, it is now becoming increasingly clear that the key to whether people will be healthy or sick, live a long life or die prematurely, is likely to be found in the characteristics of their individual common behavioral performance repertoires. These include activities such as eating, drinking, sleeping, exercising, smoking, and, significantly from an industrial health perspective, whether speed laws and other safety measures are observed to avoid accidents and violence. Clearly, the roots of these health-related performances can be readily traced to behavioral processes of obvious relevance, yet they have received surprisingly little attention from traditional medical disciplines. The expanded applications of behavior analysis to industrial health problems have broadened the scope of potential interventions. Under such circumstances, the interaction between applied medical science and behavior analysis in the promotion of industrial health promises to emerge as a major movement through the remainder of this century. It will no doubt continue into the next century as a significant approach to treatment, prevention, and maintenance of health in the work-place.

REFERENCES

Barrett, J. E. (1980, April). Behavioral pharmacology: Recent developments and new trends. *Science*, 215–218.

Benson, H., Herd, J. A., Morse, W. H., & Kelleher, R. T. (1969). Behavioral inductions of arterial hypertension and its reversal. *American Journal of Physiology, 217*, 30–34.

Blaine, J. D., Meacham, M. P., Janowsky, D. S., Schoor, M., & Bozzetti, L. P. (1976). Marihuana smoking and simulated flying performance. In M. C. Braude & S. Szara (Eds.), *The pharmacology of marihuana*. New York: Raven.

Brady, J. V., Emurian, H., & Fischman, M. (in press). Experimental neurosis: By-products of aversive control. *Proceedings of the International Pavlovian Conference of Emotions and Behavior*, Moscow, USSR.

Brady, J. V., & Griffiths, R. R. (1977). Drug maintained performance and the analysis of stimulant reinforcing effects. In E. H. Ellinwood & M. M. Kilbey (Eds.), *Cocaine and other stimulants*. New York: Plenum.

Colpaert, F. C., & Rosecrans, J. A. (Eds.). (1978). *Stimulus properties of drugs: Ten years of progress*. Amsterdam: Elsevier.

Darley, C. R., & Tinklenberg, J. R. (1974). Marihuana and memory. In L. L. Miller (Ed.), *Marijuana: Effects on human behavior*. New York: Academic Press.

Deane, G. E., & Zeaman, D. (1958). Human heart rate during anxiety. *Perceptual and Motor Skills, 8*, 103–106.

DiCara, L. V., & Miller, N. E. (1969) Transfer of instrumentally learned heart rate changes from curarized to noncurarized state: Implications for a mediational hypothesis. *Journal of Comparative and Physiological Psychology, 62*(2, Pt.1), 159–162.

Dykman, R. A., & Gantt, W. H. (1958). Cardiovascular conditioning in dogs and in humans. In W. H. Gantt (Ed.), *Physiological bases of psychiatry*. Springfield, IL: Thomas.

Emurian, H. H., Emurian, C. S., & Brady, J. V. (1978). Effects of a pairing contingency on behavior in a three-person programmed environment. *Journal of the Experimental Analysis of Behavior, 29*, 319–329.

Engel, B. T., & Gottlieb, S. H. (1970). Differential operant conditioning of heart rate in the restrained monkey. *Journal of Comparative and Physiological Psychology, 73*(2), 217–225.

Falk, J. L. (1971). The nature and determinants of adjunctive behavior. *Physiology & Behavior, 6*, 577–588.

Findley, J. D., & Brady, J. V. (1965). Facilitation of large ratio performance by use of conditioned reinforcement. *Journal of the Experimental Analysis of Behavior, 8*, 125–129.

Fischman, M. W., & Schuster, C. R. (1982). Cocaine self-administration in humans. *Federation Proceedings, 41*, 241–246.

Geller, I., Stebbins, W. C., & Wayner, M. J. (eds.). (1979). *Test methods for definition of effects of toxic substances on behavior and neuromotor function*. Fayetteville, NY: ANKHO International.

Griffiths, R. R., Bigelow, G. E., & Henningfield, J. E. (1980). Similarities in animal and human drug taking behavior. In N. K. Mello (Ed.), *Advances in substance abuse*. Greenwich, CT: Jai Press.

Harris, A. H., Findley, J. D., & Brady, J. V. (1971). Instrumental conditioning of blood pressure elevations in the baboon. *Conditioned Reflex, 6*(4), 215–226.

Harris, A. H., Gilliam, W. J., & Brady, J. V. (1976). Operant conditioning of large magnitude 12-hour heart rate elevations in the baboon. *Pavlovian Journal of Biological Science, 11*(2), 86–92.

Honig, W. K. (Ed.). (1966). *Operant behavior: Areas of research and application*. New York: Appleton-Century-Crofts.

Honig, W. K., & Staddon, J. E. R. (eds.). (1976). *Handbook of operant behavior*. Englewood Cliffs, NJ: Prentice-Hall.

Iverson, L. L., Iverson, S. D., & Snyder, S. H. (Eds.). (1977). *Handbook of psychopharmacology: Principles of behavioral pharmacology* (Vol. 7). New York: Plenum.

Kelleher, R. T., & Morse, W. H. (1968). Schedules using noxious stimuli. III. Responding maintained with response-produced electric shocks. *Journal of the Experimental Analysis of Behavior, 11*, 819–838.

Pappas, B. A., DiCara, L. V., & Miller, N. E. (1970) Learning of blood pressure responses in the noncurarized rat: Transfer to the curarized state. *Physiology & Behavior, 5*(9), 1029–1032.

Pomerleau, O. F., & Brady, J. P. (Eds.). (1979). *Behavioral medicine: Theory and practice*. Baltimore: Williams & Wilkins.

Rachlin, H. (1970). *Introduction to modern behaviorism*. San Francisco: Freeman.

Razran, G. (1961). The observable unconscious and the inferable conscious in current Soviet psychophysiology: Interoceptive conditioning, semantic conditioning, and the orienting reflex. *Psychological Reviews, 68*, 81–147.

Schuster, C. R., & Balster, R. L. (1977). The discriminative stimulus properties of drugs. In T. Thompson & P. B. Dews (Eds.), *Advances in behavioral pharmacology* (Vol. 1). New York: Academic Press.

Turkkan, J. S., & Harris, A. H. (1981). Differentiation of blood pressure elevations in the baboon using a shaping procedure. *Behavioral Analysis Letters, 1*, 97–106.

Winter, J. C. (1978). Drug-induced stimulus control. In D. E. Blackman & D. J. Sanger (Eds.). *Contemporary research in behavioral pharmacology*. New York: Plenum.

Yanagita, T. (1973). Pharmacology and the future of man. *Proceedings of the 5th International Congress of Pharmacology, San Francisco*. Basel: Karger.

2

Social Learning Principles for Organizational Health Promotion: An Integrated Approach

DAVID B. ABRAMS, JOHN P. ELDER, RICHARD A. CARLETON, THOMAS M. LASATER, and LYNN M. ARTZ

OVERVIEW

Interest in primary prevention through organizational health promotion is growing at a rapid pace fueled by various concerns, such as the rising costs of health care and the quality of life (DeLeon & Pallack, 1982; Matarazzo, 1980; Matarazzo, 1982; Pomerleau, 1979; Tanabe, 1982; Weiss, 1982). This chapter combines ideas derived from social learning theory (SLT) and from related sociological and educational disciplines. The aim is to develop a framework for organizational health promotion. SLT is a cognitive-behavioral theory that is most clearly delineated at the level of individual behavior (Bandura, 1969, 1977). A major challenge for health promotion, however, is to achieve changes at a more macro level, such as in entire

The material presented in this chapter was prepared under Grant #RO1 2362901A to Richard A. Carleton, M.D. The authors would like to thank Michael J. Follick, Ph.D. for reviewing the manuscript and for contributing to the ideas presented herein. We are particularly indebted to the chairpersons of the individual, group, and organizational level committees of the Pawtucket Heart Health Program who have put these ideas into practice. These people are, in alphabetical order, as follows: Paula A. Beaudin, M.A.; Andrea V. Ferreira, M.P.H.; Patricia M. Knisley, R.D.; Thomas Lasater, Ph.D.; Gussie S. Peterson, M.S., R.D.; Anthony Rodrigues, M.Ed.; Patricia E. Rosenberg, M.A.; Frederick C. Schwertfeger, M.A.T.; and Robert C. K. Snow, M.S. We would also like to thank the evaluation unit staff headed by Sonja M. McKinlay, Ph.D., with Sarah A. McGraw, M.A. as formative and process evaluation supervisor. Our secretaries deserve a special note of thanks, as does Anne Jaworski, a special volunteer.

organizations (Brown & Margo, 1978; Matarazzo, 1980; Stunkard 1976). What is needed is a theoretical integration to bridge the gap between micro- and macro-level principles. As we shall see, SLT offers broad-based applicability at the macro level as well as fine-grained specificity at the micro level.

The model presented herein was used as the organizational level component of the Pawtucket Heart Health Program—a community-wide multiple risk factor intervention for the prevention of coronary heart disease (Carleton, 1980). The project focuses on changing lifestyle practices (smoking, sedentary lifestyle, overeating, blood pressure management, dietary intake of cholesterol, and stress reduction) during a five-year period in a community of 72,000 people. For illustrative purposes, examples will be drawn from the Pawtucket project's approach to organizations (worksites, churches, schools).

Challenges for Bridging the Micro–Macro Gap

The ultimate goal of health promotion is, of course, behavior change, leading to changes in mediating mechanisms of chronic disease (e.g., cholesterol) and finally to changes in morbidity, mortality, and longevity. Traditional health promotion efforts in industry tend to emphasize educational technologies, such as print media campaigns or detection and screening protocols, to achieve change. For the most part, change is achieved by increasing knowledge and modifying attitudes. The Health-Beliefs Model (Rosenstock, 1966, 1974), the "PRECEDE" model (Green, Kreuter, Reeds, & Partridge, 1980), communication theory, and recent "Macro Level" SLT derivatives, such as the Stanford Heart Disease trial, typify the theoretical emphasis of some of these approaches (Maccoby, Farquhar, Wood, & Alexander, 1979).

Although the macro approach creates a motivational climate for change, disadvantages are that explicit behavioral skill training and maintenance strategies are not easy to program. Furthermore, many individuals already know about the health consequences of negative lifestyle practices (e.g., smoking) and yet they do not, cannot, or will not change. Large-scale campaigns may even have an iatrogenic effect on some individuals who become rebellious to any suggestions of change. Because macro-level programs focus primarily on antecedents (precursors) to behavior change, maintenance and generalization of lifestyle change are difficult. Few studies focus on measuring physiological mediating mechanisms, and almost none focus on endpoints. Maintenance of change throughout an entire life-time is probably necessary for primary prevention of chronic diseases.

Despite limitations, the macro approach is advantageous, because a large number of people can be reached relatively quickly and at low cost per person. There is the possibility that some individuals in organizations will either change spontaneously or seek additional help from existing resources.

It has been suggested that large-scale, less intrusive programs, designed to encourage self-directed efforts, may be more effective (Loro, Fisher, & Levenkron, 1979). Those who change on their own may be better equipped to maintain the change (Schachter, 1982). Finally, organizations are complex social networks, and they lend themselves to change techniques that transcend the simple transfer of clinic programs to organizational settings.

At the individual or micro level of analysis, there are established SLT principles for facilitating individual or small group lifestyle change (Wilson, 1980). Programs based on SLT have been successfully applied in clinical settings in diverse areas, such as smoking cessation, weight loss, stress management, and medication compliance (Franks, Wilson, Kendall, & Brownell, 1982). The major strength of the SLT approach is a direct focus on the target behaviors (rather than on knowledge and attitudes) and the provision of explicit skill training with active participation, role modeling, feedback, and reinforcement. Both modeling and active participation are believed to be the strongest mechanisms of modifying self-efficacy and behavior (Bandura, 1977).

Disadvantages of the micro approach in organizational settings are that providing individualized programs to a large number of people is either not feasible or too expensive and that maintenance of lifestyle change is also difficult to achieve. Of those who complete micro-level programs, 40 – 85 percent relapse within six months of treatment (e.g., Hunt & Bespalac, 1974; Marlatt & Gordon, 1978; Owen, 1983). No matter how good the skill training is, or how strongly motivated the individuals are, they are thrown back into a social climate and physical environment filled with negative health practices. Isolated individuals or small groups have to fend for themselves like tiny rowboats in a stormy sea of negative health practices in the rest of the environment. Upper limits exist, therefore, on even the best protocol that can be developed using an individually oriented program philosophy. Studies have yet to demonstrate unequivocally that less costly behavioral self-help groups are as effective as professionally led groups or that they can have a significant public health impact. Recently, Schachter (1982) noted that change is a lengthy process of repeated attempts, making it important to look beyond the individual and his or her single treatment episode.

Maintenance and generalization of lifestyle change are major challenges for both macro and micro approaches. Generalization is related to the issues of maintenance and to the goal of transcending the individual by achieving organizational structural change. Generalization can take three forms: (1) generalization across organizational settings (e.g., home, school, worksite); (2) generalization across people (family members, friends, colleagues at work); and (3) generalization across health practices (behaviors) can be encouraged (e.g., an individual who quits smoking should consider increasing their exercise level). Some new organizational level projects have begun to tackle problems of generalization by adopting a *wellness philosophy* (e.g., *Live for Life* at Johnson & Johnson; Wilbur, 1983). To achieve maintenance, issues of generalization must be addressed on a much larger scale than is necessary for the micro or macro approaches currently available.

Other factors must also be considered to meet the challenge of an optimal organizational health promotion protocol. Among the more important dimensions are (1) individual differences (e.g., Best & Hakstian, 1978) in the membership and (2) mediating variables that can significantly enhance or retard an intervention. Recent research suggests that individual differences in stages of change can be identified and that interventions must be matched to the stage that an individual is currently experiencing (DeClemente & Prochaska, 1982). But most of our past research using SLT protocols has been conducted in clinics with individuals who are already contemplating change. The typical recipient of a program is a motivated middle-class person who arrives at a university-based psychology department or medical school. By contrast, most members of organizations are not motivated at all. The area of marketing psychology, of "selling products" to "nonbelievers," requires a new set of intervention strategies.

Other important mediating variables include social network support systems (Berkman & Syme, 1979; McKinlay, 1980), social climate and environmental variables (Moos, 1976), and cultural/ethnic factors that moderate health seeking and health care utilization (Croog & Levine, 1972; Dressler, 1980; Hessler, Nolan, Ogbru, & New, 1975). These factors have received considerable attention in medical sociology and anthropology but are underemphasized in health psychology.

In summary, the problem of bridging the gap between micro and macro levels is as follows: at the macro level, health promotion specialists have expertise in mass media, health education, knowledge, and attitude change principles; however, they tend not to focus on psychological principles related to the specifics of skills training for behavior change or on the issue of maintenance and generalization. Micro-level psychologists, on the contrary, have expertise in individual or small group SLT principles and in the direct modification of health behavior. Neither micro nor macro models have resolved challenges in the area of maintenance. Interactions between individuals, groups, organizations, and other social systems are also under-emphasized by both micro and macro specialists. The issue of dealing with large, unmotivated populations is a challenge to behavioral health psychologists.

Having delineated several challenges, this chapter will take a first step toward developing a comprehensive model for health promotion. A clear model is important because it helps optimize treatment, facilitates replication, and encourages testable hypotheses about the process of lifestyle change. To bridge the gap between micro-level principles and macro-level approaches, the analysis will explore principles of behavior change in three qualitatively different, but interactive systems within the social structure of an organization: (1) *individual level* (cognitive, behavioral); (2) *small group level* (family, friendship circles, and worksite colleagues); (3) *organizational level*

LEVEL ╲ PHASE	PROMOTION	BEHAVIORAL SKILLS TRAINING	MAINTENANCE
INDIVIDUAL			
GROUP			
ORGANIZATION			

FIGURE 2.1. An integrated model of phases and levels of organizational health promotion.

(worksite, civic, religious, and health organizations). The interactions between and within levels 1, 2, and 3 will also be explored in a comprehensive integration of principles (see Figure 2.1). At the individual level, although genetic, biological, and physiological processes are important for a proper understanding of health and illness, they are beyond the scope of this exploration. Our goal, therefore, is to contribute to an understanding of how to achieve large-scale, organization-wide change. It is assumed (and this is a big assumption) that if lifestyle change is achieved it will lead to positive health outcomes throughout the lifespan (Follick, Abrams, Smith, Henderson, & Herbert, in press).

The major elements for health promotion must include the following: (1) promotional, marketing, or incentive systems that create a motivational climate for behavior change; (2) specific behavioral skill training using SLT principles of modeling, active participation, and self-control; and (3) programming of generalization and maintenance so that those who change remain more healthy and, most important, begin to influence others around them. Finally, the entire system must be dynamic, flexible, and responsive to changing problems and issues as they arise in the organization. It must also be data oriented with self-correcting feedback loops based on cost-effectiveness and outcome evaluation. The dynamic model should be replicable and relatively easy to evaluate using both process (e.g., program evaluation) and outcome (e.g., changes in biochemical indices of risk) measures (McGraw & Abrams, 1983).

The major departure from traditional SLT programs is the goal for change, which shifts from the individual to the entire culture of the organization. The overall goal of health promotion is to achieve organizational structural change—to start a growing *Social Movement that diffuses to others*. The health program must eventually create a *new social climate* and *physical environment* for a permanent pro-health oriented philosophy in the *majority of members* of the organization. Such a process of generalization and maintenance of change, affecting an ever-increasing number of individuals within an organization, is termed *diffusion* (Rogers & Shoemaker, 1971). In some respects, the concepts of generalization and diffusion overlap. As will become clear in subsequent sections of this chapter, diffusion is a central concept, used to bridge the micro-macro gap. In addition, by simultaneously focusing an intervention on multiple levels, new solutions to the maintenance problem will emerge.

FROM THE INDIVIDUAL TO THE ORGANIZATION

Individual Differences in Readiness to Change

Ultimately, lifestyle change must occur at the individual level. Individuals within an organization, however, will be at different points along a hypothetical continuum of "readiness to change." Readiness for change can be defined operationally as a combination of the following: (1) predisposing biological and social learning history factors; (2) beliefs and knowledge about lifestyle and health (personal outcome expectations); (3) motivation and self-efficacy expectations (cognitively formulated intentions and self-confidence to change); (4) level of skills; and (5) degree of environmental (social and inanimate) support for change. Only if and when environment pressures are exerted (external pressure) and lifestyle change is perceived by the individual as a high priority (cognitive factors) will a commitment to change be made. One goal of a health promotion program is to move individuals and their environments from low readiness to high readiness, at which point they will be able to take action and learn new skills.

Prochaska (1979) and DeClemente and Prochaska (1982) have shown that individuals who are at different stages of readiness to change need to be given different kinds of interventions. Individuals who are ready for action— i.e., high on readiness to change—can be viewed as natural resources within the population and are called *early adopters* (Rogers & Shoemaker, 1971). Early adopters will be the first ones to join behavioral skill training programs. By contrast, individuals who are low on readiness require stronger promotional campaigns (advertising, consciousness raising) and pressure or support from family or close friends. Low readiness individuals are referred to as *late adopters* (Rogers & Shoemaker, 1971). Some members of an organization, of course, will actively or passively resist change. This is their prerogative in a free and open society but an intervention must plan for these individuals as well. Some of these individuals may not avoid change out of rebelliousness or ignorance but simply because they have strong cultural beliefs about health and illness (Harwood, 1971). These beliefs could be at variance with modern Western medicine. Nevertheless, resistant, late adopter, and early adopter subcategories of individuals within an organization should be taken into account in a model of health change. A more fine-grained protocol might consider the following categories of readiness and might plan different intervention strategies to match each one: (1) *resistant,* (2) *neutral (immotive),* (3) *low readiness,* (4) *high readiness,* (5) *action,* (6) *maintenance,* and (7) *termination* (DeClemente & Prochaska, 1982).

Self-Efficacy and Ownership

Bandura (1977) asserts that self-efficacy expectations mediate change and maintenance of change. Self-efficacy is the belief that an individual has mastered the skills necessary to engage in the new behavior. Self-efficacy is best increased by means of participant modeling and successfully engaging in behaviors without attributing the skills to external aids. Thus both verbal persuasion and insight (i.e., knowledge and attitude change) are less likely to result in increases in self-efficacy compared to direct, self-guided practice of the new behavior. Self-efficacy also determines resistance to relapse. In the face of minor setbacks, the higher the self-efficacy the less likely a return to old habits. A critical determinant of self-efficacy is attribution of responsibility for the change to his or her own doing versus outside sources (Goldfried & Robins, 1982). A person who attributes change to luck, his or her doctor or psychologist, or a drug is likely to have low self-efficacy and is vulnerable to relapse when the external agents are withdrawn. Craighead, Stunkard, and O'Brien (1981) showed that a group of people who were given a weight-loss drug (fenfluramine hydrochloride) plus behavior therapy lost more weight than a behavior therapy alone group, but they relapsed more rapidly. Presumably members of the combined treatment attributed their weight change to the drug even though they practiced the same behavioral skills as the other group. Since self-efficacy did not change in the combined group, they were more vulnerable to relapse. Others have also suggested that individuals who change by their own efforts and interventions that are least intrusive may ultimately be more effective (Loro, Fisher, & Levenkron, 1979; Schachter, 1982).

Self-efficacy theory raises a critical issue for health promotion. An intensive promotional campaign, aggressively doing things *to* the target organization, creates considerable external pressure to change. Change, however, may not be attributed to the actions of the members of the organization themselves but rather to the pressure of external agents. Self-efficacy would be unaffected due to external attributions of causality (Goldfried & Robins, 1982). Consequently, the intervention would produce only short-term change, which would rapidly dissipate after the external pressures are withdrawn.

To meet the goal of organizational structural change, change should occur within the organizational membership, not as a passing fad fueled by external campaigns. Patterns that are mutually and reciprocally reinforced by other members of the organization may be preferable to those that are driven from outside sources. Perhaps a distinction should be made between *external* promotion and *internal* promotion. External promotion refers to the use of educational materials and screening protocols conducted by outsiders. Typically, these are conducted once a year and they use "canned" messages, posters, or brochures. Internal promotion refers to the use of actual behavioral changes (made by early adopter members of an organization) as the stimulus material for future promotional campaigns. This strategy capitalizes on using natural support networks and role models within the organization. External promotion can help set the stage and facilitate health change efforts. The intensity, introduction, and withdrawal of external stimuli, however, is critically important. Eventually, internal promotion via feedback based on actual progress made by members of the organization themselves should predominate in a health promotion effort. Self-efficacy theory suggests that organizational activation from within the organization itself must be explicitly addressed. The role of health psychology in this model, therefore, is clearly one of mediating and facilitating change, as co-participants with the organizational membership, not as external agents.

Thus for a successful organizational health change program, both internal and external promotional campaigns should be developed. Campaigns should be ongoing rather than one-shot, and they should be well coordinated. Consideration must be given to individual differences in readiness to change and to the goal of organizational structural change. A sense of ownership of the program by the members of the organization will probably be necessary to achieve generalization and maintenance. Efforts should be made to capitalize on the unique natural resources within an organization, such as its social networks, role models, and ability to provide organizational incentives for lifestyle change (e.g., monetary bonus or days off work for quitting smoking).

Individual Skills Training, Self-Control and Problem Solving

After promotional systems are established, then specific skills for behavior change must be taught to those who are interested—i.e., the early adopters. These skills can be taught using either programmed texts or self-instructional techniques, such as microtraining (Glasgow & Rosen, 1978; Ivey, 1978). Programmed instruction leads to the development of effective programs using standard SLT principles, such as demonstrating, shaping, rehearsal, reinforcement, practice, and feedback. Shaping, for example, allows for gradual acquisition of new skills by successive approximation and selective positive reinforcement (Skinner, 1953). Self-control and general problem-solving principles can be used as an overall guiding framework to achieve optimally effective self-instructional procedures (Kanfer & Busemeyer, 1982).

Self-control procedures supposedly operate when a previously automatic chain of events leading to an undesirable behavior is interrupted by an environmental or cognitive event (Kanfer, 1975). This change questions a person's previous responses, and it demands an examination of the self (e.g., "If George the foreman can quit smoking so can I!"). Next, a process is instituted to help an individual substitute more desirable alternatives. In general, the basic steps in self-control include the following: (1) self-assessment (or self-monitoring), (2) self-selection of goals, (3) self-evaluation, against predetermined standards, and (4) self-production of reinforcing or punishing consequences.

A pervasive component of self-control for individual behavior change consists of ongoing self-assessment or self-monitoring (e.g., Abrams & Wilson, 1979; Ciminero, Nelson, & Lipinski, 1977; Nelson, 1977). Usually, the lifestyle pattern to be changed (i.e., the target behavior) is operationally defined in terms of observable behavior and made easy to record (e.g., number of miles jogged per outing, number of cigarettes smoked, number of calories consumed). Once a baseline (the current rate before trying to change behavior) has been established, then an individual can begin to set change goals.

Often when we decide to change, we set a goal for ourselves (e.g., "I will run one mile in ten minutes"). The trick here is to help individuals select appropriate standards that are not overly demanding, especially in the beginning of behavior change programs. In general, an immediate goal ("I want to walk one mile today") is preferable to a long-term, vague, or global goal ("I want to be healthy"). Once goals are set, new skills are learned and implemented. Then an individual can evaluate progress made toward the goals. Evaluation against self-monitoring data provides feedback to refine or change one's skills in the light of new evidence. Evaluation of both skills and progress toward goals forms part of a self-correcting feedback system that continues to use the self-monitoring data until the change process is completed and the new behavior is internalized.

Behavioral self-control techniques are similar to general problem-solving protocols (Goldfried & D'Zurilla, 1969; Kanfer & Busemeyer, 1982). The steps involved include the following: (1) *assessment* of the situation and relevant facts; (2) *formulation* of the major problem(s) and the circumstances surrounding them (goal setting); (3) *brainstorming* of alternatives (generation of solutions without regard for their practicality or value); (4) *selection* and *implementation* of the most realistic alternative, which may include the need to learn new skills; (5) *evaluation* of the solution and of progress toward goals; (6) *reassessment, correction*, and *refinement*, which may involve returning to step 1 and going through the process again until the problem is solved. In this way, problems are fine-tuned, and unworkable solutions are rejected so creative new directions can be explored in their place. Research has shown that specific training in problem solving for socially oriented problems can enhance a person's ability to develop and choose effective solutions to his or her own problems (D'Zurilla & Nezu, 1982).

Beyond the Individual

Using key concepts from self-control and problem-solving theory, and principles of Organizational Behavior Modification (OBM), a framework can be developed for total organization-wide change programs. The important new development is that self-control principles can be applied to various levels of social structure, not just at the individual, micro level of analysis. In essence, self-control theory is used in an iterative and recursive manner at the individual, group, and organizational levels of social structure (see Figure 2.2). Individuals, groups, and the organization as a whole all complete "cycles" of self-control, using qualitatively different targets for change that are appropriate for that level of analysis. The general strategy is applied at the individual level (inner feedback loop in Figure 2.2) through to the organizational level (outer loop in Figure 2.2).

Just as individuals engaged in lifestyle change assess their needs, monitor key behaviors, set goals, learn new skills, evaluate progress, and use feedback to correct themselves, so too can entire organizations (through special task forces or committees). For example, an organization assesses and monitors its membership's needs, sets organizational goals, implements programs, and then evaluates, refines, and continues with its plans. Organizational assessment of needs can consist of a survey of membership interest in specific lifestyle change programs, an examination of the organization's physical environment, and its policies governing health. Organizations can then self-monitor certain organizational targets of change (e.g., the number of members who attend change programs each month, the number of new no-smoking areas created, the number of newsletter articles on lifestyle changes made by members). Organizational goal setting may

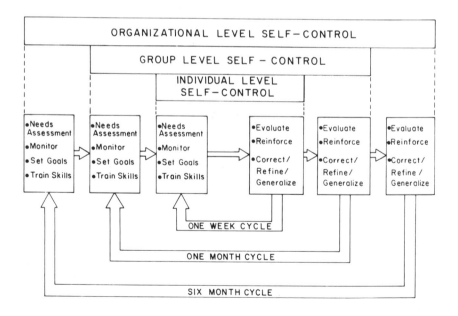

FIGURE 2.2. Self-control steps applied to the individual, group, and organizational levels of an organizational intervention strategy.

involve creating organizational incentives, changing the physical environment, and providing organizational support or facilities for healthy practices. Subgoals might include helping to set up correspondence or self-help group programs (e.g., complete four quit-smoking groups and three fitness groups within six months).

Reliable and valid data, collected on key variables, are an integral part of a self-control model at the organizational level (McGraw & Abrams, 1983). Program and organizational progress should be evaluated on an ongoing basis. Part of organizational "self-monitoring" can serve to alert the planning committees to problems and to success. Because changing organizations is complex, the timeframe for one complete organizational self-control "cycle" is rather long (e.g., once every six months). At the end of each six-month period, the entire health promotion program, its goals, achievements, and failures are evaluated, and refined, while new directions are planned for the next cycle.

Similar to establishing an organizational self-control system, group and individual self-control sequences can be implemented, each with its own goals, and even shorter timeframes for completion of each self-control "cycle" (see inner two feedback systems in Figure 2.2). In recursive-iterative fashion then, collections of individuals help to meet group goals and collections of groups help to meet organizational goals. In this comprehensive, multi-level, self-control model, problems of promotion, skills training, generalization, and maintenance can be directly addressed. In the remainder of this chapter, several selected examples of applications of multi-level interventions in individuals, groups, and organizations will be presented to illustrate the basic concepts. Emphasis will be placed on optimizing multi-level effects and on issues of generalization and maintenance.

Group Level Strategies: Competition

Most traditional behavioral medicine interventions in groups involve little more than individual programs conducted in group settings. Minimal attention has been paid to group-level contingencies and their potential power beyond individual self-control protocols. Several behavioral medicine researchers have, however, sensed the importance of group self-control factors in enhancing behavior change or maintenance. Weight-loss studies, for example, have incorporated spouse (e.g., Brownell et al., 1978) and peer (Zitter & Fremouw, 1978) support in augmenting weight-reduction procedures.

Just as in the case of individual behavior, a group as a whole can engage in group monitoring, group goals can be set, change can be implemented, and group progress can be evaluated. The group as a whole contributes to the index of change, not the individual per se. Groups provide group reinforcement, correct or refine their strategies, or eventually explore new areas. Groups can also use problem-solving skills to help generalize and maintain change by either soliciting support or by influencing others. If groups are empowered to evaluate their goals and to set new goals, then a weight-loss group, for example, can begin to investigate nutrition and cholesterol at a worksite cafeteria or in a school where their children attend classes (generalization across people and settings).

The successful use of group self-control procedures is illustrated in a recent worksite weight-loss program. Abrams and Follick (1983) transcended the traditional individual behavioral self-control approach. They had each group monitor the average weight lost by the entire group. This average weight was used as part of a group self-control index. The group averages were used as the index for a competition between groups while still maintaining support within groups and fostering cooperation. Group reinforcement was applied in the form of a "winning group" of the week—i.e., the group that had the greatest *average* weight loss from all the participating groups. All group statistics were published in the worksite newsletter without identifying any individuals. This created support from the entire organization, and several individuals who were not part of the program suddenly wanted to join (internal promotion stimuli). Using competition between groups represents a wholesale departure from traditional clinic-based programs. It can be a powerful incentive system and could also be used to reduce dropouts (e.g., by having an attendance index). Many other applications are possible. Group self-control techniques offer a level of reciprocal support and influence that is, in some ways, more than the simple sum of the individual members.

Groups actually form part of complex social networks. An understanding of social network parameters can help to optimize group techniques. Social network structures and their ultimate influence on health-seeking behaviors have been studied extensively in sociology, social psychology, and anthropology (McKinlay, 1980). These studies reveal how social network parameters (such as frequency, density, supportiveness, or range of contacts between members) can moderate health care utilization, the transmission of attitudes, and health-seeking behaviors (Cobb, 1976; McKinlay, 1980). Social network factors have, for the most part, been studied from epidemiological perspectives (e.g., Berkman & Syme, 1979). Social network parameters have not as yet been incorporated into health promotion efforts as direct targets of change. Consequently, the technology for optimizing changes in social network structures and the feasibility of doing this on a large scale remain challenges for the future. Given the power of social network structures, and combining this knowledge with behavioral psychology (Social Skills Training), may result in quantum leaps in our ability to address generalization and maintenance issues.

There is substantial literature on the use of social skills in mental-health contexts (e.g., Curran & Monti, 1982). Social skills principles offer techniques that extend the application of group approaches to health interventions. Attempts to define socially skilled behavior usually include the cost-effective social response. Socially competent individuals are those who maximize positive and minimize negative social or personal consequences through eliciting cooperation and by being verbally adept (Gough, 1965), through cost-effective social problem-solving (Goldfried and D'Zurilla, 1969), through asserting both negative and positive feeling (Wolpe, 1969), and through minimizing social anxiety (Gough, 1965; Wolpe, 1969). Training in social skills usually consists of instructions (written or spoken) with a rationale, modeling, role-playing of both problem-solving (i.e., perceptual and translational) and nonverbal and verbal (i.e., motoric) behaviors, feedback (from both self and others), and homework assignments (Ivey, 1978).

Social Skills Training offers a potentially powerful technique for enhancing the maintenance and generalization prospects for health promotion programs, especially when combined with social network theory. The level of one's interpersonal competence may predict the degree to which she or he will be able to arrange a supportive natural social environment or convince others to change their behaviors. The supportiveness of the environment, in turn, will in large part determine whether the individual is able to maintain a lifestyle change or indeed whether such a change is even initiated. In addition to these reciprocal interactions, indirect enhancement of group cohesion and increased abilities to modify more distal organizational environments for the further promotion and maintenance of healthy lifestyles may also occur to those with socially skilled repertoires. For example, groups can join together to form larger committees to influence entire organizations, or to arrange organizational events (like fun runs). Veteran group members who have successfully changed can be encouraged to volunteer to be trained as group leaders to start up and run new groups. This would allow change to take place in an exponential growth pattern within an organization, using existing resources to the fullest.

Not all social network members will be ready to support lifestyle changes actively; therefore, it is necessary to engender skills for handling those who are nonsupportive or negative about lifestyle change. Individuals must be able to select the sets of behaviors most relevant to their particular social network. A possible application of this concept to health promotion involves preparing people for dealing with negative (e.g., using extinction when someone jokes or criticizes you) as well as positive situations (e.g., using self-disclosure when attempting to convince others to become involved in your program).

One major challenge facing organizational health promotion is to reach a large number of people relatively inexpensively. This often rules out using professional psychologists as small group leaders, which has been done in most clinical settings. However, social skills training and problem-solving techniques can be used to provide self-help or "correspondence course" types of behavior change programs (Rogers, Killen, Cameron, & Weinberger, 1982). Using a structured self-help protocol, Peterson, Abrams, Elder, and

Beaudin (in press) demonstrated that a worksite self-help protocol for weight loss can be as much as three times more cost-effective (dollars per kilogram lost) than a professionally led program, even though the self-help groups had less average weight loss than the professional groups. To enable behavioral programs to meet the formidable challenge of making a large scale, cost-effective, public health impact, more research must be devoted to developing and evaluating low-cost group level change protocols.

Group self-control capitalizes on potentially powerful sources of social influence that clearly transcend traditional SLT approaches. Not only can intergroup competition be used, but specific group contingencies and social skills training can also be employed. Group self-control, especially in groups that are members of the same social networks (e.g., family, work associates), is more likely to achieve maintenance and generalization than will individuals working in isolation (Brownell et al., 1978). Group support can be used to reduce the reliance on professional change agents. The disciplines of sociology, anthropology, and behavioral psychology must collaborate to make optimal use of both social network parameters and social skills training to enhance transmission of health behaviors, to facilitate generalization and maintenance of change, and to promote diffusion (see later) along natural social network lines. By explicitly programming social support skills and assertive skills into natural worksite social networks, one may be able to quicken the pace of structural change in the organization and to reduce relapse episodes in those who do change.

Organizational Level Strategies

Organizational self-control has been well documented under the Organizational Behavior Modification (OBM) banner. However, OBM has, up to now, been devoted to issues of management, tardiness, and productivity rather than to health (Andrasik, Heimberg, & MacNamara, 1981). For our purposes, the principles of behavior change would be identical to those already employed in organizational settings, except with "health" instead of "productivity" as the goal. There are potentially powerful organizational level contingencies that are available for optimizing health promotion efforts (Abrams & Follick, 1983). To further enhance lifestyle change programs, individuals and groups may benefit from the added power of the larger organization system. To promote health and recruit members into behavior change programs, a variety of incentive systems can be used including prizes and monetary rewards (e.g., Bowles, Malott, & Dean, 1982). Sometimes savings can also accrue to the organization in the form of reduced health insurance premiums. Organizational goal setting, monitoring of programs, and consequation/feedback can also be incorporated into an *organizational self-control* approach to lifestyle change.

Performance feedback is a powerful technique for organizational behavior change. *Performance feedback* can be defined as a behavior modification strategy that is based on the objective communication (verbal or written) of past or current behavior as it relates to specifically stated organizational performance goals. Feedback may be delivered privately or publicly, and it may be addressed to individual behaviors, to aggregate group behaviors, or to aggregate total organizational behaviors. For example, feedback could be provided about each individual who quits smoking or the total number of individuals in a group, or about the percentage of quitters from the total membership of the entire organization who smoke.

The use of feedback to the entire organization acts as a public incentive, both for early adopters to maintain their nonsmoking status and for others (later adopters) who may be induced to try to quit in the future. Because people in organizations have frequent and regular contact with one another, the powerful facilitating effects of modeling and vicarious learning can be emphasized (Bandura, 1977). One effective method for providing perform-ance feedback is the use of highly visible displays that inform members about the lifestyle change of others. Some displays have provided public feedback of individual behavior and others simply provide feedback on total group performance (Collins, Komoki, & Temlock, 1979; Kreitner, Reif, & Morris, 1977). Public posting of individual or group performance has led to significant increases in desired behavior patterns (Kreitner, Reif, & Morris, 1977). Because giving public feedback about individuals can have a punitive component in certain cases, ethical considerations and informed consent must be exercised.

Moos and his colleagues (Billings & Moos, 1981; Cronkite & Moos, 1978, Moos, 1976) have studied extensively the impact of social climate and environmental settings on human behavior. They also note the interaction between environment and coping capacity. Both the physical environment and the social climate in home, work, religious, and other settings, can have a critical influence on lifestyle change efforts. At the individual level, the personal environment (e.g., an office at work, the kitchen at home) can be examined and changed to be more supportive of positive health practices. But groups and organizations as a whole can be convinced to address issues such as an overall "health policy" for their worksites, or religious or social organizations (e.g., no smoking areas, increase the availability of healthy foods in vending machines). Social skills technology (see group section) can be used to help modify the organizational social climate so that negative health practices are assertively handled and positive practices are reinforced by members of the organization. Attention to social climate and environ-mental settings can further disseminate health and optimize interventions.

One recent study used several organizational level techniques for lifestyle change with promising results (Abrams & Follick, 1983). They used organizational goal setting, public displays of group achievements, the organization's newsletter for reinforcement and feedback, as well as inter-group competition, while maintaining within-group cohesion, social network supports, and general problem-solving training to enhance generalization and maintenance of lifestyle change. The feasibility of the program and main-tenance of weight loss at six-month follow-up were demonstrated.

Self-control and OBM principles can also be extended between different organizations. For example, several organizations (e.g., industries, churches, or schools) could be encouraged to pool resources, share costs, and sponsor joint programs. They can also use cooperation (working together within their organization) and competition between organizations on a friendly basis to create additional incentives for change and maintenance. Comparing total organizational progress on a variety of aggregate indices could be achieved by using data from organizational monitoring procedures (e.g., percentage of the total membership joining new programs per month, the total number of kilograms of weight lost per month). Such competition could be publicized in local organizational newsletters and in local community newspapers just like sports teams compete in baseball or football. Each organization would follow the principles of *organizational self-control* and would monitor its own "behavior," set goals, evaluate, and refine progress to "win" the competition. A "grand prize" could be given at a large community health fair at the end of a year. If successful, the fair would itself become an annual interorganiz-ational event.

In an early report of an incentive program for stopping smoking, Rosen and Lichtenstein (1977) conducted an uncontrolled study using a financial incentive not to smoke. All employees, regardless of smoking status at pretest, received a $5 per month bonus for not smoking at work. With no other program or procedures, 58 percent of employees stopped smoking at work with 33 percent still not smoking after one-year follow-up. Stachnick and Stoffelmayr (1983) use a seven-month worksite program, with contracting, social support, 20 group meetings, and various large financial incentives including a lottery and intergroup competitions. Results were very encouraging, with more than 50 percent of all the smokers enrolling in the program and from 80 percent to 91 percent reporting abstinence at six months. Unfortunately, no biochemical verification of smoking status was used. The organization-wide incentives and the competition cannot be separated from the other procedures in terms of their independent contribution to outcome. Nevertheless, these ideas are worthy of replication and controlled investigation since they are relatively inexpensive, easy to implement, and require little professional training.

Recently Brownell, Stunkard and their colleagues reported a program using competition among three different banks in the local community (Brownell, Stunkard, McKeon, & Felix, 1983). This extends to a community level the intergroup competition principles first reported by Abrams & Follick (1983). It is noteworthy that neither Abrams and Follick (1983) nor Brownell et al. (1983) used elaborate programs, yet they have achieved encouraging results. Perhaps this indicates the potentially powerful forces operating when one taps into individual, group, and organizational psychological principles simultaneously. It is our opinion that such multi-level programs signal a major change in the field of behavioral medicine and health psychology. Especially significant are the implications of multi-level strategies for enhancing generalization and maintenance.

OPTIMIZING THE INTERACTIONS BETWEEN INDIVIDUAL AND ORGANIZATIONAL PRINCIPLES

Diffusion of the Change Process

Disseminating new patterns of health behavior between and within levels of social structure is termed *diffusion of innovation*. Basically, diffusion involves generalization across behaviors, people, and settings. The concept of diffusion is derived from the seminal works of Rogers and Shoemaker (1971) and Bandura (1977, pp. 50–55). The notion of using early adopters within an organization (i.e., people high on readiness to change) as role models and resources to encourage others to change is a key concept in this model, and the "point of departure" of the total organizational lifestyle change effort. With early adopters, new behaviors start, and if robust, they survive, diffuse to others, and endure to be integrated into the social climate—namely, the culture of the organization. It is a type of social Darwinism in which new innovations are "selected" by needs or problems. Then they either fail to blossom—that is, they become passing fads or fashions (Bandura, 1977)—or they endure on a more permanent basis because of increasing and wide-spread acceptance as viable solutions to the need or problem. In the case of adoption of new innovations, reciprocal support for lifestyle change from an increasingly larger number of individuals occurs between and within levels of social structure.

Early adopters become role models and educators to encourage others to change. This process is facilitated by the following: (1) skillful use of promotional campaigns using macro-level principles (e.g., media that highlights actual behaviors of role models), (2) social skills training and social networking (see group section), and (3) organization-wide incentives (see organizational section). Diffusion occurs along social network lines to those who have a high frequency of contact with the early adopters. Some individuals may react negatively to change, thus inhibiting diffusion.

In one direction, diffusion spreads from micro to macro levels—that is, individual to organization. Early adopter individuals link together into groups, groups link together within and between organizations, and these more healthy members of organizations in turn join together creating a growing social climate of change. In the other direction (macro to micro), diffusion is encouraged using well-timed organizational media campaigns, group level incentives (e.g., competition), and so on down to the individual levels (recall Figure 2.2). The entire process of change from micro to macro levels and vice versa can be termed the *snowball* effect of diffusion. A word of caution here: snowball effects are not always in the desirable direction; one has to look only at the rise and fall of the Third Reich as an example. In essence, the concept of diffusion, from micro to macro levels and vice versa, is an extension of reciprocal determinism from a theory of individual change to one of organizational change.

To capitalize on the principle of reciprocal determinism, it is also necessary to attempt to induce *simultaneous* changes in individuals and in their most immediate, relevant, and frequently encountered social networks and physical environments. A healthy individual, therefore, both contributes to the health supportiveness of his or her social network (e.g., family, friends, work associates) *and* is more likely to remain healthy if others in his or her social network are also changing in the same direction at the same point in time. Thus maintenance of lifestyle change is automatically enhanced if members of the family and people at the worksite (and other environments where the individual spends most of his or her time) are themselves changing and reciprocally supporting new patterns of health. Generalization and maintenance of change are thus directly addressed in this model in a different way from the individual SLT psychology perspective. Different levels of social structure can play unique roles as resources for facilitating social change. Individuals are reciprocally supported between and within levels. Thus a self-maintaining, reciprocal support network is gradually developed and strengthened for maintenance of more healthy lifestyles. This is achieved by planning multi-level interventions in organizations, groups, or individuals that naturally interact with one another in reciprocal fashion. The problem of maintaining a changed lifestyle is addressed via arranging a supportive proximal environment along with individual change.

Proximal and Distal Social Influence

A key concept for optimizing change at different levels of social structure is the distinction between proximal and distal sources of influence. Proximal and distal are defined as the two polar opposites of a continuum, which varies according to the degree of reciprocal social control exerted relative to the individual. Both the social and inanimate environment can exert varying degrees of social influence on a particular individual. The degree of influence is also a function of the social network structure relating the individuals, groups, organizations, and other sectors of the population. The more intense, intimate, or similar that individuals, groups, or organizations are to each other, and the more frequently they interact, the more proximal the social influence. The hypothesis is that for a particular target level the more proximal sources of interaction there are (with individuals who are making lifestyle changes) the more likely it is that other individuals will be persuaded to change or be able to maintain change. For example, for an individual, a proximal source of social support might be his or her family or a group of close friends or work colleagues. A distal source may be business associates in another organization.

Identifying proximal sources of social influence, relative to a specific target for change, can enhance promotion, facilitate skill training, and greatly improve generalization, maintenance, *and* diffusion from early to late adopters. To illustrate how powerful multi-level interventions in proximal social networks can be, the image of a magnifying glass concentrating the sun's rays comes to mind. Consider a hypothetical health promotion project: an individual works in a local factory with a group of fellow workers, some of whom are beginning to practice healthy behaviors. In addition, this individual's spouse is beginning to behave in a more healthy manner because of programs based at a local church where the spouse does volunteer work. One of this person's close friends is obtaining considerable support for his quit smoking program and has excellent health-related social skills. Assume also that this individual's school children are participating in a school smoking prevention and fitness program that explicitly encourages adult participation. This hypothetical person, because of proximal social influences, would be experiencing strong environmental pressures for healthy lifestyle behaviors from several proximal sources simultaneously. Thus an important optimizing strategy for health promotion and maintenance is to *impact on as many proximal sources as possible and simultaneously*. Clearly, the worksite lends itself to this kind of multilevel strategy.

Although achieving simultaneous change in several proximal sources is difficult to engineer in practice, the perception of a large number of people, groups, and organizations all changing and talking about it enhances maintenance for those who have changed. It would also have a strong impact on late adopter individuals who have not yet changed. They would begin to feel excluded from a major change in their entire social system. This is, in fact, happening to smokers in the USA today; more often than not, they are seen as the "out group." We referred to this phenomenon as a major *structural* change in the population in the introduction to this chapter, where organizational structural change was identified as the ultimate goal of a health promotion effort (Leventhal et al., 1980).

Structural Change and Critical Mass

According to Bandura (1977), the reciprocal interaction between a target level (in this case the individual and his or her proximal group and organizational environment) forms a potentially powerful self-maintaining system in which the individual (at work, at home, with friends) is both a recipient and an agent of a new fixed interaction pattern (healthy lifestyle habits). From an individual's perspective, the expected patterns of reciprocal reinforcement (or the rules that govern their behavior) are changing in the majority of people in the proximal social network. The rules of interaction then are seen as different from the old rules that used to govern the health behaviors of these social networks (in an unsupportive or negative environment). When the majority of individuals *expect* others not to smoke, to serve fruit instead of high calorie desserts to visitors at dinner parties, and so forth, then a powerful structural change has occurred.

The goal of an organizational health program is to achieve what is called a *critical mass* in order to obtain a structural change in the entire organization. A critical mass is defined as that state of an organization in which the majority of members are reciprocally supporting new health practices and extinguishing negative practices. In a sense, a new, stable "subculture" emerges, in which expectations of normal conduct involve positive lifestyle practices in a majority of the members.

After a critical mass is achieved, then maintenance of change in the face of a negative host environment is no longer a difficult issue because the social networks to which the individuals are tied are themselves becoming a healthy host environment. It would be expected practice that a couple would serve fresh fruit after dinner, that the worksite would replace coffee and donuts with orange juice and fresh fruit, and the local church would have a "health food sale" instead of a "cake sale." The diffusion process across individual, group, and organizational social network levels and the critical mass concept provides a potential solution to the longstanding problems of generalization and maintenance of change.

SUMMARY AND CONCLUSION

Optimal organizational health change is obtained by focusing simultaneously on three target levels of social structure (individual, group, organization) and three phases of change (promotion, skills training, generalization/maintenance). Promotional stimuli were used differently from those used in traditional health education campaigns. A distinction was made between internal (dynamic feedback of actual changes being made by members of the organization) versus external (static, "canned") promotional campaigns. The interaction between promotional variables and individual differences in motivation to change was discussed. Individual self-control theory was extended to apply to all three levels of social structure. Self-efficacy, social skills training, modeling, and social networks were key concepts used as building blocks in the overall model.

Principles for optimizing change within each level and between levels were then delineated using additional concepts from various areas of social science, especially using group principles and organizational level concepts. Techniques derived from social psychology (e.g., group process, cooperation-competition) were examined. Organizational Behavior Modification added significant qualitative refinements to the model. Especially important was the notion of an "organizational level" self-control cycle. Finally, diffusion theory and reciprocal determinism were integrated between phases and across levels. Diffusion was facilitated by an analysis of proximal and distal sources of social influence, social network principles, and multilevel interventions. A diffusion process leading to a critical mass and to the ultimate goal of organizational structural change was described.

When self-control is operationally defined and applied to various levels of social complexity, some new ways of looking at old problems emerge. Of particular significance is the way an integration between and within levels yields a model that directly addresses the challenging issues of the following: (1) dissemination to unmotivated individuals, (2) achievement of generalization, (3) maintenance of change, and (4) self-correction and refinement of ongoing programs. Bridging the micro-macro gap between individual, group, and organizational principles provides significantly broader options for lifestyle change. The present model can be extended to other levels of social complexity, such as whole communities (Abrams, Elder, Lasater, & Carleton, 1982), and to other problems in which primary prevention can make an impact. Psychological principles, derived from Social Learning Theory and related disciplines, can make a significant contribution to primary prevention efforts within organizational settings. However, what is needed is an integrated and systematic blueprint for optimal change rather than the isolated use of one of the techniques.

Health psychology is standing on the threshold of some major advances in theory and application. The integration of individual, group, and organizational level behavioral principles extends the traditional behavioral programs developed in clinical settings during the last 20 years. The new settings and the challenges of making a large-scale, cost-effective, public health impact are forcing behavioral scientists in new directions. We cannot assume that the traditional protocols developed by behavior therapy 20 years ago can be simply transferred to worksite settings. It is clear that a wholesale conceptual shift is required and that research must focus on evaluating setting, population, program, process, and outcome variables within this new context.

REFERENCES

Abrams, D. B., Elder, J., Lasater, T., & Carleton, R. (1982, November). *Social learning principles for community health promotion: An integration across levels.* Paper presented at the annual convention, Association for Advancement of Behavior Therapy, Los Angeles, CA.

Abrams, D. B., & Follick, M. J. (1983). Behavioral weight loss intervention at the worksite: Feasibility and maintenance. *Journal of Consulting and Clinical Psychology, 51,* 226–233.

Abrams, D. B., & Wilson, G. T. (1979). Self-monitoring and reactivity in the modification of cigarette smoking. *Journal of Consulting and Clinical Psychology, 47,* 243–251.

Andraskik, F., Heimberg, J., & MacNamara, J. (1981). Behavior modification of work and work-related problems. In M. Hersen, R. Eisler, & P. Miller (Eds.), *Progress in behavior modification.* New York: Academic Press.

Bandura, A. (1969). *Principles of behavior modification.* New York: Holt, Rinehart & Winston.

Bandura, A. (1977). *Social learning theory.* Englewood Cliffs, NJ: Prentice-Hall.

Bandura, A. (1977). Self-efficacy: Toward a unifying theory of behavioral change. *Psychological Review, 84,* 191–215.

Berkman, L. F., & Syme, L. (1979). Social networks, host resistance and mortality: A nine year follow-up of Alameda County residents. *American Journal of Epidemiology, 109,* 186–204.

Best, J. A., & Hakstian, A. R. (1978). A situation specific model for smoking behaviors. *Addictive Behaviors, 3,* 79–82.

Billings, A., & Moos, R. (1981). The role of coping responses and social resources in attenuating the stress of life events. *Journal of Behavioral Medicine, 4,* 139–154.

Bowles, M. D., Malott, R. W., & Dean, M. R. (1982). The evaluation of an incentive program used to recruit health club memberships through friend referrals. *Journal of Organization Behavior Management, 3,* 65–72.

Brown, E. R., & Margo, G. E. (1978). Health education: Can the reformers be reformed? *International Journal of Health Services, 8,* 3–26.

Brownell, K., Heckerman, C., Westlake, R., Hayes, S., & Monti, P. (1978). The effect of couples training and partner co-operativeness in the behavioral treatment of obesity. *Behavior Research and Therapy, 16,* 323–333.

Brownell, K., Stunkard, M., McKeon, P., & Felix, M. (1983). *Cooperation competition: Four studies in department stores and banks.* Paper presented at Annual Convention, American Psychological Association, Anaheim, CA.

Carleton, R. A., (1980). *The Pawtucket Heart Health Program.* National Heart, Lung and Blood Institute. (Grant # R01 HL 23629011A awarded July, 1980).

Ciminero, A. R., Nelson, R. O., & Lipinski, R. P. (1977). Self-monitoring procedures. In A. R. Ciminero, K. S. Calhoun, & H. E. Adams (Eds.), *Handbook of behavioral assessment.* New York: Wiley.

Cobb, S. (1976). Social support as a moderator of life stress. *Psychosomatic Medicine, 38,* 300–314.

Collins, R. L., Komoki, J., & Temlock, S. (1979, September). *Behavioral definition and improvement of customer service in retail merchandising.* Paper presented at the annual meeting of the American Psychological Association, New York.

Craighead, L. W., Stunkard, A. J., & O'Brien, R. (1981). Behavior therapy and pharmacotherapy for obesity. *Archives of General Psychiatry, 38,* 763–766.

Cronkite, R. C., & Moos, R. H. (1978). Evaluating alcoholism treatment programs: An integrated approach. *Journal of Consulting and Clinical Psychology, 46,* 1105–1119.

Croog, S., & Levine, S. (1972). Religious identity and response to serious illness: A report on heart patients. *Social Science and Medicine, 7,* 17–32.

Curran, J., & Monti, P., (Eds.). (1982). *Social skills training: A practical handbook for assessment and treatment.* New York: Guilford.

DeClemente, C. C., & Prochaska, J. O., (1982). Self-change and therapy change of smoking behaviors: A comparison of processes of change in cessation and maintenance. *Addictive Behaviors, 7,* 133–142.

DeLeon, P. H., & Pallak, M. S. (1982). Public health and psychology. *American Psychologist, 37,* 934–935.

Dressler, W. W. (1980). Ethno-medical beliefs and patient adherence to a treatment regimen: A St. Lucian Example. *Human Organizations, 39,* 88–91.

D'Zurilla, T., & Nezu, A. (1982). Social problem solving in adults. In P. C. Kendall (Ed.), *Advances in cognitive-behavioral research and therapy* (Vol.1). New York: Academic Press.

Follick, M., Abrams, D., Smith, T., Henderson, O., & Herbert, P. (in press). Behavioral intervention for weight loss: Acute versus long-term effects on HDL and LDL cholesterol levels. *Archives of Internal Medicine,* pp. 1571–1574.

Foreyt, J. P., Scott, L. W., & Giotto, A. M. (1980). Weight control and nutrition educational programs in occupational settings. *Public Health Reports, 95,* 127–136.

Franks, C. M., Wilson, G. T., Kendall, P. C., & Brownell, K. D. (1982) *Annual review of behavior therapy: Theory and practice* (Vol.8). New York: Guilford.

Glasgow, R. E., & Rosen, G. M. (1978). Behavioral bibliography: A review of self-help behavior therapy manuals. *Psychological Bulletin, 85*, 1–23.

Goldfried, M. R., & Robins, C. (1982). On the facilitation of self-efficacy. *Cognitive Therapy and Research, 6*, 361–379.

Goldfried, M. R., & D'Zurilla, T. (1969). A behavior analytic model for assessing competence. In L. Spielberger (Ed.), *Current topics in clinical and community psychology*, New York: Holt, Rinehart & Winston.

Gough, H. G. (1965). A validational study of the chapin social insight test. *Psychological Reports, 17*, 355–368.

Green, L. M., Kreuter, M. W., Reeds, S. G., & Partridge, K. B. (1980). *Health education planning: A diagnostic approach*. Palo Alto, CA: Mayfield.

Harwood, A. (1971). The hot-cold theory of disease. *Journal of the American Medical Association, 7.*

Hessler, R. W., Nolan, M. F., Ogbru, B., & New, P. K. M. (1975). Intra-ethnic diversity: Health care of the chronic. *American Journal of Human Organization, 34*, 253–258.

Hunt, W., & Bespalec, D. (1974). An evaluation of current methods of modifying smoking behavior. *Journal of Clinical Psychology, 30*, 432–438.

Ivey, A. E. (1978). *Microcounseling innovations, interviewing counseling, psychotherapy and psycho-education* (2nd Ed.). Springfield, IL: Thomas.

Kanfer, F. H. (1975). Self management methods. In F. H. Kanfer & A. P. Goldstein (Eds.), *Helping people change*, New York: Pergamon.

Kanfer, F. H., & Busemeyer, J. R. (1982). The use of problem solving and decision-making in behavior therapy. *Clinical Psychology Review, 2*, 239–266.

Kreitner, R., Reif, W. E., & Morris, M. (1977). Measuring the impact of feedback on the performance of mental health technicians. *Journal of Behavior Management, 1*, 105–109.

Leventhal, H., Safer, M., Cleary, P., & Gutmann, M. (1980). Cardiovascular risk modification by community based programs for lifestyle change: Comments on the Stanford Study. *Journal of Consulting and Clinical Psychology, 48*, 150–158.

Leventhal, H. J., & Cleary, P. D. (1980). The smoking problem: A review of the research and theory in behavioral risk modification. *Psychological Bulletin, 88*, 370–405.

Loro, A.D., Fisher, E. B., & Levenkron, J. E. (1979). Comparison of established and innovative weight reduction treatment procedures. *Journal of Applied Behavior Analysis, 12*, 141–155.

Maccoby, N., Farquhar, J., Wood, D., & Alexander, J. (1979). Reducing the risk of cardiovascular disease: Effects of a community-based campaign on knowledge and behavior. *Journal of Community Health, 3*, 100–114.

Marlatt, G. A., & Gordon, J. R. (1978). *Determinants of relapse: Implications for the maintenance of behavior change.* Paper presented at the 10th International Conference on Behavior Modification, Banff, Canada.

Matarazzo, J. D. (1980). Behavioral health and behavioral medicine: Frontiers for a new health psychology. *American Psychologist, 35*, 807–817.

Matarazzo, J. D. (1982). Behavioral health's challenge to academic, scientific, and professional psychology. *American Psychologist, 37*, 1–14.

McGraw, S., & Abrams, D. (1983, March). *Program evaluation in community and organizational health promotion.* Paper presented at the annual convention, Society for Behavioral Medicine, Baltimore, MD.

McKinlay, J. (1980). Social network influences on morbid episodes and the career of help seeking. In L. Eisenbert & A. Kleinman (Eds.), *The relevance of social science for medicine.* Boston: Reidel.

Moos, R. H. (1976). *The human context: Environmental determinants of behavior*, New York: Wiley.

Nelson, R. O. (1977). Methodological issues in assessment, via self-monitoring. In J. D. Cone &

R. F. Hawkins (Eds.), *Behavioral Assessment*. New York: Brunner/Mazel.

Owen, N. (1983). Behavioral medicine: Prospects and limits. In J. L. Sheppard (Ed.), *Advances in Behavioral Medicine* (Vol.3), Sydney, Australia: Cumberland College of Health Sciences.

Peterson, G., Abrams, D., Elder, J., & Beaudin, P. (in press). Professional versus self-help weight loss at the worksite: The challenge of making a public health impact. *Behavior Therapy*.

Pomerleau, O. F. (1979). Behavioral medicine: The contribution of the experimental analysis of behavior to medical care. *American Psychologist, 34,* 654–663.

Prochaska J. O. (1979). *Systems of psychotherapy; A transtheoretical analysis.* Homewood, IL: The Dorsey Press.

Rogers, E. M., & Shoemaker, F. F. (1971). *Communication of innovations: A cross-cultural approach.* New York: Free Press.

Rogers, T., Killen, J., Cameron, R., & Weinberger, E. (1982). *Community-based correspondence courses for weight control.* Symposium presented at the 16th Annual Convention of the Association for Advancement of Behavior Therapy, Los Angeles, CA.

Rosen, G., & Lichtenstein, E. (1977). An employee incentive program to reduce cigarette smoking. *Journal of Consulting and Clinical Psychology, 45,* 957.

Rosenstock, L. M. (1966). Why people use health services. *Milbank Memorial Fund Quarterly, 44,* 94–127.

Rosenstock, L. M. (1974). Gaps and potentials in health education research. *Health Education Monographs, 8,* 21–27.

Schachter, S. (1982). Recidivism and self-cure in smoking and obesity. *American Psychologist, 37,* 436–444.

Skinner, B. F. (1953). *Science and human behavior.* New York: Macmillan.

Stachnick, T., & Stoffelmayr, B. (1983). Worksite smoking cessation programs: A potential for national impact. *American Journal of Public Health, 73,* 1395–1396.

Stunkard, A. (1976, December). *Obesity and the social environment: Current status and future prospect.* Bicentennial conference on food and nutrition, Philadelphia, PA.

Tanabe, G. (1982). The potential for public health psychology. *American Psychologist, 8,* 942–944.

Weiss, S. T. (1982). Health psychology: The time is now. *Health Psychology, 1,* 81–91.

Wilbur, C. S. (1983). *Live for Life at Johnson and Johnson.* Symposium presented at the Annual Convention Society for Behavioral Medicine, Baltimore, MD.

Wilson, G. T. (1980). Cognitive factors in lifestyle change: A social learning perspective. In P. R., Davidson & S. M., Davidson (Eds.), *Behavioral medicine: Changing health lifestyles.* New York: Brunner/Mazel.

Wolpe, T. (1969). *The practice of behavior therapy.* Oxford: Pergamon.

Zitter, R. E., & Fremouw, W. (1978). Individual versus partner consequence for weight loss. *Behavior Therapy, 9,* 808–813.

3

Stress and Disease in the Workplace: A Methodological Commentary on the Accumulated Evidence

STANISLAV V. KASL

It is my intention to provide an overview of research activities included within the broad designation, "stress at work." My basic aim is to arrive at cause-and-effect interpretations, that is, statements linking adverse health outcomes to specific environmental conditions at work. Since I do not intend to deal with purely physical and chemical hazards in the workplace, I will have to address (regrettably) the meaning of the designation "stress at work." Since much of the evidence to be considered is ambiguous regarding causal interpretations, I will have to organize my commentary around methodological issues as well as the substantive findings.

A number of reviews, overviews, and compilations of articles concerning stress at work make this research area relatively easy to access (Beehr & Newman, 1978; Cooper & Marshall, 1980; Cooper & Payne, 1978, 1980; Gardell, 1982; Holt, 1982; House, 1981; Hurrell & Colligan, 1982; Kahn, 1981; Kahn, Hein, House, Kasl, & McLean, 1982; Kasl, 1974, 1978, 1984a; Kasl & Cobb, 1983; Levi, Frankenhaeuser, & Gardell, 1982; McGrath, 1976; McLean, 1979; Moss, 1981; Sharit & Salvendy, 1982; Shostak, 1980). The volumes that discuss stress reduction and stress management (Beech, Burns, & Sheffield, 1982; Brief, Schuler, & Van Sell, 1981; Cooper, 1981; Gmelch, 1982; McLean, 1981; Matteson & Ivancevich, 1982a) also tend to provide overviews of the research evidence, but these tend to be more skimpy and less balanced, since they are driven by the need to justify the stress-reduction perspective.

The above reviews do not provide a convergent view, partly because different authors have different objectives. Some reviews are primarily descriptive and informative, explaining what work is being conducted and what independent-intervening-dependent variables are typically chosen for study (e.g., Holt, 1982). Others are quite selective, choosing to teach from a few illustrative studies generalizable principles and findings (e.g., Kahn, 1981). Still others seek to organize the evidence so that they can offer a list of work environmental conditions or dimensions that are deemed stressful (e.g., Beehr & Newman, 1978). Furthermore, high variance in the reviews is also due to the variety of ways in which methodological-design issues are handled. Fundamentally, methodological criticism of quasi-experimental field research concerns probable and possible flaws, as well as probable and possible alternative explanations. Certainty about presence of spurious effects and confounding and reverse causation may be as difficult to achieve as certainty about a desired cause-and-effect relationship; thus reviewers have considerable latitude in raising methodological issues and in dismissing particular evidence as unsound. Finally, it must be acknowledged that reviewers have their own preexisting values and beliefs and convictions, which influence the way they handle the same empirical material; for instance, the wide discrepancy in interpreting the literature on stress and ischaemic heart disease seen in a recent exchange (Boman, 1982; Tennant, 1982).

It is likely that in future reviews of the stress at work areas will have to restrict themselves to specific and cicumscribed content areas, such as machine pacing (Salvendy & Smith, 1981) or shift work (Tasto, Colligan, Skjei, & Polly, 1978), if they are to remain effective and accurate guides through the past and beacons to the future. For example, the general stress at work research evidence appears to contain many contradictory results but most of the reviews barely manage to convey that problem. This problem occurs in such reviews because it is quite difficult to keep track of the inconsistencies: to pin down the extent of contradictory evidence of the specific dimensions or hypotheses involved, and of the possible reasons for nonreplication, such as noncomparable methodologies. Consequently, reviewers shy away from the issue of consistency. But this is a disservice to the field: the probing of inconsistencies is one essential feature of evaluating evidence and moving the field forward.

Since I have contributed to this glut of reviews and overviews, I feel a particularly strong need to integrate, to build on the past work, to move beyond, to say something new. Unfortunately, such a sentiment may be virtuous but not entirely appropriate. The typical reader will not stop here to consult all the previous reviews before proceeding; thus reviewing only the most recent studies would lead to an incomplete picture for the less than fully informed reader. Furthermore, if the new evidence does not justify new conclusions or new perspectives, then "saying something new" is hardly helpful. For example, the Toronto conference on occupational stress (Jick &

Burke, 1982; Payne, Jick, & Burke, 1982) identified a number of issues worthy of our attention in the coming decade: (1) acute versus chronic effects; (2) the role of objective conditions versus subjectively appraised conditions; (3) the difference between psychological and physiological indicators of strain; (4) the role of coping; and (5) the work–family–society interface. But these issues have been with us for some time already. No matter; these issues have persisted, have remained unresolved, and it would be inappropriate merely to drop them to say something new.

It is my intention in this chapter to focus, first of all, on conceptual and methodological issues, which I believe have been persistent problems, or sources of controversy, during the last decade. Second, and to a lesser extent, I wish to discuss recent formulations or research findings, which I believe are opening up promising areas or providing new information. Third, I wish to examine the link between the stress at work research and intervention studies, stress reduction, and health promotion at the worksite.

THE PERSPECTIVE OF THIS CHAPTER

The perspective, which guides my commentary and criticisms, might be best characterized as that of *occupational epidemiology*. This perspective argues that our primary goal is to identify linkages between exposure to some aspect of the (objective) work environment and adverse (or possibly, positive) health outcomes. We should design our studies, and conduct our analyses, to optimize our chances of demonstrating that such linkages represent, in fact, cause-effect relationships. Additional effort, used to design the study (such as additional data collection or data analysis), should be in the service of the primary goal (e.g., eliminate self-selection biases) or two additional goals: (1) identify variables, which explain additional variance in the adverse health outcomes and which, thereby, indicate differential reactivity to the environmental exposure; and (2) provide information on the underlying mechanisms involved in the overall association. Obviously, the two additional goals are well intertwined: advancing one is likely to advance the other. This perspective would be equally valid if I were to review intervention studies at the worksite, rather than those dealing with stress at work.

I intend to use the above perspective as a rock-bottom basis for thinking about various conceptual and methodological issues and evaluating diverse studies. I recognize that other perspectives could be articulated and defended. I espouse this position only for the "stress and disease in the workplace" domain; it is likely to be much less suitable for the general stress and disease area, and it is certainly unsuitable for a "psychology of stress" orientation.

STRUGGLES CONCERNING DEFINITIONS OF STRESS

The term *stress* has been used several fundamentally different ways:

1. As an *environmental* condition, susceptible to *objective* definition and measurement. Clearly, this meaning of the term stress is most relevant to the occupational epidemiology perspective enunciated above; many investigators might be inclined to use the term "stressor" in this context (Elliott & Eisdorfer, 1982).

2. As a *subjective perception* or *appraisal* of an objective environmental condition. This use of the term is, above all, characteristic of studies in which the putative independent variable is some subjective description of the work setting; the notion that "stress is subjective" may be implied or explicitly affirmed. However, we need to distinguish between studies in which the subjective measure is the sole independent variable from those in which it is paired with some objectively established environmental condition, which is theorized to be the stimulus for the perception. Later, I will argue that much mischief has been created by the use of subjective formulations and subjective measures; admittedly, the amount and nature of the mischief does depend on whether the design includes the assessment of the objective environmental condition as well.

3. As a *response* or *reaction*. This includes a variety of both proximal and distal outcomes, including changes in affective states, biological parameters (e.g., neuroendocrine levels), as well as incidence of specific disease states. Again, two versions of this use of the term stress may be recognized. One is that certain states or diseases (e.g., elevated cortisol levels, demoralization and nonspecific distress, ulcers) are seen as stress reactions or stress diseases and their mere presence is a sufficient reason for invoking the term stress, even though the antecedents of this state or disease have not been established or even examined. A good illustration of this use of the term is seen in the procedure of inviting workers into a "stress" reduction program solely on the basis of elevated scores on some symptom checklist. The second version uses the term stress under the same conditions as the first, except the response or reaction is linked to specific antecedents. Here, the worksite "stress" reduction program would be offered to workers with similarly elevated symptoms but only to those who work in certain demanding jobs and have described their jobs in certain ways.

4. As a *complex relational* term linking environmental characteristics and personal characteristics, or more specifically, as an excess of environmental demands beyond the individual's capacity to meet them. This is potentially an attractive use of the term stress since, at its best (when both components of the relation are measured objectively), it is fully responsive to the above proposed occupational epidemiology perspective: study an environmental condition and add to it variables, which will account for differential reactivity. However, when the measures involved are subjective ones, then we are simply once again confronting the potential mischief of such an approach (alluded to previously under the second use of the term stress and discussed later), and maybe increasing such mischief because two subjective measures are paired together, as in the Person-Environment Fit formulation (French, Caplan, & Van Harrison, 1982).

It should be recognized that in addition to disagreement on general use of the term stress—as a stimulus, as a response, and as some relation between the two—the field also lacks consensus on defining criteria within each of these usages. For example, the unique defining characteristics of those environmental conditions to be thought of as "stressors" in the workplace have never been agreed on; that is, different authors have proposed different lists of specific work dimensions, which they nominated for inclusion (Kasl, 1978). Similarly, when stress is used as a response term, we again have no consensus on the unique set of responses that would, singly or collectively, represent criteria for presence of stress. Baum, Grunberg, and Singer, (1982) recommend the use of multiple measures: self-reports, behavioral-performance, psychophysiological, and neuroendocrine. They do this primarily on methodological grounds, but the conceptual grounds are equally compelling. Of course, the list of potential indicators of stress response remains enormous and the concept is thereby hardly elucidated or narrowed down.

The use of stress as a relational term has considerable endorsement in the stress at work research domain (Holroyd & Lazarus, 1982; *Journal of Human Stress*, 1975; Kasl, 1978; McGrath, 1970; Shirom, 1982). The concept of excess environmental demands on a person's resources has intuitive appeal. However, since none of the authors using this concept has attempted to provide additional defining criteria or additional precision (what demands, what resources, how much excess), or has tried to supply guidelines for operationalization, one has to infer that this is about as far as we can go with the construct. Tinkering with the concept of stress remains irresistible and an occupational hazard. For example, Schuler (1982) defines stress as "perceived dynamic state about something important." But such a novel offering to the stress field strikes one as bizarre: after 25 years of myriad definitions, one must ask just what the gain in a new definition would be.

I believe that the fundamental dilemma is that in the vernacular the central notion of stress is embodied by the saying "one person's meat is another person's poison." This makes stress completely idiosyncratic and subjectivistic; it is also circular in the sense that something can't be stressful unless it has bad consequences. The problem for scientific theory construction is that we have not yet figured out how to take this vernacular concept and refine it so that it becomes useful in our research. Instead, we have become addicted to its high marquee value and its rich surplus meaning. In our research we need to avoid circularity and to define stress cleanly either as a stimulus condition (so that we can study its consequences) or as a response (so that we can study its antecedents). But such "cleaner" conceptualizations leave us dissatisfied because we feel that something has been left out: on the stimulus side, in the process itself, or on the response side. Hence the vacillation, lack of clarity, absence of consensus.

THE STRESS PROCESS: FORMULATIONS THAT CAUSE PROBLEMS

In this section I would like to argue that the currently dominant theoretical formulations regarding stress and coping have not adequately served investigators in occupational epidemiology, those concerned with stress and disease in the workplace. This problem is not necessarily caused by formulations that are flawed. Rather, they are unsuitable models for decisions about research design and methodology and, to a lesser extent, for formulations of hypotheses and research questions. The dominant theoretical positions I am talking about are as follows: (1) the general theory of stress-coping-adaptation of Lazarus (e.g., Holroyd and Lazarus, 1982; Lazarus, 1978; Lazarus and DeLongis, 1983); and (2) The Institute for Social Research (ISR) model, developed specifically for the work setting (e.g., Caplan, Cobb, French, Harrison, & Pinneau, 1975; House, 1981; Kahn, 1981).

In an earlier review (Kasl, 1978), I argued that the convergence of theoretical opinion on the need for a subjective and idiographic (ipsative) approach to stress had led to "a self-serving methodological trap which has tended to trivialize a good deal of research on work stress or role stress: The measurement of the "independent" variable . . . and the measurement of the "dependent" variable . . . are sometimes so close operationally that they appear to be simply two similar measures of a single concept" (p.13). In this chapter I wish to offer a metastasized and updated version of this criticism. It should be noted that the recent criticisms of the stressful life events methodology as a way to measure stress, the putative independent variable in stress-disease research (e.g., Dohrenwend & Dowrenwend, 1981; Kasl, 1983, 1984b; Schroeder & Costa, 1984), represent developments of parallel methodological concerns that are equally applicable to the issue at hand.

Let me begin the argument with fictitious case histories of two 25-year-old men who lost their blue-collar jobs because of a permanent plant shutdown. Both found difficulties in finding a new job. The first man appraised his situation as a belated opportunity to return to school. He talked to his wife and found her supportive of the idea. He borrowed money, enrolled in a community college program, and obtained basic training in computers. He then found a new job with a company that found his new skills useful and was willing to train him further. Two years after the plant closing, this man had a strong sense of mastery, high self-esteem, and felt supported by his wife. The second man became depressed by his inability to find a new job. He appraised his situation as one where his previous skills were no longer useful and his self-esteem plummeted. He still went through the motions of looking for a job but his discouragement grew. He did not find his wife supportive, he was irritable, and he coped by withdrawing and by drinking. Two years after the plant closing, he was unemployed, depressed, drinking heavily, and had little to do with his wife or friends.

The basic questions in my argument are as follows: What is the appropriate unit of observation, the molar unit of behavior, that should guide us to define the outcome(s)? How many research questions, how many hypotheses do we have here? My argument is that only a single outcome variable exists and that the various steps in the process are intimately linked to each other, since they are stages in one single evolving process. The primary research questions are as follows: How many become like the first man and how many like the second? (Of course, this can be rephrased to imply a continuous outcome variable.) And why the differential outcome? Unfortunately, the dominant theoretical positions have permitted (perhaps encouraged) investigators to ignore these primary research questions. Instead, the investigators divide the molar unit of behavior into trivially small components and begin to study the relationships between these components. How is appraisal related to coping? What kind of coping leads to what kind of secondary reappraisal? How is coping related to sense of mastery? How is sense of mastery related to self-esteem? How is social support related to depression?

I suspect that I have overstated the case to make the point. The process of stress-coping-adaptation is perhaps not as homogeneous and the linkages among the components and stages are not as uninteresting as I suggested (see, for example, the results reported by Pearlin, Lieberman, Menaghan, & Mullan, (1981) in their examination of the consequences of various disruptions in the work role). Furthermore, the point is primarily applicable when the components and stages are all intrapsychic processes. When there is a new step involving new person-environment interaction (such as in the fictitious case history above, when the new job unexpectedly terminates), then our molar behavior becomes unsuitably large. Nevertheless, the main point is that—from the occupational epidemiology perspective outlined earlier—the study of relationships among closely linked or overlapping intrapsychic steps in a single evolving outcome is a secondary priority.

The role of theoretical formulations in directing research questions and study designs should not be underestimated. For example, Lazarus has argued (e.g., Holroyd & Lazarus, 1982; Lazarus, 1978; Lazarus & DeLongis, 1983) that we must not view coping as a static trait—in fact, that we should deemphasize trait and moderator variables in general—and that we must adopt an ipsative approach in which intraperson variability across situations becomes our focus. The consequences of this formulation for occupational stress research may be quite unhealthy. One of our most powerful research design solutions to causal ambiguities is the prospective assessment of risk factors and moderators. However, if coping is time-event-place specific, we are forced to measure it only after the fact. Such retrospective accounts of coping will be quite suspect and the (reverse) influence of outcomes on such accounts could be considerable. Similarly, the ipsative strategy of intra-person variability may be too oblique a way to attempt achieving our goal, namely, understanding normative differences in health status outcomes. The danger that such ipsative information will never link up with normative health outcomes is considerable (e.g., Rose, Hurst, & Herd, 1979; Rose, Jenkins, Hurst, Herd, & Hall, 1982; Rose, Jenkins, Hurst, Kreger, Barrett, & Hall, 1982).

The biggest single influence of the dominant stress formulation has been to simultaneously encourage and justify subjective measurements of environmental exposure, the use of subjective perceptions or appraisals of an objective environmental condition. In the next section I wish to elaborate on some of the problems this influence has caused in trying to interpret the accumulating evidence in occupational stress epidemiology.

SOME PERSISTENT PROBLEMS WITH SUBJECTIVE MEASURES OF ENVIRONMENTAL EXPOSURE

In terms of the perspective of this chapter, such subjective measures are usefully introduced into a design if (1) they add explanatory variance to the overall association between objective environmental exposure and disease outcome, and/or (2) they point to underlying mechanisms involved. However, these measures have limitations and, depending on the specifics of the rest of the design, they can be considerable: cross-sectional or retrospective designs, using only a subjective measure of exposure and depending on self-report to measure the outcome (e.g., symptom checklist), combine design characteristics that enhance the weaknesses of the subjective measures of exposure.

It is possible to identify several different types of confounding, which represent serious alternative explanations to the one interpretation of interest to the investigator, namely, (putative) effects of exposure on outcome.

1. Overlap in content: both the measure of exposure and of outcome reflect the same broad concept, same broad area of content.
2. Influence of a personal characteristic or trait: this trait influences both the subjective perceptions and the outcome, thereby creating a spruious association between the latter two.
3. Influence of a shared response set: both the measures of exposure and of outcome are subject to stable response tendencies (particularly

social desirability, tendency to complain), which create an association between the two.

4. The presence of the outcome (e.g., a painful and/or serious health condition) alters the perception and/or the reporting of the exposure; this may be a direct effect of the disease or an indirect effect via the initiation of a cognitive process of attribution and reattribution.

Several comments should be made about this list. First, the list covers potential sources of confounding, but the actual presence of such confounding in a particular study may be difficult to pin down securely. Second, the types listed have quite imprecise boundaries and separating one from another may not be easy. For example, the association between descriptions of one's job as "hectic and demanding" and the reporting of non-specific symptoms of distress may be interpreted as an effect of a trait on both measures (neuroticism, maladjustment, low ego strength, etc.) or as an effect of a response tendency (complaining versus denial) that both measures have in common. Finally, it cannot be argued that objective measures of exposure are automatically immune from these sources of confounding; rather, it can be argued that subjective measures are more often and more extensively vulnerable to such confounding. When objective measures of exposure are used (e.g., holding a specific job or working with a specific machinery), the confounding possibilities involve primarily self-selection, that is, those variables that influence both the assignment to exposure and the outcome variable.

It is also worth noting that prospective study designs—a longitudinal study of an initially healthy cohort on whom exposure is assessed and who are then followed for the development of a target condition—do not remedy three of the four possible problems. Only the fourth, dealing with the effects of disease (reverse causation), is remedied by the prospective design. This is worth remembering, since prospective and longitudinal designs have been touted as the panacea for our problems with causal interpretations.

A few examples from the current research literature can illustrate more concretely some of these methodological concerns. In a cross-sectional study of stress among medical technologists (Matteson & Ivancevich, 1982b), the authors used a four-item measure of overall work stress (e.g., "There is a great deal of strain involved in working in this organization.") in order to divide their subjects into two groups. They then proceeded to show that subjects high on this general stress measure also report more nonwork stress, health complaints, psychiatric symptoms, and so on. It is difficult to become excited by such analyses and such findings; certainly we can be dealing with overlap in content, influence of a general trait (neuroticism-distress), a shared response set, and influence of outcomes on perceptions. Some small amount of interest might exist in seeing whether this occupational group reports more work stress, but the authors specifically reject the possibility of making such across-job comparisons.

Anytime we wish to increase our association between measures of exposure and measures of outcome, all we have to do is take some items from the latter and include them with the former. For example, the Lazarus team has reported (DeLongis, Coyne, Dakof, Folkman, & Lazarus, 1982; Kanner, Coyne, Schaefer, & Lazarus, 1981) that their daily hassles scale is related more strongly to psychological and somatic health symptoms than the more traditional measure of stressful life events. That is hardly surprising, considering the items included in the hassles scale: fear of rejection, nightmares, not enough personal energy, physical illness, being lonely, and so forth.

When information on a large number of variables is collected, avoiding content overlap between predictors and outcomes becomes, in part, a matter of establishing thoughtfully our data analyses and making careful choices about what we label as independent or dependent variables. For example, Cooper and Melhuish (1984) collected rich and valuable information on their subjects, male and female senior executives, including physical and laboratory data. The pool of variables designated as dependent variables were factor analyzed and one of the four factors obtained was labeled "poor physical fitness": it included data on overweight, pulse rate, and results of the Harvard step test. In a stepwise multiple regression, all their designated independent variables were included. Because they included such items as "lack of exercise" and "heavy smoke," these came in as powerful "predictors" of poor fitness for both men and women. This type of data analysis strategy obscures the (presumably) primary objective of the study: the role of occupational stress in various health outcomes, including fitness.

The issues become somewhat more complex when the normally ambiguous cross-sectional association between a subjective measure of exposure and a particular outcome such as self-report of symptoms, yields differential results, such as for different variables or different subgroups of subjects. For example, Pepitone-Arreola-Rockwell, Sommer, Sassenrath, Rozee-Koker, & Stringer-Moore, (1981) found that among women in university nonfaculty positions, a subjective index of job stress was associated with an index of symptoms of poor health (e.g., headaches, sleep and digestive problems, allergies) but not with symptoms specifically concerning menstrual dysfunction. Such differential results can lead to more complex scenarios of confounding versus causation. For example, we could argue that it is too implausible to believe that the influence of an underlying trait, or a shared response set, or an attribution process, would be differential for different classes of symptoms and that, therefore, the pattern of results does suggest more strongly a cause-effect relation. Of course, such an argument may be based only on one's faith in parsimony.

Sometimes, the differential results between a subjective measure of exposure and a self-reported outcome are obtained for different subgroups of subjects. Most pertinent for job stress research are subgroups defined by occupational categories. For example, Gentry and Parkes (1982) found that intensive care unit (ICU) nurses were more likely to report stresses due to workload and issues of death and dying that were non-ICU nurses; however, no differences in psychological strain (anxiety, depression, hostility) were observed. Similarly, in a study of stresses among supervisory police officers (Cooper, Davidson, & Robinson, 1982), different correlates of an overall index of mental health were observed at different organization levels: for junior- and middle-level officers, the unique predictors involved reports of long working hours and lack of available personnel, while for senior-level officers, the unique predictor involved concerns regarding good community relations. Results such as these are certainly suggestive of differences in causal dynamics. However, these findings would be much stronger if (1) the groups did actually differ in average levels of outcomes, rather than just in correlational patterns of associations; and (2) we had objective measures of the specific work dimensions so that we could show that the differential associations with outcomes apply to them as well. Without such additional evidence, it is difficult to dismiss an attribution process, which mocks causal stress dynamics; individuals with higher levels of distress look for environmental sources that would explain their higher levels, and different jobs simply lend themselves to selecting different attributions.

Studies that have examined the hypothesis that high levels of social support will buffer the effects of job stress typically work with subjective measures of stress, strain, and social support. Their aim is to demonstrate a differential slope of association between job stress and strain, steeper for people low on support and weaker or zero slope for those high on support. The major studies in this area (Blau, 1981; House & Wells, 1978; Karasek, Triantisa, & Chaudhry, 1982; LaRocco, House, & French, 1980; LaRocco & Jones, 1978; Winnbust, Marcelissen, & Kleber, 1982) find some support for the proposition; however, the effects are quite weak and sporadic (i.e., involving varying combinations of stress-strain-support indicators). The issue here is what types of confounding remain plausible when such a differential pattern is obtained. One possibility is that reporting high support from family and friends is cognitively consistent with reporting stresses but few symptomatic consequences. Conversely, feeling unsupported is consistent with reporting more strain at higher levels of stress. At low levels of perceived stress, however, no differential demands for cognitive consistency are generated by differential levels of social support. Of course, one will need longitudinal designs and triangulation of measurements for all three constructs (stress, strain, support) before we can begin to examine seriously such a speculation.

I trust the reader will recognize that this diatribe against subjective measures of exposure has its boundaries. Certainly, such subjective measures are perfectly acceptable risk factors for disease. For example, in a prospective study of angina pectoris in Israel, it was found that the men who reported problems with supervisors or co-workers, such as "being hurt" or "not being appreciated by them," were at greater risk for later angina (Medalie &

Gouldbourt, 1976; Medalie, Snyder, Groen, Neufeld, Goldbourt, & Riss, 1973). These items remained as predictors in the multiple logistic model, even after established biomedical risk factors were entered. In this study, the longitudinal design and the different methodologies for assessing the risk factor (self-report) and the outcome (clinical examination) argue for a causal interpretation. However, we don't know what kind of a risk factor we have identified. Its links to actual behavior of co-workers or supervisors is unclear. In fact, we don't know that such items represent any kind of a "reaction" (however subjective) to an as-yet-undetermined environmental stimulus; they may simply be manifestations of a stable trait of anxiety-neuroticism, a demonstrated risk factor for angina (Jenkins, 1976, 1982).

It should also be recognized that my argument is not against subjective measures of exposure per se, but rather against (1) their complete substitution for objective indicators of exposure and (2) the insufficient recognition of the various kinds of mischief these measures can cause. Clearly, objective measures of exposure should be paired with suitably commensurate subjective indicators. This technique does not guarantee interpretable or useful findings or absence of confounding, but it does promise we will know more than if only one set of measures is used. It is, however, still important to keep track of the total pattern of findings and interpret them carefully.

If the objective measure of exposure is confirmed as a risk factor for a disease outcome, and the subjective measue is associated with both the objective measure and the disease outcome (and the latter two associations are stronger than the first), then we have a strong suggestion of a disease impact of environmental exposure, operating primarily through the intervening or mediating step of subjectively appraised exposure. It is, of course, still possible that a self-selection process has introduced a confounding that mimics a causal association. For example, workers who do not get along with others may drift into afternoon or rotating shifts; they will be higher on marital and family conflict (the outcome), and they may report more disruption of social and leisure activities due to the shift (the subjective exposure). These are plausible findings (e.g., Mott, Mann, McLaughlin, & Warwick, 1965) and yet would be altogether spurious due to self-selection, which is particularly bothersome in shiftwork studies (Colligan, 1980).

When the objective and subjective measures of exposure are essentially unrelated to each other—a not unusual situation in the occupational stress literature (e.g., Kasl, 1978; Payne & Pugh, 1976)—and only the subjective measure is related to disease outcome, then we have a number of possibilities: (1) we have not identified properly the objective environmental exposure for which the subjective measure does act as a link in the causal chain; (2) the subjective measure is a true risk factor but it has no linkages to any actual environmental exposure; and (3) we are only dealing with one of the four types of confounding, which were introduced earlier in this section.

When the objective and subjective measures of exposure are reasonably well related to each other, but only the latter relates to the outcome, we have somewhat of a puzzle, only half of the picture that we need to claim a causal process and an intervening variable. The results of a recent British study of mental health effects of aircraft noise (Jenkins, Tarnapolsky, & Hand, 1981; Tarnapolsky, Watkins, & Hand, 1980; Watkins, Tarnapolsky, & Jenkins, 1981) illustrate this puzzle nicely. The objective measure of exposure, The Noise and Number Index (NNI), was substantially related to the subjective measure, being bothered and annoyed. However, only the latter was associated with levels of various symptoms, with use of psychotropic drugs, and so on. In fact, within each level of annoyance, higher symptom levels and higher use of drugs were observed in the lower NNI residential areas. The one interpretation that makes sense is that the subjective measure of exposure was causally heterogeneous: it reflected both the influence of an objective environmental stimulus as well as the manifestation of a trait with no specific environmental anchors. Since other manifestations of that trait involve symptoms and drug use, an association between the subjective measure and outcome was naturally observed; however, the objective environmental condition is not implicated in the causal schema.

The results of this study lead us to a general concern. How much can we take symptoms reflective of a particular *trait* (anxiety, irritability, etc.) and transform them into a *state* measure, which, moreover, by the magic of wording the items differently becomes a suitable subjective measure of a specific environmental exposure? For example, we can change "How often do you feel irritated or annoyed?" to "How often do you feel irritated or annoyed by . . . the noise from aircraft (or whatever)?" But can our subjects really screen out the trait component of that item and then subtract other environmental influences on their state, or are we giving our subjects a convenient opportunity to attribute their symptoms and feelings to a particular source?

STUDY DESIGN CHOICES IN OCCUPATIONAL STRESS RESEARCH

Thus far the methodological comments have concentrated on conceptual and measurement issues, particularly the need to have objective data on exposure. Issues of study design came in tangentially, that is, only insofar as the comments were contingent on type of design used. In this section, I wish to discuss some additional issues that concern general study design. However, I will concentrate only on design issues that do not presume an unrealistic list of design options available to the typical investigator.

The classical and highly serviceable design in occupational medicine is simple enough: establish differences in disease-specific morbidity and/or mortality by occupation and place of work, and then search for environmental agents in the workplace, the exposure to which might explain these differences. This strategy works well when certain conditions are optimized: (1) self-selection into occupations (e.g., health status, personal characteristics) and company selection policies are minimal or do not produce confounding; (2) type and extent of exposure can be pinpointed and quantified; (3) detection and ascertainment of cases and noncases is complete and without bias (e.g., not contingent on seeking or receiving treatment); (4) latency between exposure and detection is short; (5) the disease is rare and (preferably) unicausal. The story of angiosarcoma of the liver and polyvinylchloride (e.g., Creech & Johnson, 1974) illustrated these optimal conditions admirably.

In the stress at work area, apparently similar strategies of research are often less serviceable. For example, Colligan, Smith, and Hurrell (1977) used admission records of Tennessee community mental health centers to compute rates for 130 major occupations; US census data for the state provided the numerator for each occupation. Some of the occupations with high rates were health technicians, waiters and waitresses, practical nurses, inspectors, musicians, and so forth. This is certainly a crude first step, and it would be relatively easy to list some of the numerous biases involved in such an approach, including the differences between illness and illness behavior, the many influences on using that specific source of care, drifting in and out of occupations, and so on. In a somewhat similar approach, Schuckit and Gunderson (1973) examined US Navy records and computed rates of "first" (in the Navy) psychiatric hospitalization by occupation. Within such a closed system, one would guess that the hospitalization data would have fewer inherent weaknesses than the rates derived from the community mental health centers. Nevertheless, this study failed to provide any leads about possible stressful aspects of the jobs with high rates. Instead, it was found that men in the jobs with high rates were older, of lower education, of lower social-class origin, and more likely to be divorced or single. Since these are characteristics the men brought with them to the jobs, and since these are well-established correlates of treated or untreated mental health problems, one can conclude that this whole approach only identified self-selection factors. A more recent report using US Navy hospitalization records examined a wider spectrum of diagnoses (Hoiberg, 1982). Although certain occupational groups did tend to have consistently elevated rates (e.g., hospital corpsmen, mess management specialists), it is difficult to see what promising job stress dimensions, if any, have been thus identified.

The strategy of searching for occupational differences in morbidity or mortality in one data source and then examining other sources to see what psychosocial-environmental dimensions distinguish jobs with high and low rates remains a popular approach, particularly for the study of coronary heart disease (Alfredsson, Karasek, & Theorell, 1982; French & Caplan, 1970; Karasek, Theorell, Schwartz, Pieper, & Alfredsson, 1982; Sales & House, 1971). The low cost of this strategy will no doubt continue to make it attractive, despite any methodological reservations that might be identified. The salient feature might be mentioned here: not all differences observed between the jobs with high and low rates of disease are necessarily involved in the causal process, but we have no way of knowing which differences are relevant. Since this is an ecologic strategy, possibly none of the group differences will lead to an ultimate identification of individual risk factors.

There is no way to escape in this section a discussion of prospective and longitudinal designs, however formidable and idealized these may appear to those investigators with limited resources and only temporary access to a study population. Too much evidence exists from the general stress and disease research area suggesting that findings from cross-sectional and retrospective studies may not replicate in longitudinal and prospective designs; for instance, the apparently plentiful evidence on stressful life events from retrospective designs and the meager evidence from prospective studies (Kasl, 1983, 1984c).

The previous section on subjective measures of exposure identified four sources of confounding and noted that prospective designs only help with one, the influence of outcome (i.e., disease) on subjective retrospective accounts of risk factors. Two additional benefits of the prospective design transcend measurement issues. One benefit is that the risk factors may change as a result of the disease; for example, the blood pressure levels of individuals after a heart attack may be quite different from what they were for the ten years before. This is not a problem of biased measurement; this is an issue of actual change. The second benefit is that longitudinal designs permit a better hold on the issue of selective attrition due to mortality, institutionalization, and loss to follow-up.

The benefits of ruling out the influence of disease on measurement can be considerable. For example, the evidence regarding myocardial infarction (MI) reveals that (1) higher levels of anxiety and distress are a consequence of MI but not its antecedents (Jenkins, 1976, 1982) and (2) the general public believes that "stress, worry, tension, pressure" are the most important cause of heart attacks (Marmot, 1982; Shekelle & Lin, 1978). Because of this evidence, retrospective accounts of heart attack victims about their premorbid life circumstances, including work stresses, are strongly suspect, both because of the likely "search for meaning" as well as due to the distress of the disease. International data on patients with ischemic heart disease and controls reveal that European men are more likely to ascribe retrospectively the stresses to the job situation, and American men to family conflict (Orth-Gomer, 1979; Siegrist, Dittmann, Rittner, & Weber, 1982). Without additional data we can't tell if this difference is an important etiological clue or reflects only cultural influences on the attribution process and lay explanations of etiology.

Differences in results from cross-sectional versus prospective study designs are not necessarily due only to the impact of disease, but may also apply to "silent" conditions as well. For example, Jenkins, Somervell, & Hames (1983) recently compared cross-sectional and longitudinal correlates of blood pressure in a group of supermarket employees. Some measures of distress, such as anxiety and hostility, were found to be positively related to blood pressure but only in cross-sectional analyses. A related measure of distress (i.e., the Reeder stress scale) was related to *lower* blood pressure increase over time, while showing a nonsignificant positive association cross-sectionally. It is difficult to interpret these results; however, the red flag regarding differential results from cross-sectional and longitudinal data is certainly there.

In order not to discredit the cross-sectional methodology excessively, it is worth examining the circumstances under which a discrepancy in results from cross-sectional versus prospective analyses should be accepted at face value (i.e., reflecting a true difference in the etiological picture), rather than starting a search for methodological explanations. The circumstances concern either historical time or passage of time in the life of the person (e.g., stage of life cycle or stage of adaptation to a particular exposure). For example, the data from the Evans County cardiovascular study (Cassel, 1971) showed an initial cross-sectional (i.e., prevalence) association, with white males in lower social strata having lower rates of CHD than white males from middle and upper strata. But in the propsective follow-up, the longitudinal (i.e., incidence) association was no longer observed. The interpretation that this is a true historical change is highly defensible (Morgenstern, 1980).

Similarly, within the life of a person, a risk factor may be linked to disease only during a certain age span or stage of adaptation, but not at other times. This phenomenon would also lead to discrepant cross-sectional and longitudinal findings, which would represent a true difference. Several variations on this phenomenon may be recognized as follows:

1. **A Survival Effect.** Cigarette smoking ceases to be a risk factor for CHD in the elderly (Pooling Project Research Group, 1978); this is presumably because of selective survival of the less vulnerable, plus the fact that few people start smoking as mature adults.

2. **Insufficient Length of Exposure.** Smoking among young men may not be linked to lung cancer simply because the exposure had not been of sufficient duration.

3. **Age-linked Vulnerability.** Infection with E-B virus in young children almost never leads to clinical infectious mononucleosis, while among adolescents, significant numbers who experience such infection develop clinical mono (Evans, 1978).

4. **"Saturation" Effect.** Perhaps being higher on a trait such as submissiveness (Julius, 1981) leads to a slightly faster increase in blood pressure during adolescence and adulthood, but once moderately elevated levels are reached, submissiveness is unrelated to further increases.

In the above examples of reasons for true differences between cross-sectional and longitudinal results, the common requirement is that the two types of study designs being compared not cover the same historical time or life-span time (maturation, duration of exposure). This is, in fact, a quite common occurrence, since most often the cross-sectional results derive from the initial wave of data collection of a prospective study. Furthermore, in such instances, the noncomparability of time is always in one direction: the initial cross-sectional results (if unbiased) reflect the end product of etiological dynamics, which took place earlier and in a somewhat younger cohort.

Designing a prospective study is not necessarily as straightforward as taking a representative cross-sectional sample of community adults and simply adding to it some follow-up data collection—current practice notwithstanding (e.g., Billings & Moos, 1982; Eaton, Regier, Locke, & Taube, 1981; McFarlane, Norman, Streiner, & Roy, 1983; Pearlin, Lieberman, Menaghan, & Mullan, 1981; Williams, Ware, & Donald, 1981). In a prospective design, the cohort has been examined at the optimal point (developmental stage, environmental exposure) for (1) detecting the dynamics of the transition from exposure to overt disease and (2) describing the most "representative" characteristics of the etiological process of disease development. On the other hand, when the cohort is *not* examined because something extraordinary has happened to it, or is expected to happen, then the longitudinal follow-up may reveal very little. This result would be particularly evident when the etiological dynamics have already played themselves out (with respect to a particular exposure or a particular outcome) and nothing more is happening. Then our most rigorous analysis of the longitudinal data, adjusting for initial levels of health status, will yield next to nothing. For example, there is considerable evidence from the mental health literature on blue-collar workers in assembly-line jobs (e.g., Caplan et al., 1975; Chinoy, 1955; Kasl, 1974; Kornhauser, 1965) that adaptation to a boring, monotonous job occurs rather early in their careers, that the primary mode of adaptation is not expecting work to be a meaningful human activity, and that those unable to adapt "successfully" are likely to drift out of such jobs. Under these circumstances, a longitudinal study of middle-aged blue-collar workers might fail to reconstruct any one of the dynamics that impact on assembly-line work.

The investigator who uses a prospective study design is under special obligation to be thoughtful about establishing an adequately probing design, which is most likely to detect the causal dynamics of interest. This is necessary because of the long-established habit of resolving inconsistencies in results from a variety of designs by giving more attention to the findings from the prospective study. The strong design in the occupational stress research area would have the following characteristics.

1. The cohort is picked up prior to exposure.
2. The environmental condition (exposure) is objectively defined and measured.
3. Self-selection into exposure conditions is minimized by exploiting opportunities for "natural experiments," changes in the work setting

that lead to highly comparable groups of exposed and unexposed individuals.

4. Potential confounding variables, primarily biological risk factors and initial health status, are assessed and their influence monitored in analysis.

5. The period of follow-up is adequate to take the cohort through crucial periods of adaptation and disease causation.

6. Mediating processes are studied, and vulnerability factors that interact with exposure are included.

This list of characteristics is, of course, unrealistic. The investigator most often does not have the freedom to make such choices, and knowledge about the etiology of the particular disease may not be sufficient to make all the right choices, even if the options were there. Nevertheless, the list is an invitation to look beyond the "ordinary" prospective study design and look for "extraordinary" research opportunities. A Belgian study of job stress and CHD in two banks, one of which had undergone changes in work setting and work demands, is a marvelous example of such an opportunity (Kittel, Kornitzer, & Dramaix, 1980).

Investigators must retain a strong interest in the nuances of different prospective study designs. It is not sufficient to have one undifferentiated picture of what a prospective design is, and then hope that inconsistencies from case-control or cross-sectional studies will be resolved by reference to findings from an undifferentiated class of prospective designs. That leaves no room for explaining differences in other results all based on prospective data. For example, the research on Type A behavior as a risk factor for CHD (The Review Panel, 1981) reveals considerable variability in results, all based on prospective designs. Three major studies have confirmed its status as a risk factor (Brand, Rosenman, Sholtz, & Friedman, 1976; French-Belgian Collaborative Group, 1982; Haynes & Feinleib, 1982), but several other, more recent results have been all negative (Matthews, 1985). They include two clinical trials, the Multiple Risk Interventional Trial and the Aspirin Myocardial Infarction Study, a study of Finnish men, and The Honolulu Heart Program studying Japanese American men. Furthermore, among the three studies reporting positive findings, some variability exists regarding end points to which type is or is not associated. In Framingham data, it was related to angina but not to uncomplicated MI, while in the Belgian study, it was related to MI and sudden death but not angina; in the WCGS data, it was related to all three end points. Furthermore, in Framingham, Type A was a significant risk factor only for men in white-collar jobs (versus blue-collar) and for housewives (versus working women). Clearly, such a variation in findings will not be explained until we have a better understanding of the nuances that exist in prospective designs. If Type A is an interactive risk factor, jointly with specific parameters of the work setting (e.g., Chesney, Sevelius, Black, Ward,

Swan, & Rosenman, 1981; Davidson & Cooper, 1980; Matteson & Ivancevich, 1982c; Rose, Jenkins, & Hurst, 1978), and if the consequences of such inter-action are modified by possibilities of self-selection in the occupational structure, then we may have truly complicated dynamics for which the ordinary prospective study design may be inadequately probing.

BIOLOGICAL MECHANISMS AND A PARADIGM FOR JOB STRESS

When we examine the evidence for a relationship between job stress and a specific disease outcome, such as cardiovascular disease, we find a fair amount of suggestive evidence, but it does not add up to an impressive and coherent picture. The prospective studies continue to show a mixed picture: no relationship (e.g., Kornitzer, Kittel, Dramaix, & deBacker, 1982), some association with angina only (e.g., Aro, 1981), and some relationships with cardiovascular outcomes measured in less than an optimal way (Karasek, Baker, Marxer, Ahlbom, & Theorell, 1981; Karasek, Theorell, Schwartz, Pieper, & Alfredsson, 1982). The Framingham data show relatively little: reporting more job promotions during the previous 10 years appears to be a risk factor, but only for blue-collar male workers (Haynes, Feinleib, & Kannel, 1980). However, the general literature on job mobility and CHD has shown little (e.g., Berkanovic & Krochalk, 1977; Lehman, Schulman, & Hinkle, 1967). Among Framingham women, having a nonsupportive boss and decreased job mobility appear to be risk factors, but only among clerical workers (Haynes & Feinleib, 1980). Among women with higher education, these same two factors also increase the risk for CHD in the husbands of these women (Haynes, Eaker, & Feinleib, 1983). Results from other research, concerned with depression and psychosomatic symptoms, confirm the possibility of adverse effects of wives' job stresses on husbands' health (Billings & Moos, 1982a). Stressful life events are rarely examined prospectively in relation to CHD (Haney, 1980). One such study (Theorell & Floderus-Myrhed, 1977) showed that an index of "workload" increased the risk for MI. Unfortunately for simple interpretations, the index included items that reflected both increases and decreases in responsibility and complaints both about too little and too much responsibility. Furthermore, those dying of MI were actually lower on the index, while only MI survivors were higher, compared to controls. Occupational differences in CHD mortality and morbidity continue to be of considerable interest (Carruthers, 1980; Guralnick, 1963; King, 1970; Russek, 1962), but lack of prospective designs and absence of even minimal controls on biomedical risk factors make such differences, for the purpose of identifying job stress factors, extremely tenuous. Thus only pinpointed comparisons of occupational groups, such as those associated with high and low levels of responsibility (priests and brothers) within two orders of monks (Caffrey, 1969), will lead to reasonable interpre-tations. Prospective studies have identified other risk factors, such as hostility

(Barefoot, Dahlstrom, & Williams, 1983; Shekelle, Gale, Ostfeld, & Paul, 1983) and exhaustion at the end of the day and inability to relax (Jenkins, 1982), but these have no established anchors in the work environment. Of course, there is considerable CHD research involving retrospective case-control design (see for reviews: Boman, 1982; Haney, 1980; Jenkins, 1976, 1982), but their contribution to knowledge is extremely tentative. There is an increasing recognition of the methodological problems involved, but proposed solutions, such as using the MMPI Lie Scale in statistical adjustments (Siegrist, Dittmann, Rittner, & Weber, 1982) or adjusting for standard risk factors (Orth-Gomer & Ahlbom, 1980), do not adequately tackle the underlying problems.

Overall, then, the job stress-CHD evidence is quite fragmentary, and neither specific work environment dimensions nor the essential pathogenic process have been clearly illuminated. And yet, the job stress-disease relationship has been most extensively examined for CHD, compared to other physical health outcomes.

In this section I wish, first, to examine some of the research on biological mechanisms that might be involved in the job stress-disease link and, then, to suggest a paradigm for job stressors.

Studies of biological mechanisms in stress-disease associations tend to fall into two classes: (1) Those primarily seen as indicators of stress, such as corticosteroids or catecholamines; and (2) those established as risk factors for a specific disease, such as blood pressure and total serum cholesterol for CHD. Two issues arise from using the first class. One is that we don't really know how such endocrine variables link up with disease outcomes, or even other risk factors, such as indicators of immunological functioning. Neither identifying situations that produce elevated endocrine levels nor identifying individuals who are responders will enable us to predict a disease outcome. For example, results from the Air Traffic Controllers Study showed that neither high average levels of cortisol nor responding to increased workload with increases in cortisol, was predictive of physical or psychiatric morbidity during the course of the study (Rose et al., 1978, 1979, & 1982). The second issue is that on neither conceptual nor empirical grounds is it acceptable to make the strict equation, reactivity = presence of a stressor and non-reactivity = absence of a stressor. In a recent review of the evidence on endocrine responses to stress, Rose (1980) notes many instances in which the endocrine responses (especially cortisol), though sensitive to acute stressors, undergo extinction when the individuals are reexposed to those stressors, or are exposed to chronic stressors. However, it is not clear if this should be interpreted, conceptually, that the stressor ceased to be stressful and, empirically, that the risk of disease was reduced. Instead, it could be that such endocrine variables have their own patterns of reactivity and simply are not suitable for monitoring the consequences of chronic exposures. These points must be remembered when looking at the association between stress at work and endocrine reativity.

Two other issues also arise from the use of the second class of variables, established risk factors. One issue again concerns chronic reactivity but is slightly different. Specifically, considerable evidence exists regarding acute stressors (usually in the laboratory) and transient changes in risk factors, such as blood pressure (Shapiro, 1978) or plasma lipids (Dimsdale & Herd, 1982). However, there is little basis for making the extrapolation that if such stressors became chronic, the consequences for the risk factors would be equally chronic. For example, in a study of male blue-collar workers who lost their jobs because of a permanent plant shutdown, blood pressure and serum cholesterol went up acutely as the men became unemployed but came down again, even for these men who remained unemployed for a long period (Kasl & Cobb, 1980). And cross-sectional associations between blood pressure and cholesterol and indicators of stable job pressures usually yield nonsignificant correlations across a variety of occupations (Caplan et al., 1975; French et al., 1982), something we would not expect if chronic exposure led to chronic elevations of risk factors. The second point is that the risk of a specific disease, such as CHD, can be elevated because of some trait or environmental exposure even after one adjusts for the established risk factors. In fact, the typical analyses from Framingham or with respect to Type A have shown just that (e.g., Brand et al., 1976; Haynes et al., 1980, 1983). This means, of course, that the environmental exposure does not (only) act through the established risk factors, and that, therefore, the absence of an association between environmental exposure and a risk factor does not rule out elevated risk for the disease itself.

A consideration of various studies dealing with job stress and biological mechanisms (stress indicators, disease risk factors) suggests a number of interpretations and conclusions:

1. Chronically dangerous work environments do not appear to be chronically arousing; individuals adapt to them, especially as they acquire skills that reduce the danger (e.g., Bourne, Rose, & Mason, 1967; Dutton, Smokensky, Leach, Lorimer, & Hsi, 1978; Pinter, Tolis, Gnyda, & Katsarkas, 1979).

2. When a dangerous work environment demands attention, vigilance, and concentration, and when high levels of responsibility (for people or equipment) are attached to the work tasks, then habituation or adaptation to chronic exposure may not take place (e.g., Bourne, Rose, & Mason, 1968; Roman, 1963; Rubin, 1974; Smith, 1967).

3. Boring and monotonous activities on the job are not, by themselves, associated with arousal (Thackray, 1981), though adverse mental health consequences may be observed (Kasl, 1978).

4. Repetitive activities that are machine-paced, however, will be generally arousing (e.g., Khaleque, 1981; Salvendy & Smith, 1981; Wilkes, Stammerjohn, & Lalich, 1981). In particularly arousing jobs,

such as those of saw-mill workers, work is paced by machine, is highly repetitive yet involves skilled judgments made at short intervals, and is highly constraining with the need to maintain the same posture (Frankenhaeuser & Gardell, 1976; Gardell, 1976; Johansson, Aronsson, & Lindstrom, 1978). The combination of high job demands and narrow latitude with respect to meeting those demands may be generically pathogenic (Karasek et al., 1981, 1982); this generalization, however, may not apply to jobs where wider latitude is arousing because of the implication of greater individual responsibility (Rose et al., 1982).

5. Physiological arousal may also be increased by payments on a piece-work basis (pay proportional to output), which may increase the pace and the constraints, and reduce relaxing social interaction among workers (Levi, 1974; Timio & Gentili, 1976).

6. The magnitude of arousal during working hours and the speed of unwinding from work (return to baseline levels) in the home setting may be influenced by a variety of other factors, such as difficult rush-hour commuting to and from work (Lundberg, 1976; Stokols & Novaco, 1981), recency of a vacation (Johansson, 1976), and working overtime, particularly unwanted overtime among female workers (Frankenhaeuser, 1979; Rissler, 1977). The speed of unwinding from work is likely to influence the spillover effects of work stressors on leisure activities (Frankenhaeuser, 1977).

7. Job involvement may be the best umbrella concept for those influences from stable traits and attitudes that feed into a work orientation, which enhances arousal effects of the various work dimensions (e.g., Rose et al., 1978). Behavior Type A (The Review Panel, 1981) may be a good example of one of these influences.

Evidence, such as that reviewed above, allows one to develop a paradigm for pathogenic conditions in the work setting involving "stress" (Kasl, 1981). The basic assumption in this paradigm is that we are trying to identify work environmental conditions that lead to chronic reactivity and, eventually, to disease outcomes. The following characteristics suggest themselves: (1) the stressful work condition is enduring rather than acute, though it is most likely to be intermittent rather than constant; (2) failure to meet the demands of the work setting has serious consequences (e.g., high level of responsibility for lives of others, expensive equipment, or profits) and disengagement is difficult; (3) habituation and adaptation to the chronic situation is difficult because of environmental constraints and because of the need to maintain some form of involvement, vigilance, or arousal; (4) there is a "spillover" of the effects of the work role on other areas of functioning (e.g., family, leisure) so that the daily impact of the demanding job situation becomes cumulative and health-threatening, rather than being daily defused or erased, when it would be expected to have minimal long-term impact.

The job of the air traffic controller appears to exemplify well the above paradigm, and the stress perspective has been applied to it vigorously (e.g., Crump, 1979). The early results with respect to health impact (Cobb & Rose, 1973) looked more impressive than the findings of the carefully conducted longitudinal study of Rose et al. (1978). However, the latter is more suitable for describing the short-term dynamics of stress-arousal-adaptation than for documenting long-term health effects and for comparing this group to other occupations. Another job, which may fit the paradigm, is that of sea pilots, and their cardiovascular health has received some attention in the European literature recently (Cook & Cashman, 1982; Mundal, Erikssen, & Rodahl, 1982; Zorn, Harrington, & Goethe, 1977). An old report on the health of train dispatchers before the advent of electronic equipment (McCord, 1948) would also fit into the picture here.

The above paradigm is merely a reflection of the already expressed occupational epidemiology perspective, which characterizes this chapter, and of the emphasis on stress as an objective environmental condition. It is primarily intended to help investigators identify work environments that are likely to have health consequences. The emphasis is on work conditions to which it is difficult to adapt (hence, the expectation of chronic reactivity), but it takes only a first step in that direction. For example, we know little about the stress effects of knowledge that one was exposed to a carcinogen, such as asbestos (Selikoff & Hammond, 1979), in one's work setting. However, the major limitation is in the notion of "spillover" effects, the continuation of the impact of work in nonwork settings. This research area is quite limited and thus the formulation remains quite vague. The social support literature does contain some studies that examine jointly the work and the family settings, but most of the research concerns effects of the "stressed" worker on other members of the family, rather than how the family setting facilitates or hinders the defusion of job stressors after one is off the job (Cox, 1980; Gardell, 1982; Kasl & Wells, 1985; Levi, Frankenhauser, & Gardell, 1982).

INTERVENTIONS AT THE WORKSITE

Since disease risk factors identified from passive observation (quasi-experimental field research) seldom fully deserve the label "cause," one looks toward intervention studies as the only practical opportunity for full experimental control. When the experimentally induced modification of a risk factor is followed by reduced risk for disease, then we have powerful supplementary evidence. It is for this reason that I wish to consider the intervention studies. A secondary purpose is that such intervention studies may tell us something new about the dynamics of the work environment, even if they do not specifically show that a reduction in some particular work stressor has health benefits.

My conclusion from a sampling of this research literature is that it is completely without relevance to either my primary or my secondary purpose; perhaps it has merit from other perspectives.

Consider, for example, the health promotion activities (for recent reviews, see: Brennan, 1982; Fielding, 1982, 1984; Pomerleau, 1983). One learns what they are, what they cost, and what the benefits are. But the specific setting in which they were carried out seems to be quite unimportant. In general, the preventive health behavior literature pays minimal attention to the setting (and its dynamics) in which the intervention is done (e.g., Kirscht, 1983), though specific settings, such as the family (Baranowski, Nader, Dunn, & Vanderpool, 1982) or elderly housing (Pickard & Collins, 1982) are sometimes analyzed for their advantages as sites for intervention. Comparative conceptual analyses of advantages and disadvantages of different settings for conducting intervention or health promotion are rare, and comparative evaluative research is even rarer. The worksite is seen as a good place for health promotion, but aside from the issue of cost-effective access to target individuals and vague references to social support systems in the workplace (e.g., Andreoli & Guillory, 1983; Statement from . . . , 1984), a reader will not learn much about the interplay of workplace dynamics and health promotion dynamics. For example, Fielding (1982) reviews the effectiveness of various programs, but one only learns that, for example, weight reduction is more difficult to achieve than smoking cessation, not if weight reduction at the worksite achieves results different from, say, a clinic-based program. Only in the area of high blood pressure control have there been some specific attempts to compare programs at different sites. One study (Logan, Milne, Achber, Campbell, & Haynes, 1982) claimed to show the superiority of worksite intervention over community-based program, but this was only because the latter was no program at all, just referral to "usual care." When properly designed, comparisons may not show the superiority of worksite blood pressure programs (e.g., Alderman & Davis, 1980). There do seem to be some nonspecific effects of worksite health promotion programs on job satisfaction and attitudes (e.g., Blair, Collingwood, Reynolds, Smith, Hagan, & Sterling, 1984; Fielding, 1984), which presumably would not obtain for programs in other settings.

Stress management programs at the worksite (for overviews, see Murphy, 1984; Newman & Beehr, 1979) also appear to be irrelevant to the stress at work perspective of this chapter. One reason for this is that such programs typically do not address the environmental source of stress at work, but only address the issue of workers's "distress" (broadly defined). Environmental changes are rare; among them, introduction of flextime is a common modification that is examined (e.g., Hicks & Klimoski, 1981; Krausz & Freibach, 1983; Narayanan & Nath, 1982), even though the stress at work literature does not particularly identify this as a source of stress. Typically, the benefits are only on absenteeism, possibly on job satisfaction. Some interventions, such as participation in decision making, are ambiguous manipulations and unless they are accompanied by some actual redesign of work (e.g., Wall & Clegg, 1981), they are best viewed as managing the response side of the stressor-outcome relationship (e.g., Jackson, 1983). A second reason for the irrelevance of these interventions (to the perspective of this chapter) is that, with rare exceptions (e.g., Ganster, Mayes, Sime, & Tharp, 1982), the interventions deal with nonspecific "distress" (psychological or biological) without being concerned with its origins, (i.e., its ties, if any, to stress at work). The lack of concern with origins of the distress is both in the sense of selection of subjects for the program as well as in the sense of the actual content of the intervention. This approach is particularly curious in such areas as modification of Type A behavior (Friedman, Thoresen, & Gill, 1981; Roskies, Spevack, Surkis, Cohen, & Gilman, 1978; Thoresen, Telch, & Eagleston, 1981), where the conceptualization of this construct—a predisposition triggered by an appropriate environmental challenge—would seem to demand that the intervention be heavily concerned with environmental stimuli.

In passing, one might note that evaluative reseach in this area frequently exhibits methodological weaknesses, which further compromise the usefulness of this literature. These include the following: lack of attention to statistical regression (Seamonds, 1982), absence of comparison groups or adequate follow-up (Pauly, Palmer, Wright, & Pfeiffer, 1982), and inattention to artifacts, such as pretreatment anticipation effects creating the appearance of a subsequent treatment effect (Drazen, Nevid, Pace, & O'Brien, 1982). Particularly disturbing is the practice of analyzing treatment effectiveness not on those to whom treatment was offered, but only on those who showed high participation or compliance (e.g., Bertara & Cuthie, 1984; Durbeck, Heinzelmann, Schacter, Haskell, Payne, Moxley, Nemiroff, Limoncelli, Arnoldi, & Fox, 1972); this introduces a strong self-selection bias that mocks program effectiveness.

My conclusion is this: People are obese, they smoke, and they are tense. The origins of these are overwhelmingly preemployment. It is appropriate to help people with such issues, and one place to do it is the worksite. But how this has illuminated the stress-at-work dynamics is too difficult to discern.

CONCLUDING REMARKS

In this chapter I have tried to offer a coherent perspective, that of occupational epidemiology, which argues that the first and most important goal is to identify causal linkages between exposure to objective environmental conditions at work and health-disease outcomes. From this perspective I derived both a framework for reviewing the substantive findings and for developing a paradigm for the environmental conditions of interest, as well as for offering criticism of existing theoretical formulations and of research methodology. At the center of my argument is a strong objection to the use of subjective measures of exposure; they are seen as precipitating numerous problems, above all confounding and triviality. At their best, such measures are valid risk factors but of undetermined environmental anchors; hence, their problemmatic relevance to the occupational epidemiology perspective. The central argument has one major qualification: when the study design is fully responsive to the primary goal of occupational epidemiology (as stated above), then subjective measures of exposure cause no harm—if our interpretations are careful—and may help us identify additional vulnerability factors or underlying biobehavioral mechanisms.

REFERENCES

Alderman, M. H., & Davis, T. K. (1980). Blood pressure control programs on and off the work-site. *Journal of Occupational Medicine, 22*, 167–170.

Alfredsson, L., Karasek, R., & Theorell, T. (1982). Myocardial infarction risk and psychosocial work environment: An analysis of the male Swedish working force. *Social Science and Medicine, 16*, 463–467.

Andreoli, K. G., & Guillory, M. M. (1983). Arenas for practicing health promotion. *Family and Community Health, 5*, 28–40.

Aro, S. (1981). Stress, morbidity, and health-related behavior. *Scandinavian Journal of Social Medicine, 25* (Suppl.), 1–130.

Baranowski, T., Nader, P. R., Dunn, K., & Vanderpool, N. A. (1982) Family self-help: Promoting changes in health behavior. *Journal of Communication, 32*, 161–172.

Barefoot, J. C., Dahlstrom, W. G., & Williams, R. B., Jr. (1983). Hostility, CHD incidence, and total mortality: A 25-year follow-up study of 255 physcans. *Psychosomatic Medicine, 45*, 59–63.

Baum, A., Grunberg, N. E., & Singer, J. E. (1982). The use of psychological and neuroendocrino-logical measurements in the study of stress. *Health Psychology, 1*, 217–236.

Beech, H. R., Burns, L. E., & Sheffield, B. F. (1982). *A behavioral approach to the management of stress.* Chichester, England: Wiley.

Beehr, T. A., & Newman, J. E. (1978). Job stress, employee health, and organizational effective-ness: A facet analysis, model, and literature review. *Personnel Psychology, 31*, 665–699.

Berkanovic, E., & Krochalk, P. C. (1977). Occupational mobility and health. *Advances in Psychoso-matic Medicine, 9*, 123–139.

Bertera, R. L., & Cuthie, J. C. (1984). Blood pressure self-monitoring in the workplace. *Journal of Occupational Medicine, 26*, 183–188.

Billings, A. G., & Moos, R. H. (1982). Stressful life events and symptoms: A longitudinal model. *Health Psychology, 1*, 99–117.

Billings, A. G., & Moos, R. H. (1982a). Work stress and the stress-buffering roles of work and family resources. *Journal of Occupational Behavior, 3*, 215–232.

Blau, G. (1981). An empirical investigation of job stress, social support, service length, and job strain. *Organizational Behavior and Human Performance, 27*, 279–302.

Blair, S. N., Collingwood, T. R., Reynolds, R., Smith, M., Hagan, R. D., & Sterling, C. L. (1984). Health promotion for educators: Impact on health behaviors, satisfaction, and general well-being. *American Journal of Public Health, 74*, 147–149.

Boman, B. (1982). Psychosocial stress and ischaemic heart disease: A response to Tennant. *Australian and New Zealand Journal of Psychiatry, 16*, 265–278.

Bourne, P. G., Rose, R. M. & Mason, J. W. (1967). Urinary 17-OCHS levels. Data on seven helicopter ambulance medics in combat. *Archives of General Psychiatry, 17,* 104–110.

Bourne, P. G., Rose, R. M., & Mason, J. W. (1968). 17-OCHS levels in combat. *Archives of General Psychiatry, 19,* 135–140.

Brand, R. J., Rosenman, R. H., Sholtz, R. I., & Friedman, M. (1976). Multivariate prediction of coronary heart disease in the Western Collaborative Group Study compared to the findings of the Framingham study. *Circulation, 53,* 348–355.

Brennan, A. J. J. (Ed.). (1982). Worksite health promotion. *Health Education Quarterly, 9* (Suppl.), 5–91.

Brief, A. P., Schuler, R. S., & Van Sell, M. (1981). *Managing job stress.* Boston: Little, Brown.

Caffrey, B. (1969). Behavior patterns and personality characteristics related to prevalence rates of coronary heart disease in American monks. *Journal of Chronic Diseases, 22,* 93–103.

Caplan, R. D., Cobb, S., French, J. R. P., Jr., Harrison, R. V., & Pinneau, S. R., Jr. (1975). *Job demands and worker health.* DHEW Publication No. (NIOSH) 75–106. Washington, DC: U.S. Government Printing Office.

Carruthers, M. (1980). Hazardous occupations and the heart. In C. L. Cooper & R. Payne (Eds.), *Current concerns in occupational stress,* pp.3–22. Chichester, England: Wiley.

Cassel, J. C. (1971). Summary of major findings of the Evans County cardiovascular studies. *Archives of Internal Medicine, 128,* 890–895.

Chesney, M. A., Sevelius, G., Black, G. W., Ward, M. M., Swan, G. E., & Rosenman, R. H. (1981). Work environment, Type A behavior, and coronary heart disease risk factors. *Journal of Occupational Medicine, 23,* 551–555.

Chinoy, E. (1955). *Automobile workers and the American dream.* Garden City, NY: Doubleday.

Cobb, S., & Rose, R. M. (1973). Hypertension, peptic ulcer, and diabetes in air traffic controllers. *Journal of the American Medical Association, 224,* 489–492.

Colligan, M. J. (1980). Methodological and practical issues related to shiftwork research. *Journal of Occupational Medicine, 22,* 163–166.

Colligan, M. J., Smith, M. J., & Hurrell, J. J., Jr. (1977). Occupational incidence rates of mental health disorders. *Journal of Human Stress, 3*(3), 34–39.

Cook, T. C., & Cashman, P. M. N. (1982). Stress and ectopic beats in ships' pilots. *Journal of Psychosomatic Research, 26,* 559–569.

Cooper, C. L. (1981). *The stress check.* Englewood Cliffs, NJ: Prentice-Hall.

Cooper, C. L., Davidson, M. J., & Robinson, P. (1982). Stress in the police service. *Journal of Occupational Medicine, 24,* 31–36.

Cooper, C. L., & Marshall, J. (Eds.). (1980). *White collar and professional stress.* Chichester, England: Wiley.

Cooper, C. L. & Melhuish, A. (1984). Executive stress and health. Differences between men and women. *Journal of Occupational Medicine, 26,* 99–104.

Cooper, C. L., & Payne, R. (Eds.). (1978). *Stress at work.* Chichester, England: Wiley

Cooper, C. L., & Payne, R. (Eds.). (1980). *Current concerns in occupational stress.* Chichester, England: Wiley.

Cox, T. (1980). Repetitive work. In C. L. Cooper & R. Payne (Eds.), *Current concerns in occupational stress,* pp. 23–41. Chichester, England: Wiley.

Creech, J. L., & Johnson, M. N. (1974). Angiosarcoma of liver in the manufacture of polyvinyl-chloride. *Journal of Occupational Medicine, 16,* 150–151.

Crump, J. H. (1979). Review of stress in air traffic control: Its measurement and effects. *Aviation, Space, and Environmental Medicine, 50,* 243–248.

Davidson, M. J., & Cooper, C. L. (1980). Type A coronary-prone behavior in the work environment. *Journal of Occupational Medicine, 22,* 375–383.

DeLongis, A., Coyne, J. C., Dakof, G., Folkman, S., & Lazarus, R. S. (1982). Relationship of daily hassles, uplifts, and major life events to health status. *Health Psychology, 1*, 119–136.

Dimsdale, J. E., & Herd, J. A. (1982). Variability of plasma lipids in response to emotional arousal. *Psychosomatic Medicine, 44*, 413–430.

Dohrenwend, B. S., & Dohrenwend, B. P. (Eds.). (1981). *Stressful life events and their contexts.* New York: Prodist.

Drazen, M., Nevid, J. S., Pace, N., & O'Brien, R.M. (1982). Worksite-based behavioral treatment of mild hypertension. *Journal of Occupational Medicine, 24*, 511–514.

Durbeck, D. C., Heinzelmann, F., Schacter, J., Haskell, W. L., Payne, G. H., Moxley, R. T., Nemiroff, M., Limoncelli, D. D., Arnoldi, L. B., & Fox, S. M. (1972). The National Aeronautics and Space Administration—U.S. Public Health Service health evaluation and enhancement program. *American Journal of Cardiology, 30*, 784–790.

Dutton, L. M., Smolensky, M. H., Leach, C. S., Lorimer, R., & Hsi, B. P. (1978). Stress levels of ambulance paramedics and fire fighters. *Journal of Occupational Medicine, 20*, 111–115.

Eaton, W. W., Regier, D. A., Locke, B. Z., & Taube, C. A. (1981). The epidemiologic catchment area program of the National Institute of Mental Health. *Public Health Reports, 96*, 319–325.

Elliott, G. R., & Eisdorfer, C. (Eds.) (1982). *Stress and human health.* New York: Springer.

Evans, A. S. (1978). Infectious mononucleosis and related syndromes. *American Journal of Medical Sciences, 276*, 325–329.

Fielding, J. E. (1982). Effectiveness of employee health improvement programs. *Journal of Occupational Medicine, 24*, 907–916.

Fielding, J. E. (1984). Health promotion and disease prevention at the worksite. *Annual Review of Public Health, 5*, 237–265.

Frankenhaeuser, M. (1977). Job demands, health and well-being. *Journal of Psychosomatic Research, 21*, 313–321.

Frankenhaeuser, M. (1979). Psychoneuroendocrine approaches to the study of emotion as related to stress and coping. In H.E. Howe, & R.A. Dienstbier (Eds.), *Nebraska symposium on motivation*, pp. 123–161. Lincoln: University of Nebraska Press.

Frankenhaeuser, M., & Gardell, B. (1976). Underload and overload in working life: outline of a multidisciplinary approach. *Journal of Human Stress, 2*, 35–46.

French, J. R. P., Jr., & Caplan, R. D. (1970). Psychosocial factors in coronary heart disease. *Industrial Medicine and Surgery, 39*, 383–397.

French, J. R. P., Jr., Caplan, R. D., & Van Harrison, R. (1982). *The mechanisms of job stress and strain.* Chichester, England: Wiley.

French-Belgian Collaborative Group. (1982). Ischemic heart disease and psychological patterns. *Advances in Cardiology, 29*, 25–31.

Friedman, M., Thoresen, C. E., & Gill, J. J. (1981). Type A behavior: Its possible role, detection, and alteration in patients with ischemic heart disease. In J. W. Hurst (Ed.), *Heart update V*, pp. 81–100. New York: McGraw-Hill.

Ganster, D. C., Mayes, B. T., Sime, W. E., & Tharp, G. D. (1982). Managing organizational stress: A field experiment. *Journal of Applied Psychology, 67*, 533–542.

Gardell, B. (1976). *Job content and quality of life.* Stockholm: Prisma.

Gardell, B. (1982). Scandinavian research on stress in working life. *International Journal of Health Services, 12*, 31–41.

Gentry, W. D., & Parkes, K. R. (1982). Psychological stress in intensive care unit and non-intensive care unit nursing: A review of the past decade. *Heart and Lung, 11*, 43–47.

Gmelch, W. H. (1982). *Beyond stress to effective management.* New York: Wiley.

Guralnick, L. (1963). Mortality by occupation and cause of death. *Vital Statistics—Special Reports* (Vol. 53, Nos. 3, 4, and 5). Washington, DC: U.S. Government Printing Office.

Haney, C. A. (1980). Life events as precursors of coronary heart disease. *Social Science and Medicine, 14A*, 119–126.

Haynes, S. G., Eaker, E. D., & Feinlieb, M. (1983). Spouse behavior and coronary heart disease in men: Prospective results from the Framingham Heart Study. I. Concordance of risk factors and the relationship of psychosocial status to coronary incidence. *American Journal of Epidemiology, 118*, 1–22.

Haynes, S. G., & Feinleib, M. (1980). Women, work and coronary heart disease: Prospective findings from the Framingham Heart Study. *American Journal of Public Health, 70*, 133–141.

Haynes, S. G., & Feinleib, M. (1982). Type A behavior and incidence of coronary heart disease in the Framingham Heart Study. *Advances in Cardiology, 29*, 85–95.

Haynes, S. G., Feinleib, M., & Kannel, W. B. (1980). The relationship of psychosocial factors to coronary heart disease in the Framingham study. III. Eight-year incidence of coronary heart disease. *American Journal of Epidemiology, 111*, 37–58.

Hicks, W. D., & Klimoski, R. J. (1981). The impact of flextime on employee attitudes. *Academy of Management Journal, 24*, 333–341.

Hoiberg, A. (1982). Occupational stress and illness incidence. *Journal of Occupational Medicine, 24*, 445–451.

Holroyd, K. A., & Lazarus, R. S. (1982). Stress, coping, and somatic adaptation. In L. Goldberger & S. Breznitz (Eds.), *Handbook of stress*, pp. 21–35. New York: Free Press.

Holt, R. R. (1982). Occupational stress. In L. Goldberger & S. Breznitz (Eds.), *Handbook of stress*, pp. 419–444. New York: Free Press.

House, J. S. (1981). *Work stress and social support*. Reading, MA: Addison-Wesley.

House, J. S., & Wells, J.A. (1978). Occupational stress, social support, and health. In A.A. McLean, G. Black, & M. Colligan (Eds.), *Reducing occupational stress: Proceedings of a conference*, 8–29. DHEW (NIOSH) Publication No. 78–140. Cincinnati, OH.

Hurrell, J. J., & Colligan, M. J. (1982). Psychological job stress. In W.N. Rom (Ed.), *Environmental and occupational medicine*, pp. 425–430. Boston: Little, Brown.

Jackson, S.E. (1983). Participation in decision making as a strategy for reducing job-related strain. *Journal of Applied Psychology, 68*, 3–19.

Jenkins, C. D. (1976). Recent evidence supporting psychologic and social risk factors for coronary disease. *New England Journal of Medicine, 294*, 987–994, 1033–1038.

Jenkins, C. D. (1982). Psychosocial risk factors for coronary heart disease. *Acta Medica Scandinavica, 660* (Suppl.), 123–136.

Jenkins, C. D., Somervell, P. D., & Hames, C. G. (1983). Does blood pressure usually rise with age? . . . Or with stress? *Journal of Human Stress, 9*(3), 4–12.

Jenkins, L., Tarnapolsky, A., & Hand, D. (1981). Psychiatric admissions and aircraft noise from London airport: Four-year, three-hospitals' study. *Psychological Medicine, 11*, 765–782.

Jick, T. D., & Burke, R. J. (1982). Occupational stress: Recent findings and new directions. *Journal of Occupational Behavior, 3*, 1–3.

Johansson, G. (1976). Subjective wellbeing and temporal patterns of sympathetic adrenal medullary activity. *Biological Psychology, 4*, 157–172.

Johansson, G., Aronsson, G., & Lindstrom, B. P. (1978). Social psychological and neuroendocrine stress reactions in highly mechanized work. *Ergonomics, 21*, 583–599.

Journal of Human Stress. (1975). Editorial. *Journal of Human Stress, 1*, 3.

Julius, S. (1981). The psychophysiology of borderline hypertension. In H. Weiner, M. A. Hofer, & A. J. Stunkard (Eds.), *Brain behavior, and bodily disease*, pp. 293–303. New York: Raven.

Kahn, R. L. (1981). *Work and health*. New York: Wiley-Interscience.

Kahn, R. L., Hein, K., House, J., Kasl, S.V., & McLean, A. A. (1982). Report on stress in organizational settings. In G.R. Elliott & C. Eisdorfer (Eds.), *Stress and human health*, (pp. 81–117). New York: Springer.

Kanner, A. D., Coyne, J. C., Schaefer, C., & Lazarus, R. S. (1981). Comparison of two modes of stress measurement: Daily hassles and uplifts versus major life events. *Journal of Behavioral Medicine, 4*, 1–39.

Karasek, R., Baker, D., Marxer, F., Ahlbom, A., & Theorell, T. (1981). Job decision latitude, job demands, and cardiovascular disease: A prospective study of Swedish men. *American Journal of Public Health, 71*, 694–705.

Karasek, R. A., Theorell, T. G. T., Schwartz, J., Pieper, C., & Alfredsson L. (1982). Job, psychological factors and coronary heart disease. *Advances in Cardiology, 29*, 62–67.

Karasek, R. A., Triantis, K. P., & Chaudhry, S. S. (1982). Co-worker and supervisor support as moderators of associations between task characteristics and mental strain. *Journal of Occupational Behavior, 3*, 181–200.

Kasl, S. V. (1974). Work and mental health. In J. O'Toole (Ed.), *Work and the quality of life*, pp. 177–196. Cambridge: MIT Press.

Kasl, S. V. (1978). Epidemiological contributions to the study of work stress. In C.L. Cooper & R. Payne (Eds.), *Stress at work*, pp. 3–48. Chichester, England: Wiley.

Kasl, S. V. (1981). The challenge of studying the disease effects of stressful work conditions. *American Journal of Public Health, 71*, 682–684.

Kasl, S. V. (1983). Pursuing the link between stressful life experiences and disease: A time for reappraisal. In C.L. Cooper (Ed.), *Stress research: Issues for the eighties*, pp. 79–102. Chichester, England: Wiley.

Kasl, S. V. (1984a). Chronic life stress and health. In A. Steptoe & A. Mathews (Eds.), *Health care and human behavior*, pp. 41–75. London: Academic Press.

Kasl, S. V. (1984b). When to welcome a new measure (editorial). *American Journal of Public Health, 74*, 106–108.

Kasl, S. V. (1984c). Stress and health. *Annual Review of Public Health, 5*, 319–341.

Kasl, S. V., & Cobb, S. (1980). The experience of losing a job: Some effects on cardiovascular functioning. *Psychotherapy and Psychosomatics, 34*, 88–109.

Kasl, S. V., & Cobb, S. (1983). Psychological and social stresses in the workplace. In B.S. Levy & D.H. Wegman (Eds.), *Occupational health*, pp. 251–263. Boston: Little, Brown.

Kasl, S. V. & Wells, J. A. (1985). Social support and health in the middle years: Work and the family. In S. Cohen & S.L. Syme (Eds.), *Social support and health*, pp. 175–198. New York: Academic Press.

Khaleque, A. (1981). Performance and strain in short-cycled repetitive work. *International Archives of Occupational and Environmental Health, 48*, 309–317.

King, H. (1970). Health in the medical and other learned professions. *Journal of Chronic Diseases, 23*, 257–281.

Kirscht, J. P. (1983). Preventive health behavior: A review of research and issues. *Health Psychology, 2*, 277–301.

Kittel, F., Kornitzer, M., & Dramaix, M. (1980). Coronary heart disease and job stress in two cohorts of bank clerks. *Psychotherapy and Psychosomatics, 34*, 110–123.

Kornhauser, A. (1965). *Mental health of the industrial worker*. New York: Wiley.

Kornitzer, M., Kittel, F., Dramaix, M., & deBacker, G. (1982). Job stress and coronory heart disease. *Advances in Cardiology, 29*, 56–61.

Krausz, M., & Freibach, N. (1983). Effects of flexible working time for employed women upon satisfaction, strains and absenteeism. *Journal of Occupational Psychology, 56*, 155–159.

LaRocco, J. M., House, J. S., & French, J. R. P., Jr. (1980). Social support, occupational stress, and health. *Journal of Health and Social Behavior, 21,* 202–218.

LaRocco, J. M., & Jones, A.P. (1978). Co-worker and leader support as moderators of stress-strain relationships in work situations. *Journal of Applied Psychology, 63,* 629–634.

Lazarus, R. S. (1978). A strategy for research on psychological and social factors in hypertension. *Journal of Human Stress, 4*(3), 35–40.

Lazarus, R. S., & DeLongis, A. (1983). Psychological stress and coping in aging. *American Psychologist, 38,* 245–254.

Lehman, E. W., Schulman, J., & Hinkle, L. E., Jr. (1967). Coronary deaths and organizational mobility. *Archives of Environmental Health, 15,* 455–461.

Levi, L. (1974). Stress, distress, and psychosocial stimuli. In A.A. McLean (Ed.), *Occupational stress,* pp. 31–46. Springfield, IL: Thomas.

Levi, L., Frankenhaeuser, M. & Gardell, B. (1982). Report on work stress related to social structures and processes. In G.R. Elliott & C. Eisdorfer (Eds.), *Stress and human health,* pp. 119–146. New York: Springer.

Logan, A. G., Milne, B. J., Achber, C., Campbell, W. A., & Haynes, R. B. (1982). A comparison of community and occupationally provided anti-hypertensive care. *Journal of Occupational Medicine, 24,* 901–906.

Lundberg, U. (1976). Urban commuting: Crowdedness and catecholamine excretion. *Journal of Human Stress, 2,* 26–32.

McCord, C. P. (1948). Life and death by the minute. *Industrial Medicine, 17,* 377–382.

McFarlane, A. H., Norman, G. R., Streiner, D. L., & Roy, R. G. (1983). The process of social stress: Stable, reciprocal, and mediating relationships. *Journal of Health and Social Behavior, 24,* 160–173.

McGrath, J. E. (1970). A conceptual formulation for research on stress. In J. E. McGrath (Ed.), *Social and psychological factors in stress,* pp. 10–21. New York: Holt, Rinehart & Winston.

McGrath, J. E. (1976). Stress and behavior in organizatins. In M.D. Dunnette (Ed.), *Handbook of industrial and organizational psychology,* pp. 1351–1395. Chicago: Rand McNally.

McLean, A.A. (1979). *Work stress.* Reading, MA: Addison-Wesley.

McLean, A.A. (1981). *How to reduce occupational stress.* Lexington, MA: Addison-Wesley.

Marmot, M.G. (1982). Hypothesis-testing and the study of psychosocial factors. *Advances in Cardiology, 29,* 3–9.

Matteson, M. T., & Ivancevich, J. M. (1982a). *Managing job stress and health.* New York: Free Press.

Matteson, M. T., & Ivancevich, J. M. (1982b). Stress and the medical technologist: I. A general overview. *American Journal of Medical Technology, 48,* 163–168.

Matteson, M. T., & Ivancevich, J. M. (1982c). Type A and B behavior patterns and self-reported health symptoms and stress: Examining individual and organizational fit. *Journal of Occupational Medicine, 24,* 585–589.

Matthews, K. A. (1985). *Assessment of Type A, anger, and hostility in epidemiological studies of cardiovascular disease.* In A. M. Ostfeld & E. D. Eaker (Eds.), *Measuring psychosocial variables in epidemiological studies of cardiovascular disease,* pp. 153–183. NIH Publication No. 85–2270. Washington, DC: US Government Printing Office.

Medalie, J. H., & Goldbourt, U. (1976). Angina pectoris among 10,000 men. II. Psychosocial and other risk factors as evidenced by a multivariate analysis of a five-year incidence study. *American Journal of Medicine, 60,* 910–921.

Medalie, J. H., Snyder, M., Groen, J. J., Neufeld, H. N., Goldbourt, U., & Riss, E. (1973). Angina pectoris among 10,000 men: 5-year incidence and univariate analysis. *American Journal of Medicine, 55,* 583–594.

Morgenstern, H. (1980). The changing association between social status and coronary heart disease in a rural population. *Social Science and Medicine, 14A,* 191–201.

Moss, L. (1981). *Management stress.* Reading, MA: Addison-Wesley.

Mott, P.E., Mann, F. C., McLaughlin, Q., & Warwick, D. P. (1965). *Shift work: The social, psychological, and physical consequences.* Ann Arbor: University of Michigan Press.

Mundal, R., Erikssen, J., & Rodahl, K. (1982). Latent ischemic heart disease in sea captains. *Scandinavian Journal of Work, Environment, and Health, 8,* 178–184.

Murphy, L. R. (1984). Occupational stress management: A review and appraisal. *Journal of Occupational Psychology, 57,* 1–15.

Narayanan, V. K., & Nath, R. (1982). A field test of some attitudinal and behavioral consequences of flextime. *Journal of Applied Psychology, 67,* 214–218.

Newman, J. E., & Beehr, T. A. (1979). Personal and organizational strategies for handling job stress: A review of research and opinion. *Personnel Psychology, 32,* 1–43.

Orth-Gomér, K. (1979). Ischemic heart disease and psychological stress in Stockholm and New York. *Journal of Psychosomatic Research, 23,* 165–173.

Orth-Gomér, K., & Ahlbom, A. (1980). Impact of psychological stress on ischemic heart disease when controlling for conventional risk indicators. *Journal of Human Stress, 6*(1), 7–15.

Pauly, J. T., Palmer, J. A., Wright, C. C., & Pfeiffer, G. J. (1982). The effect of a 14-week employee fitness program on selected physiological and psychological parameters. *Journal of Occupational Medicine, 24,* 457–463.

Payne, R., Jick, T.D., & Burke, R. J. (1982). Whither stress research?: An agenda for the 1980s. *Journal of Occupational Behavior, 3,* 131–145.

Payne, R., & Pugh, D. S. (1976). Organizational structure and climate. In M.D. Dunnette (Ed.), *Handbook of industrial and organizational psychology,* pp. 1125–1173. Chicago: Rand McNally.

Pearlin, L. I., Lieberman, M. A., Menaghan, E. G., & Mullan, J. T. (1981). The stress process. *Journal of Health and Social Behavior, 22,* 337–356.

Pepitone-Arreola-Rockwell, F., Sommer, B., Sassenrath, E. N., Rozee-Koker, P., & Stringer-Moore, D. (1981). Job stress and health in working women. *Journal of Human Stress, 7*(4), 19–25.

Pickard, L., & Collins, J. B. (1982). Health education techniques for dense residential settings. *Educational Gerontology, 8,* 381–393.

Pinter, E. J., Tolis, G., Gnyda, H., & Katsarkas, A. (1979). Hormonal and free fatty acid changes during strenuous flight in novices and trained personnel. *Psychoneuroendocrinology, 4,* 79–82.

Pomerleau, O. F. (Ed.). (1983). The University of Connecticut Symposium on employee health and fitness. *Preventive Medicine, 12,* 597–719.

Pooling Project Research Group. (1978). Relationship of blood pressure, serum cholesterol, smoking habit, relative weight and ECG abnormalities to incidence of major coronary events. Final report of the pooling project. *Journal of Chronic Diseases, 31,* 201–306.

Review Panel on Coronary Prone Behavior and Coronary Heart Disease. (1981). Coronary-prone behavior and coronary heart disease: A critical review. *Circulation, 63,* 1199–1215.

Rissler, A. (1977). Stress reactions at work and after work during a period of quantitative overload. *Ergonomics, 20,* 13–16.

Roman, J. A. (1963). Cardiorespiratory functioning in flight. *Aerospace Medicine, 34,* 322–337.

Rose, R. M. (1980). Endocrine responses to stressful psychological events. *Psychiatric Clinics of North America, 3*(2), 251–276.

Rose, R. M., Hurst, M. W., & Herd, J. A. (1979). Cardiovascular and endocrine responses to work and the risk for psychiatric symptomatology among air traffic controllers. In J.E. Barrett, R.M. Rose, & G.L. Klerman (Eds.), *Stress and mental disorder,* pp.237–240. New York: Raven.

Rose, R. M., Jenkins, C. D., & Hurst, M. W. (1978). *Air Traffic Controller Health Change Study.* Boston University School of Medicine: A report to the FAA, Contract No. DOT-FA72WA-3211.

Rose, R. M., Jenkins, C. D., Hurst, M., Herd, J. A., & Hall, R. P. (1982). Endocrine activity in air traffic controllers at work: II. Biological, psychological, and work correlates. *Psychoneuroendocrinology, 7,* 113–123.

Rose, R. M., Jenkins, C. D., Hurst, M., Kreger, B. E., Barrett, J. & Hall, R. P. (1982). Endocrine activity in air traffic controllers at work. III. Relationship to physical and psychiatric morbidity. *Psychoneuroendocrinology. 7.* 125—134.

Roskies, E., Spevack, M., Surkis, A., Cohen, C., & Gilman, S. (1978). Changing the coronary-prone (Type A) behavior pattern in a nonclinical population. *Journal of Behavioral Medicine, 1,* 201–216.

Rubin, R. T. (1974). Biochemical and endocrine response to severe psychological stress. In E.K.E. Gunderson & R.H. Rahe (Eds.), *Life stress and illness,* pp. 227–241. Springfield, IL: Thomas.

Russek, H. I. (1962). Emotional stress and coronary heart disease in American physicians, dentists, and lawyers. *American Journal of the Medical Sciences, 243,* 716–725.

Sales, S. M., & House, J. (1971). Job dissatisfaction as a possible risk factor in coronary heart disease. *Journal of Chronic Diseases, 23,* 861–873.

Salvendy, G., & Smith, M. J. (Eds.). (1981). *Machine pacing and occupational stress.* London: Taylor & Francis.

Schroeder, D. H., & Costa, P. T., Jr. (1984). Influence of life event stress on physical illness: Substantive effects or methodological flaws? *Journal of Personality and Social Psychology, 46,* 853–863.

Schuckit, M. A., & Gunderson, E. K. E. (1973). Job stress and psychiatric illness in the U.S. Navy. *Journal of Occupational Medicine, 15,* 884–887.

Schuler, R. S. (1982). An integrative transactional process model of stress in organizations. *Journal of Occupational Behavior, 3,* 5–19.

Seamonds, B. C. (1982). Stress factors and their effect on absenteeism in a corporate employee group. *Journal of Occupational Medicine, 24,* 393–397.

Selikoff, I. J., & Hammond, E. C. (Eds.). (1979). Health hazards of asbestos exposure. *Annals of The New York Academy of Sciences, 330,* 1–814.

Shapiro, A.P. (1978). Behavioral and environmental aspects of hypertension. *Journal of Human Stress, 4*(4), 9–17.

Sharit, J., & Salvendy, G. (1982). Occupational stress: Review and appraisal. *Human Factors, 24,* 129–162.

Shekelle, R. B., Gale, M., Ostfeld, A. M., & Paul, O. (1983). Hostility, risk of coronory heart disease, and mortality. *Psychosomatic Medicine, 45,* 109–114.

Shekelle, R. B., & Lin, S. C. (1978). Public beliefs about causes and prevention of heart attacks. *Journal of the American Medical Association, 240,* 756–758.

Shirom, A. (1982). What is organizational stress? A facet analytic conceptualization. *Journal of Occupational Behavior, 3,* 21–37.

Shostak, A. B. (1980). *Blue-collar stress.* Reading, MA: Addison-Wesley.

Siegrist, J., Dittmann, K., Rittner, K., & Weber, J. (1982). The social context of active distress in patients with early myocardial infarction. *Social Science and Medicine, 16,* 443–453.

Smith, H. P. R. (1967). Heart rate of pilots flying aircraft on scheduled airline routes. *Aerospace Medicine, 38,* 1117–1119.

Statement from the National High Blood Pressure Education Program. (1984). High blood pressure control in the workplace. *Journal of Occupational Medicine, 26,* 222–226.

Stokols, D., & Novaco, R. W. (1981). Transportation and well-being: An ecological perspective. In I. Altman, J. Wohlwill, & P. Everett (Eds.), *Transportation Environments.* New York: Plenum.

Tarnapolsky, A., Watkins, G., & Hand, D. J. (1980). Aircraft noise and mental health. I. Prevalence of individual symptoms. *Psychological Medicine, 10,* 683–698.

Tasto, D., Colligan, M. J., Skjei, E. W., & Polly, S. J. (1978). *Health Consequences of Shiftwork*. DHEW Publication No. NIOSH 78–154. Washington, DC: U.S. Government Printing Office.

Tennant, C. (1982). Psychosocial stress and ischaemic heart disease: An evaluation in the light of the diseases' attribution to war service. *Australian and New Zealand Journal of Psychiatry, 16,* 31–36.

Thackray, R.I. (1981). The stress of boredom and monotony: A consideration of the evidence. *Psychosomatic Medicine, 43,* 165–176.

Theorell, T., & Floderus-Myrhed, B. (1977). "Workload" and risk of myocardial infarction—A prospective psychosocial analysis. *International Journal of Epidemiology, 6,* 17–21.

Thoresen, C. E., Telch, M. J., & Eagleston, J. R. (1981). Approaches to altering the Type A behavior pattern. *Psychosomatics, 22,* 472–482.

Timio, M., & Gentili, S. (1976). Adrenosympathetic overactivity under conditions of work stress. *British Journal of Preventive and Social Medicine, 30,* 262–265.

Wall, T. D., & Clegg, C. W. (1981). A longitudinal field study of group work redesign. *Journal of Occupational Behavior, 2,* 31–49.

Watkins, G., Tarnapolsky, A., & Jenkins, L. M. (1981). Aircraft noise and mental health. II. Use of medicines and health care services. *Psychological Medicine, 11,* 155–168.

Wilkes, B., Stammerjohn, L., & Lalich, N. (1981). Job demands and worker health in machine-paced poultry inspection. *Scandinavian Journal of Work, Environment, and Health, 7* (Suppl. 4), 12–19.

Williams, A. W., Ware, J. E., Jr., & Donald, C. A. (1981). A model of mental health, life events, and social supports applicable to general populations. *Journal of Health and Social Behavior, 22,* 324–336.

Winnbust, J. A. M., Marcelissen, F. H. G., & Kleber, R. J. (1982). Effects of social support in the stressor-strain relationship: A Dutch sample. *Social Science and Medicine, 16,* 475–482.

Zorn, E. W., Harrington, J. M., & Goethe, H. (1977). Ischemic heart disease and work stress in West German sea pilots. *Journal of Occupational Medicine, 19,* 762–765.

4

Within-Group Designs: An Alternative to Traditional Control-Group Designs

JUDITH L. KOMAKI and MARJANE JENSEN

The complexity of the health and safety issues confronting workers and management today is matched only by the bewildering number of alternative methods that have been proposed to handle those issues. Proponents of each method herald the organizational benefits, the improved quality of working life, and the increased morale their method would realize if it were enacted. How to select judiciously among the available choices, the topic of this chapter, has become a critical task for the resource-conscious consumer. If one can distinguish effective means from fads before launching a new system, one has a greater likelihood of achieving success and avoiding unwarranted risks.

CLIMATE OF UNCERTAINTY

Unfortunately, the painful lack of evaluative data about the various methods for solving health and safety problems makes it difficult for any decision to rest on a sound empirical basis. Instead, the methods are assumed

Thanks to our colleagues for providing, such a rich diversity of field experiments showing that it is, indeed, feasible to document meaningful health and safety improvements in work settings; to Jerry Busemeyer for his expert guidance on the statistical analysis of time-series data; and last, but not least, to the editors for their cogent comments and warm encouragment.

to be naturally beneficial, or their effects are only cursorily examined, with consumers accepting testimonials from nonneutral parties as primary evidence. Goldstein (1980), in a comprehensive review of training in work organizations, acknowledges this unfortunate situation, noting that most "decisions are based upon anecdotal trainee and trainer reactions" (p. 238).

CONFOUNDING FACTORS

Even when more precise information is provided, few guidelines exist to aid in discriminating among choices. How does one separate fact from fantasy in claims such as the following:

Significantly less sick leave
$2 million annual savings
Stress-related illnesses reduced 30 percent
Accidents reduced by half

The most common pitfall is to believe that the claims are directly and solely attributable to the particular modifications advanced as their cause. This leap of faith is especially tempting when the proposed change fits one's own predilections, when the results produced are exactly what one would like to see, and when immediately comprehensible before-and-after comparisons are provided.

Although evidence based on before-and-after comparisons is surely better than none at all, it is not sufficient to support a verdict that the plan being credited is truly responsible for the positive results described. Before any improvements can be attributed to a particular change, other plausible alternatives must be ruled out. Let us say that information was collected about safety performance for a certain department in a certain factory; a program was introduced to reduce accidents, and accidents dropped by half. Given this information, one might want to conclude that the program was responsible for the dramatic improvement. After further probing, however, one may discover that accidents for all departments were on the decline during the same period in the same factory. This calls into question whether the program or some other event was responsible for the decrease in accidents.

Several factors, other than instituted changes, can explain why certain results occur (Campbell & Stanley, 1963):

History
Maturation
Statistical regression

Events other than the experimental program (*history*) might make the difference: for example, environmental change, the threat of an economic recession, a technological advance, the provision of new procedures, increased pressure from top management, or a new influx of trained personnel.

Sometimes improvements reflect changes that occurred merely through the passage of time, such as a worker becoming more expert at a particular task or becoming acclimated to the work environment (*maturation*).

In cases in which employees have been selected because they have done relatively poorly on a pretest measure, improvements sometimes reflect the effect of *regression artifacts*; that is, the tendency for individuals with extreme scores to move toward the mean of the population from which they were selected. Such pseudo-shifts have nothing to do with the improvement program under evaluation, but their effects can make the program seem more successful than it is in fact.

ESTABLISHED EVALUATION METHOD

How then can one rule out such potential influences? How does one definitely determine whether a particular program resulted in the desired improvements?

The established way is to use between-group or control-group designs. In the simplest case, people are assigned by lot to one of two groups. One is labeled the treatment group; the other, the control group. People in the treatment group are exposed to the program, whereas those in the control group are not. Other potential factors have been effectively ruled out because the random assignment of people to groups essentially equalizes the groups on all dimensions except the treatment; it is just as likely that people in one group will experience a significant event (history) or will become more practiced at a task (maturation) as will people in any other group. If improvements occur in the treatment group and not in the control group, one can confidently say that the program was responsible for the improvements. For a further description of these designs and potentially confounding effects, refer to the classic text by Campbell and Stanley (1963).

Although control-group designs are widely acknowledged to be the way to rule out alternative hypotheses, suitable control groups are difficult to arrange in most work settings. Rarely, outside of laboratories, can one randomly assign people to different groups with the purpose of exposing only one of the groups to a change. People in production typically stay in production; people in marketing stay in marketing. Seldom can individuals from two such departments be reassigned by lot so that some in marketing are now in production and vice versa, and people in newly formed Group 1 are exposed to the new policy change, whereas people in newly formed Group 2 are not. Despite the importance of precise evaluation, presently constituted organizational groups tend, rightfully, to be left intact. Consequently, control-group designs are rarely used in work settings.

RECOMMENDED ALTERNATIVE TO CONTROL-GROUP DESIGNS

Fortunately, a valuable alternative exists. Called within-group or single-case designs, these experimental designs are frequently used by people in the area of behavior analysis to assess the effectiveness of changes introduced in applied settings (Barlow & Hersen, 1984; Kazdin, 1982). Recently, researchers have recommended their use in work settings (Komaki, 1982; Luthans & Davis, 1982).

Instead of comparing a treatment group with a control group, comparisons are made within the group under study; that is, the group serves as its own control. The advantage is that one can draw sound conclusions about the effectiveness of changes without having to create a control group.

Another characteristic of within-group designs is the continuous assessment of performance over time. Information is typically collected on a daily basis or at least once a week. The advantage is that one can examine program effects over an extended period of time.

In the remainder of this chapter, the two most widely known and suitable within-group designs—reversal and multiple-baseline–are described, and the underlying rationale behind them explained. Examples are presented, demonstrating how these designs have been and can be used in work settings to improve decision making about health and safety issues.

REVERSAL DESIGN

In its simplest form, the reversal design includes three phases:

1. **Baseline (A).** In this phase, information is collected regularly during a period of time prior to the institution of any program.
2. **Intervention (B).** The program or intervention is introduced, and information continues to be collected regularly during a period of time.
3. **Reversal (A).** During this phase one returns to baseline conditions, discontinuing the program introduced previously, while still collecting data.

Rationale

Performance levels are examined to see if they fluctuate with the different conditions the design establishes and measures. If performance markedly improves when the intervention is in effect and worsens during the reversal phase, then one can say that the improvements resulted from the intervention and not from other variables.

The more times one can demonstrate that performance changes as the intervention is introduced and reversed, the more convincingly one can argue that the changes were caused by the intervention. A highly recommended extension of the reversal design includes a baseline (A), intervention (B), reversal (A), and then a reintroduction of intervention (B) phase. The reintroduction of the intervention phase strengthens the claim that the instituted modification caused the observed change.

Example of ABAB Reversal Design

An ABAB reversal design was used to determine whether an intervention was responsible for the improvements in the performance of a game room attendant (Komaki, Waddell, & Pearce, 1977). Initially, the owner of the game room had threatened to discharge the attendant; however, he agreed to keep the attendant on if his performance showed "visible" improvement. Data were collected: the percentage of time the attendant spent working was assessed three to five times each week for a total of 27 observation sessions during a seven-week period. Using a contingent pay sysem and a clarification of job duties, attempts to upgrade on-the-job performance were made during the intervention phase (B). During the reversal and reintroduction of intervention phases, the contingent pay system was discontinued (A) and reintroduced again (B).

Figure 4.1 shows the percentage of time the game room attendant spent

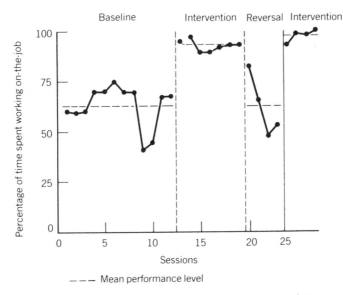

FIGURE 4.1. Example of ABAB reversal design. Shows percentage of time a game room attendant spent working on the job in a seven-week period during four experimental conditions: baseline (12 sessions in a three-week period), intervention (seven sessions in a two-week period), reversal (four sessions in one week), and reintervention (four sessions in one week). (From Komaki, Waddell, & Pearce, 1977. Copyright 1977 by Academic Press. Reprinted by permission.)

working during the four conditions. The attendant averaged 63 percent, sometimes dropping as low as 40 percent, during baseline. During intervention, his performance increased dramatically with a mean performance level of 93 percent. When conditions were reversed, the attendant's performance reverted to baseline levels, averaging only 62 percent. However, during the reintroduction of the intervention phase, the attendant's mean performance level again increased to more than 90 percent.

Because performance dramatically increased when the intervention was in effect, declined to baseline levels when the treatment was discontinued, and then increased again when the experimental modification was reintroduced, it was concluded that the intervention was responsible for the evident improvements. *History* was dismissed as a possible factor because it was not likely that another significant event would occur and then recur in precise time with the introduction and reintroduction of the intervention. The results were also not attributable to *maturation*. If maturation were the primary reason, then one might expect performance to increase gradually as time passed. However, there was a sharp drop during the reversal phase, and it was only when the intervention was reintroduced that the percentage of time spent working improved.

Regression artifacts were also dismissed because regression effects (once again, the tendency of extreme scores to move toward the mean) would be likely to appear in any series of repeated measurements and not just after the introduction and reintroduction of the intervention. In fact, because regression effects are usually a negatively accelerated function of elapsed time—that is, more likely to occur after the second, third, and fourth observations than after the twelfth, thirteenth, and fourteenth, or the nineteenth, twentieth, and twenty-first (the first and second intervention points)—it is even less likely that regression effects were responsible for the improvements during the two intervention phases. *The Hawthorne effect* (Adair, 1984; Roethlisberger & Dickson, 1939) was similarly ruled out. If the Hawthorne effect accounted for the changes in performance, then one would assume that the attendant's performance would be maintained during the last three phases when he was aware of the experimental manipulation. However, during the reversal phase his performance declined to baseline levels, and it was only when the reinforcement contingencies were reinstated that his performance improved.[1]

---

[1] Selection, another possible source of confounding, is not a problem with reversal or multiple designs, because comparisons are not made between groups but rather within groups. As a result, one need not be concerned with biases that can be introduced when selecting people for comparison groups. Likewise, testing is usually not a problem because it primarily has reference to personality or paper-and-pencil tests, a relative rarity in any repeated measures design. Differential mortality can be a problem if subjects drop out of the intervention phases more or less often than they do from the baseline or reversal phases. In assessing the results, one should be aware of this potential source of confounding.

Essential Feature

The *reversal phase*, as the name would imply, is crucial to the reversal design. Without the reversal phase, one is left with an AB or a before-and-after comparison design. With an AB design, all one can say is that changes occurred. One cannot make causal statements about the changes and the intervention because one cannot rule out potentially confounding factors, such as the rate of unemployment, new hires, or pending policy revisions.

Thus to set up a reversal design, one should do as follows:

Collect baseline data before the change.

Continue to collect data while the change is in effect.

Revert to baseline conditions at least once.

To restore baseline conditions, the active components of the program should either be discontinued or removed. If the program involves supervisors providing daily feedback, then the supervisors would not provide workers information about their performance. If the program consists of posting job openings at different locations, then the job openings would not be posted. The intentional removal or discontinuation of a program's active components constitutes a reversal. This should not be confused with a situation in which the intervention appears, for unknown reasons, to have lost its effectiveness.

Changing Phases

An important decision one must make when using the reversal design is *when* to introduce the next phase. How does one know when to stop collecting baseline data and begin the intervention? How much data should one collect during the intervention phase before removing the treatment and returning to baseline conditions? Are these phase changes a matter of convenience, or is there some criterion for the decision?

The *stability* of performance on the dependent variable is the major criterion. Stable performance has three characteristics. First, there is no trend in the mean level of performance; that is, if one graphs the performance data, the points will form a more or less horizontal line. Second, there is constant variability around that mean level within each phase; that is, the data points are all approximately the same distance from that line throughout the phase. The third characteristic is low variance; in essence, the distance between the data points and the mean level line is relatively small.

Obtaining stable performance data prior to changing phases is important for two reasons. First, it is easier to detect any change in performance across phases. If performance during a phase is fairly steady with only relatively small but consistent fluctuations around the mean level, then a change in performance after a phase change is easier to identify. If, however, the data are extremely variable, showing large fluctuations around the mean level, or if the variability changes in some unsystematic pattern during the course of the phase, then a change in performance could be lost in the fluctuations in the data. Second, if the data are stable, there is less chance that one will erroneously claim there is a change in performance across phases when there is none. The most obvious example of this phenomenon occurs when there is an increasing trend in the baseline data, the intervention is designed to improve performance, and performance continues to improve during the intervention phase. A less obvious example occurs when there is a decreasing trend in the baseline data, the intervention is designed to improve performance, and performance improves during the intervention phase. Normally, one would conclude that the treatment caused the change. However, it is possible that the natural course of the behavior is cyclic—first deteriorating for a time, then improving for a time, and then again deteriorating for a time. The timing of the intervention could have serendipitously coincided with the natural cycle. To avoid interpretation problems such as these, one should change phases only when the data are stable.

How does one determine stability? One way to assess stability is to inspect the graph visually, which Gottman (1981) calls the *interocular test* of significance. If the mean level of the data appears to be fairly horizontal or flat, the data points appear to cluster closely around the mean level, and the pattern of the clustering appears to be relatively uniform from the beginning to the end of the phase, then one assumes the data are stable. Unfortunately, the interocular test has been found to be misleading (Jones, Weinrott, & Vaught, 1977). The existence of a significant trend or nonconstant variance may not be readily apparent. Moreover, a short series of three or four consecutive data points, which are increasing or decreasing, may appear to be a trend when, in fact, they are not significant.

As a result, statistical analysis procedures are recommended when assessing trends, evaluating whether the variability is constant within each phase, and deciding whether the variability is sufficiently small. The statistical procedures should take into account the fact that the data are collected repeatedly over time on the same person or group of people and the possibility that the data may be autocorrelated—that is, recent values of the dependent variable may be partially a function of, and therefore can be predicted from, past values. (For further discussion of the unique properties of time-series data, refer to the section on statistical analyses of time series data in this chapter.)

To determine whether there is a *significant trend in the data*, a linear (autoregressive) or nonlinear (ARIMA) model is fit to the data to assess whether the data are autocorrelated. If the data are found to be autocorrelated, a least-squares estimation technique is used, which accounts for the autocorrelation in the data. The slope of the series is then estimated, and a specialized *t-test*, which also considers the autocorrelation, is calculated to determine if the slope is significantly different from zero. If it is not significantly different from zero, then one can conclude that no slope exists.

To determine whether the *variance of the data series is constant*, one first examines the beginning, middle, and end of the series in each phase. In essence, one divides the data into two or more chunks of at least twenty data points each. (Twenty or more data points per chunk are recommended as a rule of thumb because of the instability of variance estimates based on smaller samples.) Lastly, one calculates and compares the variance of each chunk. If the variance estimates of the separate chunks are approximately equal, then one can conclude the variance of the series is constant. No statistical tests, per se, have been developed to date for determining whether the variance estimates are significantly different from one another. (The traditional test [F-max test] for homogeneity of variance is inappropriate for comparing variances of time-series data because of the autocorrelation.) Despite the lack of a definitive test, at least two advantages exist for calculating the variance of chunks within each phase. One can make more informed judgements based on two sources of information—the graphic representation *and* the variance estimates. Secondly, one can report the basis on which the judgment was made.

To determine whether the *variance in the data is sufficiently small*, one calculates the variance for the phase as a whole. One then considers the estimated magnitude and variability of the next phase. For example, if under baseline conditions the average performance score is 35 percent, with a range of 15 to 50 percent, this variability might be quite acceptable if one expects the intervention to improve average performance to 75 percent. In contrast, if the intervention effect is expected to be much smaller, for instance, improving average performance to only 45 percent, then this effect might be obscured by the new relatively great variability in the data. Consequently, the decision about how small the variability should be will necessarily differ from one situation to another.

What should one do if the data are unstable? The standard answer is to continue collecting data under the present conditions until the data stabilize. However, what if one cannot expand the data collection time? Several alternatives exist, depending on the source of instability. If there is no trend, the variance is constant, and the only problem is that there is too much variability in the data, one could aggregate data points. By averaging three to five data points, the variability is reduced. (One should remember, however, that aggregating the data reduces the number of data points in the resulting series.) If the instability in the data is the result of an increasing trend in the baseline phase, one can add reversal phases to the reversal design. For example, if performance, which has shown a continual increase through the baseline (A) and the first intervention (B) phase, drops abruptly and stabilizes in the second baseline (A) phase, improves during the second intervention (B) phase, drops again in the third baseline (A), and so on, one can then conclude that the treatment was effective in improving performance. If the source of instability is a decreasing trend in the baseline data such that any apparent intervention effect could be the result of the intervention coinciding with natural cycles in performance, one might consider adding a baseline and using a multiple-baseline design. If the intervention effect is replicated, that alternative plausible hypothesis is ruled out. It is unlikely that each introduction of the treatment would coincide with the increasing portion of a cycle. Hence, if performance increases after treatment in each baseline and not before, causal conclusions can still be drawn about the effects of the program.

In summary, the decision to move to the next phase of the design rests on a careful analysis of the stability of the data. Stable performance is characterized by a lack of trend, constant range of variability across the phase, and variability that is small relative to the magnitude of the expected effect. Instead of relying on the *interocular test*, it is recommended that the decision be based on a more refined statistical analysis of the data.

Use of Reversal Designs

With Large Groups.   Although reversal designs are sometimes referred to as single-case, single-subject, or $n = 1$ designs, they are not restricted to individuals. The reversal design can be used in situations with a large number of personnel, for example:

80 employees in a small manufacturing plant (Carlson & Hill, 1982)

100 nonprofessional staff of an institution (Ford, 1981)

In a study involving an even larger group, the absenteeism rate of 215 workers in a large manufacturing and distribution center was modified by Pedalino and Gamboa (1974). Using an ABA reversal design, an incentive system was introduced after 32 weeks of baseline conditions. After the incentive system had been in place for 16 weeks, it was discontinued. Absenteeism declined when the program was in effect and increased again during the reversal phase. The authors concluded that the intervention was responsible for the significant reduction in absenteeism.

With Various Behaviors.   The reversal design can be used to evaluate changes in a variety of behaviors. Some health-related behaviors that have been studied are:

Dietary compliance (Magrab & Papadopoulou, 1977)

Weight loss (Mann, 1972)

Cigarette smoking (Jason & Liotta, 1982)

Some of the studied behaviors have been related to work performance:

Number and type of sales-client contacts (Anderson, Crowell, Sucec, Gilligan, & Wikoff, 1982; Gupton & LeBow, 1971)

Frequency of field-staff telephone calls to main office (Kreitner & Golab, 1978)

Charting of client progress in a university clinic (Fredericksen, Richter, Johnson, & Solomon, 1981)

Machine preparation time (Frost, Hopkins, & Conard, 1981)

Absenteeism (Durand, 1983)

Attendance at, and performance in, weekly meetings were the behaviors concentrated on in a study conducted in a residential facility for the mentally retarded (Hutchison, Jarman, & Bailey, 1980). An ABAB design, as displayed in Figure 4.2, was used to assess the effects of public posting of feedback on the attendance, tardiness, and performance (percentage of agenda items completed) of 10 staff members. After 9 weeks of baseline, the public posting condition was instituted for all three behaviors; and after 10 weeks (involving 10 meetings), the public posting was discontinued. This reversal period lasted for 3 meetings and then the posting procedure was reintroduced. Although the intervention had little effect on tardiness, both attendance and performance increased during the intervention phases and declined during

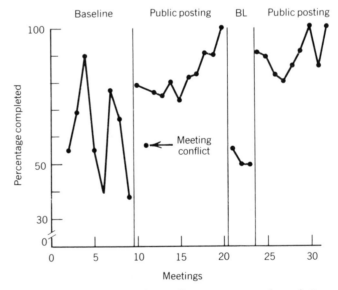

FIGURE 4.2. Example of ABAB reversal design. Shows percentage of agenda items completed during 30 weekly meetings. The intervention was introduced after meeting 9, withdrawn after meeting 19, and reintroduced after 3 more meetings. (From Hutchinson, Jarman, & Bailey, 1980. Copyright 1980 by Sage Publications, Inc. Reprinted by permission.)

the second baseline. Figure 4.2 displays the results for only the performance data (percentage of agenda items completed). Performance increased, decreased, and increased again during the first intervention, reversal and second intervention phases, respectively. It was concluded that the effect of public posting on performance was clearly demonstrated.

Figure 4.3 illustrates the ABAB reversal design used to study the effect of a group response-cost procedure on cash shortages in a family-style restaurant (Marholin & Gray, 1976). When it was found that employees regularly ended the day with cash shortages, a group response-cost policy was instituted, by which cash shortages were subtracted from the six employees' salaries on days in which the shortage exceeded one percent of total daily sales. Following 5 days of baseline, the response-cost procedure was introduced for 12 days, removed for 3 days, and then reintroduced for the final 21 days of the experiment. The magnitude of daily shortages sharply decreased when the group response-cost procedure was in effect, and increased when the procedure was removed. Therefore, it was concluded that the group response-cost was responsible for the reduction in cash shortages.

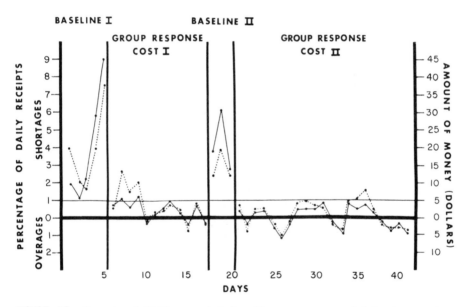

FIGURE 4.3. Example of ABAB reversal design. Shows percentage of daily shortages by restaurant employees during a 40-day period. The program was introduced on day 6, removed on day 18, and reintroduced on day 21. (From Marholin & Gray, 1976. Copyright 1976 by the Society for the Experimental Analysis of Behavior, Inc. Reprinted by permission.)

By Taking Advantage of Interruptions. In general, the reversal phase is planned. Sometimes disruptions occur, which result in the intervention being temporarily interrupted. Instead of regretting these interruptions, one can

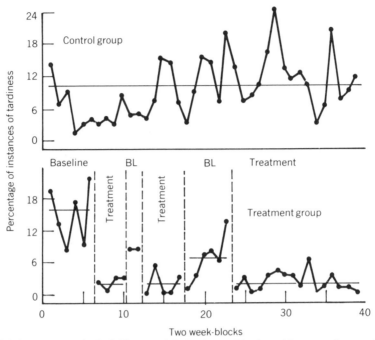

FIGURE 4.4. Example of ABABAB reversal design in combination with a control group. Upper graph presents percentage of instances of tardiness of control group during 77-week period. Lower graph presents same for treatment group. The incentive procedure was introduced after 12, 24, and 46 weeks, whereas the reversal phases occurred after 20 and 34 weeks. (From Hermann, de Montes, Dominquez, de Montes, & Hopkins, 1973. Copyright 1973 by the Society for the Experimental Analysis of Behavior, Inc. Reprinted by permission.)

take advantage of them and incorporate them into one's evaluation strategy. For example, in a study aimed at increasing attendance at self-help meetings (Miller & Miller, 1970), the reversal phase coincided with the period of time that the person responsible for implementing the intervention and obtaining the supplementary reinforcers discontinued her attendance at the meetings due to major surgery. Similarly, in the Pedalino and Gamboa study on absenteeism (1974) referred to earlier, evidence of the efficacy of the program was provided when unanticipated circumstances resulted in no incentive program being implemented for three weeks.

In Combination with a Control Group. Although the reversal design alone permits one to assess the efficacy of a program, a control group provides further confirmation of its efficacy. For example, Hermann and his associates (1973) used an ABABAB reversal design, in combination with a control group, to evaluate the effectiveness of an incentive procedure on the punctuality of workers in a manufacturing company (refer to Figure 4.4). Twelve chronically late male workers were assigned, in order of their appearance on the company payroll, to either a treatment or a control group.

For the treatment group, baseline and intervention conditions were rotated a total of three times. After 12, 24 and 46 weeks the workers were told they would receive a monetary bonus for every day they arrived on time. During return-to-baseline conditions (after 20 and 17 weeks), each worker was told that the payment for punctuality would be suspended. As the lower graph of Figure 4.4 shows, the percentage of time the workers were tardy fluctuated with the different conditions. Performance markedly improved when the treatment was in effect, resulting in late-arrival averages of only 2.5 percent, 1.8 percent, and 2.0 percent during the three intervention conditions. In contrast, during the baseline and reversal phases, the mean percentage of tardy arrivals was 15 percent, 8 percent, and 6.5 percent, respectively. On the other hand, the control group was not told about the incentive procedure and, as the upper graph of Figure 4.4 shows, did not experience the performance fluctuations shown by the treatment group. This provides additional confirmation of the effects of the incentive procedure.

In another example, 16 intact groups of retail clerks were randomly assigned to either an experimental or control group (Luthans, Paul, & Baker, 1981). In the experimental group, the effect of contingent reinforcement on five categories of work performance was investigated in an ABA design. Baseline measures were collected for four weeks, the contingent reinforcement was implemented for the next four weeks, and during the final four weeks the intervention was withdrawn. The control group remained under baseline conditions for the entire 12 weeks.

Both of these studies demonstrate how the addition of a control group, while *not* necessary to ensure sound conclusions with the reversal design, provides supplementary evidence. Where they are feasible, control groups are highly recommended additions.

Advantages and Limitations

The reversal design provides an attractive alternative to control-group designs, because one can infer cause-effect relationships from them without a traditional control group. The main drawback of the reversal design is that it requires a return to baseline conditions. Sometimes this is impossible to do: for instance, training situations in which people learn new information or acquire new skills. Other situations arise in which it is impractical to return to baseline conditions. Once progress is being made in desired directions, people implementing programs are sometimes reluctant to return to baseline, where conditions are likely to worsen and perhaps interfere with future performance. Even people who institute reversal phases sometimes cut them short. One department manager, after seeing the performance of his group worsen, felt that he could not afford to let performance return to baseline levels, so he restored the feedback program after only 11 days, despite the fact the performance had not yet declined to baseline levels (McCarthy, 1978).

The reversal design is also not a wise choice when harmful or dangerous conditions are involved. One would not want to remove safety guards or reinstall faulty equipment, for example. When problems exist with reverting to baseline conditions, one should consider the multiple-baseline design discussed in the next section.

MULTIPLE-BASELINE DESIGN

The multiple-baseline design is highly recommended. It does not require a reversal phase. It does not require a control group. Nevertheless, it can be used to draw causal conclusions about program efficacy.

As its name indicates, the multiple-baseline design involves the collection of data on two or more baselines. The baselines can be Behaviors A and B of a single group, the performance of Groups A and B, the performance of People A and B, or the performance of a single group in Settings A and B. The second characteristic of this design is that the intervention is introduced in a staggered or stepwise fashion across the baselines. In a multiple-baseline design across groups, for example, the treatment is first introduced with one group after eight weeks. When the desired change has occurred and the data are stable, the treatment is begun with the second group after 16 weeks. Again, following an observed change, the treatment is introduced with the next group, and so on, until the intervention has been introduced with all of the groups.

Rationale

To assess whether the intervention is effective, one makes within-group comparisons between baseline and intervention conditions. One examines whether performance changes after the intervention is introduced in the first group. Then, one assesses whether other groups that have *not* yet received the treatment continue at their baseline rates. Lastly, one examines whether the same change from baseline to intervention is replicated, that is, the same effect occurs at different times for the different baselines. If performance improves during, and not before, the intervention, and this result occurs each time the treatment is introduced, then one can conclude that the experimental manipulation is responsible for the changes.

The rationale relies on the probabilities of events other than the intervention occurring at the same time and in the same order. It is possible that an event other than the intervention might occur at time x for baseline A. However, it becomes less likely that the same event would occur at time y for baseline B. It is even less likely that the event would have the same effect on baseline A and on baseline B. As these events diminish in the likelihood of their occurrence, the explanation that the intervention is the event that caused the change becomes the most plausible.

Example of a Multiple-Baseline Design Across Groups

A multiple-baseline design across groups was used to assess whether a behavioral safety program was effective with personnel in two accident-ridden departments of a wholesale bakery (Komaki, Barwick, & Scott, 1978). The two departments were as follows:

1. Wrapping department, where packaged goods were bagged, sealed, labeled, and boxed.

2. Makeup department, where ingredients were mixed and the dough placed into pans.

Data were collected repeatedly over time. Information was collected in both groups about workers' safety performance four times a week during 25 weeks.

The intervention was introduced in a staggered manner to each department. After 5½ weeks, the intervention was introduced in the wrapping department. After 13½ weeks, it was introduced in the makeup department, as shown in Figure 4.5. The intervention consisted of a 30-

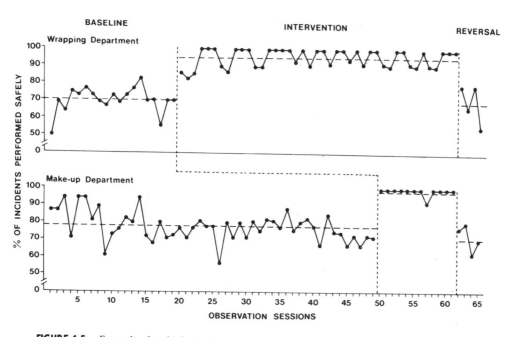

FIGURE 4.5. Example of multiple-baseline design across groups. Shows percentage of incidents performed safely in two departments of a bakery during a 25-week period. The safety program was introduced in the wrapping department after 5½ weeks and in the makeup department after 13½ weeks. (From Komaki, Barwick, & Scott, 1978. Copyright 1978 by the American Psychological Association. Reprinted by permission.)

minute performance clarification session and regular feedback regarding the percentage of incidents performed safely.

Figure 4.5 shows that, during baseline, safety performance in the wrapping department was more or less stable: 70 percent of the tasks were performed safely. Within the first week of the intervention phase, the percentage of incidents performed safely increased dramatically. After the second week, the wrapping department was regularly obtaining scores of at least 90 percent. In contrast, the makeup department, which remained under baseline conditions during the same period (sessions 20–49), showed no improvements, and performance remained at its baseline rate. After the program was introduced in the makeup department (between sessions 49 and 50), however, scores immediately rose to 100 percent and, with one exception, remained at this level.

It was concluded that the changes in performance were a function of the experimental manipulation because performance improved only after, and not prior to, the introduction of each intervention. After sessions 19 and 49, respectively, safety performance in the wrapping and makeup departments could have increased, decreased, or remained the same. However, in each case, increases followed the introduction of the intervention.

History was ruled out as an alternative explanation for the results. Although an extraneous event could have coincided with the introduction of the treatment in both departments and had a similar effect on the safety performance, this was unlikely. The outside event not only would have had to occur on two different occasions, but would also have had to coincide in timing, order, and effect. *Maturation* was also dismissed. If processes operating as a function of the passage of time or practice effects were responsible for the change, then one would expect that performance would steadily increase. However, performance in the two departments did not gradually improve throughout the study. Rather, it improved only after the intervention was introduced. The improvements were not likely to be a function of *regression artifacts*, because regression effects would be seen in any repeated measurements and not just after the experimental manipulations.

Essential Features

An integral feature of the multiple-baseline design is that *two or more baseline measures* are taken at the same time. In the above study, data were collected in two departments of the bakery. Doing this made it possible to assess the effect of the program not once, but twice. When the same effect occurs each time the program is introduced, one's conclusions gain credibility; thus one can argue more convincingly that improvements must be attributed to the program. Because of the importance of replicating the results, one baseline is not sufficient. In general, the more baselines one can obtain, the better, and the more certain the conclusions one can draw (Barlow & Hersen, 1984).

Another crucial feature of the multiple-baseline design is the *introduction of the intervention at staggered intervals*. In the study above, the program was introduced after 19 sessions for the first department, and 49 for the second. By introducing the program at different points in time, one can not only assess whether performance changes but also determine when performance changes. If the program is introduced at different times and performance changes only after the program is implemented, then one can more assuredly rule out alternative hypotheses. It is unlikely that an extraneous event would coincide and have a similar effect in the first and second departments at approximately the same times in exactly the same order as the intervention. Similarly, it is unlikely that practice effects would measurably influence the behavior at time one in the first group and at time two in the second group. In contrast, if the program is introduced simultaneously in both departments, then the situation is similar to the before-and-after design with all of its attendant problems. Outside events, practice, and regression effects become more plausible explanations in such a case. By staggering the treatment and demonstrating that performance improves only when the experimental modification is introduced, one can rule out these potential sources of confusion.

A third essential feature of the multiple-baseline design is that the *same intervention must be introduced* across the baselines. In the above study, the intervention was the same for both departments. Both departments received clarification and feedback in essentially the same manner. By introducing the same intervention again and again, one can determine whether the results are replicable. The more times one can demonstrate that a given effect occurs with a given intervention, the more confident one can be about the conclusions of program effectiveness.

Changing Phases

The decision to move from the baseline phase to the intervention phase in the multiple-baseline design is based on the criterion of data stability. To assess whether performance is stable, one must determine that (1) there is no significant trend in the data, (2) the variance is constant across the beginning, middle, and end of the phase, and (3) the variance in performance is small relative to the estimated magnitude of the treatment effect. The rationale and procedures for analyzing these three aspects of stability in the multiple-baseline design are identical to those discussed earlier with respect to the reversal design.

Types of Multiple-Baseline Designs

Across Groups.   Among within-group designs, the multiple-baseline *across groups* design is probably the most suitable to work settings because it makes use of existing groups and administrative units. The above study (Komaki et al., 1978) used separate departments, wrapping and makeup, within a single company. In recent research, the groups have consisted of the following:

1. **Sections of a police department:** Traffic section; patrol section; canine section (Larson, Schnelle, Kirchner, Carr, Domash, & Risley, 1980).

2. **Living units within a long-term care institution:** Hall E; Hall O; Hall C (Panyan, Boozer, & Morris, 1970).

3. **Production departments of an industry:** Screen printing; heat sealing; cutting and assembly; packing (Sulzer-Azaroff & de Santamaria, 1980).

4. **Work shifts within a factory department:** Morning shift; afternoon shift; night shift (Zohar & Fussfeld, 1981).

5. **Groups of university laboratory facilities:** 12 labs; 11 labs; 7 labs (Sulzer-Azaroff, 1978).

6. **Classrooms within a school:** Grade 2; Grade 3; Grade 4; Grade 5 (Maher, 1982).

This last grouping technique was used by Swain, Allard, and Holborn (1982) in their study of the effectiveness of a competitive game for improving the toothbrushing skills of school children. As seen in Figure 4.6, their two groups consisted of:

Grade one students
Grade two students

After six days of collecting baseline data, using disclosing tablets and a rating measure of oral hygiene, the authors introduced the "Good Toothbrushing Game" to the grade one students, while grade two students remained under baseline conditions. In grade two, the game was introduced seven days later. Oral hygiene scores improved (in this case, declined) in each grade, but only after the game was introduced. It was concluded that the game was an effective and practical approach to dental hygiene within the classroom setting.

Even larger groups have been used. Kempen and Hall (1977) assessed the absenteeism rates of six to eight thousand hourly employees at two plants of a manufacturing company:

Plant A
Plant B

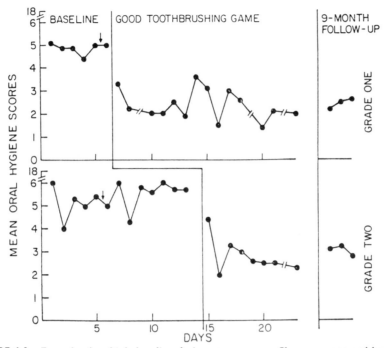

FIGURE 4.6. Example of multiple-baseline design across groups. Shows average oral hygiene scores for children in two school grades. The intervention was introduced in grade one after 6 days of baseline and in grade two after 14 days of baseline. (From Swain, Allard, & Holborn, 1982. Copyright 1982 by the Society for the Experimental Analysis of Behavior, Inc. Reprinted by permission.)

As indicated in Figure 4.7, baseline rates in Plant A were obtained for 34 months, and the implementation of an Attendance Management System followed. In Plant B, a similar system was introduced almost nine months later. Absenteeism decreased considerably in both plants following the introduction of the system. Because comparisons are made within rather than between groups in the multiple-baseline *across groups* design, it is not necessary to ensure, through random assignment, that the groups are more or less equivalent. In fact, the more the groups differ from one another in such factors as age, educational background, training, job description, supervision, and job task, the more that can be said for the generality or the external validity of the results.

Across People

The multiple-baseline design can also be arranged *across people*. It is similar to the multiple-baseline design across groups, except baselines are collected

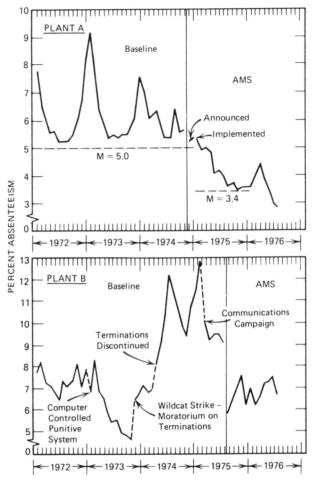

FIGURE 4.7. Example of multiple-baseline design across groups. Shows absence rates for hourly employees at experimental plants during a five-year period from 1972 to 1976. Program was introduced in Plant A in late 1974, and in Plant B in mid-1975. (From Kempen & Hall, 1977. Copyright 1977 by Behavioral Systems, Inc. Reprinted by permission.)

on different individuals instead of entire groups. The people have been individuals or parts of groups as varied as the following:

1. Uncooperative children during dental treatments (Stokes & Kennedy, 1980).
2. Married couples trying to quit smoking (Lichstein & Stalgaitis, 1980).
3. Direct-care staff of an institution (Kissel, Whitman, & Reid, 1983).
4. Managerial staff of a residential institution (Havel, Martin, & Koop, 1982).

5. Factory department supervisors (Sulzer-Azaroff & de Santamaria, 1980).

6. Teachers or teacher supervisors (Cooper, Thompson, & Baer, 1970; Cossairt, Hall, & Hopkins, 1973; Van Houten & Sullivan, 1975; Fulton & Malott, 1981).

7. School psychologists in a school district (Maher, 1981).

The multiple-baseline design *across people* was used by Hall and Hursch (1981) to evaluate a time-management training program. The people were four university faculty and staff identified as:

The nurse
The chemist
The physicist
The forester

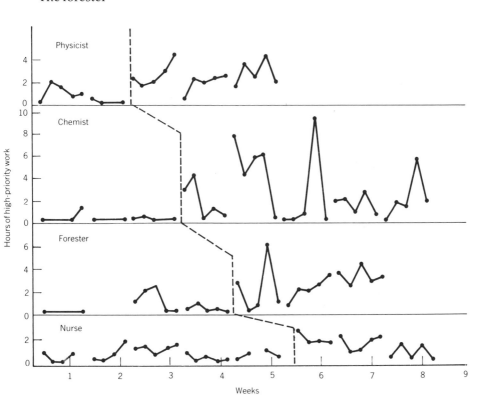

FIGURE 4.8. Example of multiple-baseline design across people. Shows total number of hours of high-priority work spent each week by four participants during an eight-week period. Training was initiated with the first participant after 2 weeks of baseline, with the second, third, and fourth participants after 3, 4, and 5 weeks, respectively. (From Hall & Hursch, 1981. Copyright 1982 by Haworth Press, Inc. Reprinted by permission.)

The time-management training consisted of a self-instructional manual, schedule planning, and weekly meetings with the investigator. It was intended to increase the time each person spent on high-priority tasks. Figure 4.8 shows that after two weeks of collecting baseline data, training was initiated with the physicist. Once a change had been observed, the chemist was exposed to the procedure, followed by the forester, and finally, the nurse. Figure 4.8 also displays the effect of training across the four people. Each person increased the number of hours spent on high-priority tasks as a result of the training program. Moreover, the four participants expressed satisfaction with the program and rated their own effectiveness as being higher after learning the time-management program.

Ten direct-care staff in a state center for the developmentally disabled were the target of a study by Burgio, Whitman, and Reid (1983). Figure 4.9 shows the manner in which data were collected about the behavior of the 10 individual staff members and the way in which the self-management procedures (self-monitoring, standard setting, self-evaluation, and self-reinforcement) were taught to three small groups of personnel (two groups of four and one group of two) at three different times. The purpose of the training was to increase the frequency of interactions between the staff and the residents. The results show that each person did interact more frequently with residents after learning the procedures than they did before. Given these results and the acceptance of the program by the staff, it was concluded that the program was an effective means for improving staff performance.

Across Behaviors

The multiple-baseline design can also be arranged *across behavior*. In this design variation, two or more activities, rather than two or more groups or people, are initially selected to be measured throughout the duration of the study.

A variety of behaviors can be targeted for change using the multiple-baseline design *across behaviors*. Some behaviors in the recent literature have been health-related:

1. **Cigarette smoking:** Puff frequency; puff duration (Fredericksen & Simon, 1978).
2. **Compliance behaviors in a regimen for diabetics:** Injecting daily insulin; testing urine periodically; exercising regularly (Schafer, Glasgow, & McCaul, 1982).

And some targeted behaviors have been descriptive of on-the-job performance:

1. **Time-management behaviors:** Arriving at work punctually; working a greater percentage of the time (Lamal & Benfield, 1978).
2. **Teaching behaviors:** Giving instructions; prompting correct responses; giving verbal praise (Page, Iwata, & Reid, 1982).

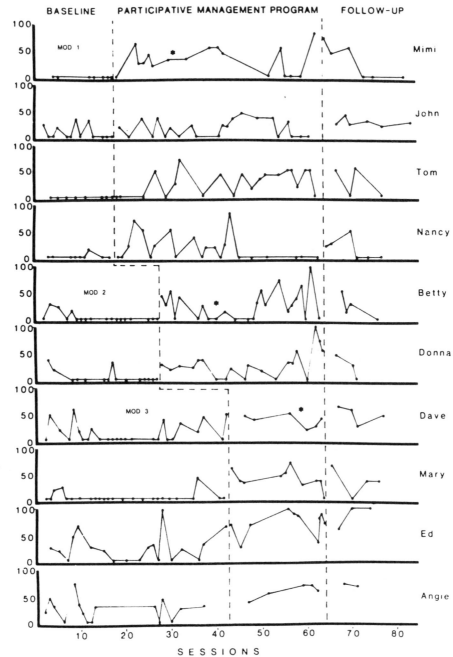

FIGURE 4.9. Example of multiple-baseline design across people. Shows percentage of observation sessions during which study participants interacted with institution's residents. The intervention was introduced to three small groups (modules) of participants at three different times. (From Burgio, Whitman, & Reid, 1983. Copyright 1983 by the Society for the Experimental Analysis of Behavior, Inc. Reprinted by permission.)

Lovett, Bosmajian, Frederiksen, and Elder (1983) employed the multiple-baseline design *across behaviors* to improve the performance of the professional staff of a community mental health center. The behaviors targeted for improvement were the amount and specificity of information recorded in clients' charts:

Recording specific information in intake summaries

Documenting treatment plans

Recording progress notes

Figure 4.10 illustrates that, after a baseline period of 7 weeks, the interven-

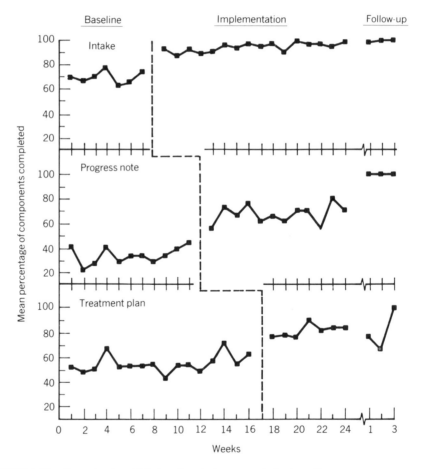

FIGURE 4.10. Example of multiple-baseline design across behaviors. Shows mean percentage of information components completed in three sections of client charts during a 25-week period and a 3-week follow-up period. The program was introduced after 8, 12, and 17 weeks, respectively, for the behaviors in the three sections of the charts. (From Lovett, Bosmajian, Frederiksen, & Elder, 1983. Copyright 1983 by Association for the Advancement of Behavior Therapy. Reprinted by permission.)

tion (simplified procedures and supervisory feedback) was introduced for the first chart section of intake summaries. At weeks 12 and 17, respectively, the intervention was introduced for the treatment plans and progress notes sections. The average percentage of items completed in each section increased only after the intervention was implemented for that section. The authors concluded that "procedural alterations and performance feedback systems can be effectively used to enhance the quality of documentation produced by an entire service unit of a community mental health center" (Lovett et al., 1983, p. 175).

In a study designed to improve the performance of grocery store clerks, Komaki, Waddell, and Pearce (1977) focused on three behaviors:

Remaining in the store

Assisting customers

Stocking merchandise

The intervention, which consisted, among other components, of extra time off with pay contingent upon performance, was introduced after 18, 24, and 30 observations, respectively, for the three behaviors. As can be seen in Figure 4.11, performance increased only after the treatment began for that

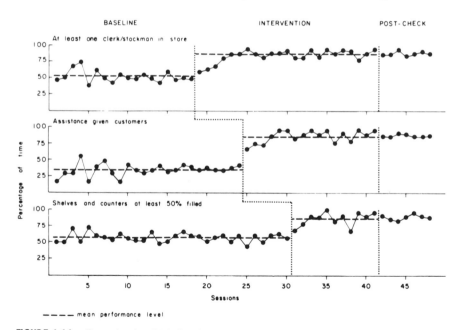

FIGURE 4.11. Example of multiple-baseline design across behaviors. Shows percentage of time the three target actions were performed by grocery personnel during a 12-week period of time. The first intervention took place after 18 sessions (3 weeks), the second after 24 sessions (4 weeks), and the third after 30 sessions (3 weeks). Post checks were carried out during the last 4 weeks. (From Komaki, Waddell, & Pearce, 1977. Copyright 1977 by Academic Press. Reprinted by permission.)

behavior. The authors concluded that, "in terms of the actual operation of the store, the increase meant that customers were being approached and served immediately, shelves and counters were rarely empty, and the occasions when at least one worker was not in the store were reduced to the times when both were needed in the storeroom to receive merchandise" (Komaki et al., 1977, p. 342).

Across Settings.   The multiple-baseline design across settings can be used when one is interested in assessing the performance of a single individual or group in different situations or circumstances, for example:

1. **Direct care staff can be assessed:**   As they teach communication skills; as they teach gross-motor skills (Page, Iwata, & Reid, 1982).
2. **Fast-food restaurant employees can be monitored as they interact with customers:**   At the cash register; in the dining area (Komaki, Blood, & Holder, 1980).

In a recent study (Kirchner, Schnelle, Domash, Larson, Carr, & McNees, 1980), the residential burglary deterrent effects of a helicopter patrol procedure were investigated in four separate areas in the Nashville metropolitan area:

High density area I
High density area II
Low density area I
Low density area II

The police chief decided how long the helicopter would remain in a particular area but was asked, for evaluation purposes, to leave the helicopter in one area for at least 10 days. Figure 4.12 shows that the helicopter procedure resulted in a lower average number of daily burglaries during helicopter patrol periods for the two high-density areas. For the two low-population density zones (not shown), however, the program did not deter burglaries. Thus it was concluded that the program was effective in one type of area but not in another.

Using Multiple-Baseline Designs to Evaluate Training Programs

Multiple-baseline designs are particularly appropriate for assessing the effects of training programs, a common activity in work settings, because they do not require a return to baseline conditions. For example, one could use a multiple-baseline design across groups, taking trainees from two sections of a plant, collecting baseline information on both groups, training the first group, and then later training the second group.

Another possibility is to use a multiple-baseline design across behaviors.

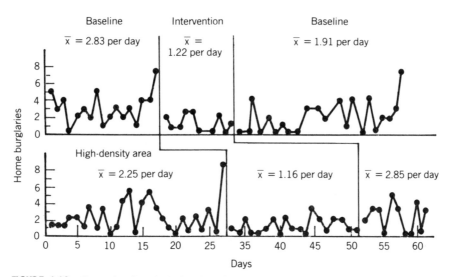

FIGURE 4.12. Example of multiple-baseline design across settings. Shows number of home burglaries in two high-density areas during a two-month period. Helicopter patrol was introduced in the first and second areas after 2½ and 4 weeks, respectively. (From Kirchner, Schnelle, Domash, Larson, Carr, & McNees, 1980. Copyright 1980 by the society for the Experimental Analysis of Behavior, Inc. Reprinted by permission.)

Miller and Weaver (1972) did this. First, they divided the content area into five subsections and devised a series of comprehensive tests on each subsection:

Research methods
Reinforcement
Stimulus control
Conditioned reinforcement
Aversive control

For 14 consecutive weeks, students took a different version of each of the five tests. After one week of baseline, the teaching package for the first sub-section of the course was introduced. Two weeks later, as indicated in Figure 4.13, materials were presented for the next subsection, and so on, until the instructional package had been introduced for all five subsections. When they found that scores for the last two subsections had not improved as dramatically as they had on the previous three sections, they revised those sections of the teaching package.

Using a similar approach, Van den Pol, Reid, and Fuqua (1983) evaluated a peer-training program for three safety-related skills. Experienced direct-care

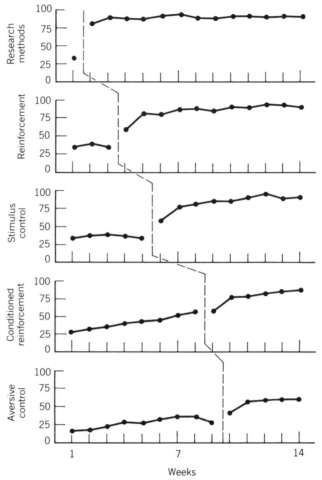

FIGURE 4.13. Example of multiple-baseline design across bahaviors used to evaluate a training program. Shows average scores on each subsection of an achievement test during 14 weeks. Teaching package was introduced on each of the five subsections at approximately 3-week intervals. (From Muller & Weaver, 1972. Reprinted by permission.)

staff of an institution for the developmentally disabled taught new staff members the appropriate steps for:

Responding to fires

Managing aggressive attacks

Assisting residents during convulsions.

Training was implemented sequentially for each skill. Figure 4.14 displays the results of training across the three skills for four new staff members. These results indicate that the peer-training program was effective in teaching new staff the necessary skills.

Advantages and Limitations

As can be seen, the multiple-baseline design is extremely versatile. One distinct advantage is that it is not necessary to return to baseline conditions. As a result, the multiple-baseline design can be used when evaluating

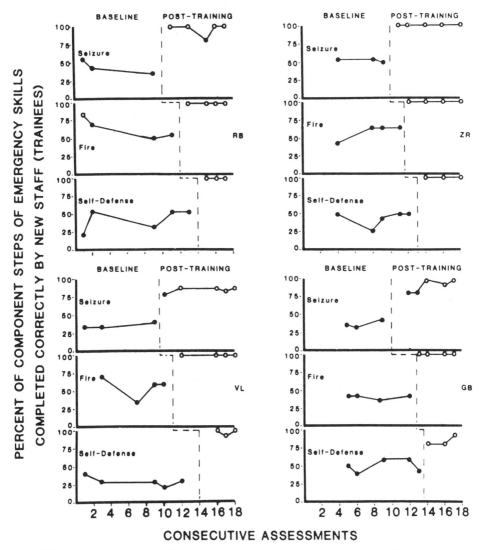

FIGURE 4.14.   Example of multiple-baseline design across behavior used to evaluate a training program. Shows percentage of steps of emergency skills completed correctly by four trainees during 18–20 consecutive assessments. Training was introduced sequentially for the three skills. All trainees received the training in the same staggered sequence, but at slightly different times. (From Van den Pol, Reid, & Fuqua, 1983. Copyright 1983 by the Society for the Experimental Analysis of Behavior, Inc. Reprinted by permission.)

techniques involving instructions, the clarification of desired performance, or the learning of new information skills. It can also be used when participants are concerned about returning to baseline conditions that are potentially harmful or dangerous.

Because the multiple-baseline design includes the collection of data on two or more baselines, it is almost always necessary to plan for it in advance, particularly when one does not use already existing archival records. On the other hand, the staggering of the techniques is relatively easy to arrange. Different groups can almost always be scheduled to receive the intervention at different points in time. One rationale, which has been successfully used, is to suggest that the program be introduced to one group on an experimental basis; if the program works, then it can be introduced to the next group; and, if there are problems, they can be solved *before* all the groups receive the treatment. When arranging a multiple-baseline design across behaviors to evaluate a training program, one can explain that introducing the components individually ensures that trainees have mastered each behavior before being allowed to proceed to the next. Reasons such as these for the staggering of treatments typically are well received, and they pave the way for the arrangements necessary for the implementation of the multiple-baseline design.

The main limitation to the multiple-baseline design is that one must be careful to select behaviors, settings, and people or groups that are *independent* of one another so that an intervention in one will not affect the others. When setting up a multiple-baseline design across behaviors, one should avoid closely related behaviors, for instance, carrying out variations of the same accounting method or maintaining similar pieces of equipment. Similarly, when the intervention is easily communicated, one should avoid groups in which the people in the first group are likely to communicate with people in the remaining groups.

BEHIND THE SCENES

The choice and execution of one's experimental design is critical to the quality of the conclusions one can draw. Unfortunately, discussions of these behind-the-scenes decisions rarely appear in print to guide researchers. An example from Komaki (1985) will be used here to illustrate how a multiple-baseline design across groups was chosen and implemented to evaluate a preventive maintenance (PM) program within the US Marine Corps. (See Figure 4.15.)

Choosing Among the Design Alternatives

The choice of experimental designs is affected by a number of factors such as the following:

The feasibility of randomly assigning people to groups and making available the treatment to only one group

The potential for regularly collecting information on the area(s) of interest

The possibility and suitability of reverting to baseline conditions

The ability to control the timing of the introduction of the program

The likelihood of identifying independent baselines

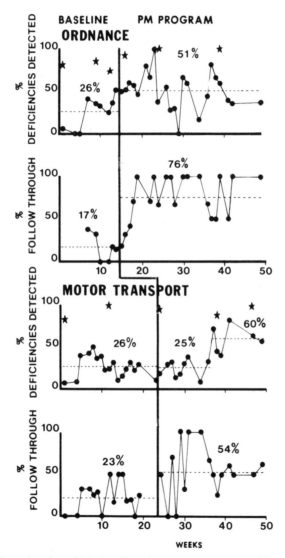

FIGURE 4.15. Example of multiple-baseline design across groups. Shows maintenance performance in two areas, detecting deficiencies and follow-through, in two sections of a heavy artillery Battery of the U.S. Marine Corps during an 11-month period. PM program was introduced after 14 and 22 weeks. (From Komaki, 1985.)

In making a decision about the design to use in evaluating the effectiveness of the PM program in the Marine Corps, the feasibility of randomly assigning people to groups was determined first. Between-group designs were quickly rejected because random assignment to groups was not feasible. Two intact groups, Ordnance and Motor Transport, already existed; personnel could not be assigned by lot to one or the other group.

A nonequivalent control group design was also ruled out, because if the program was introduced to the Ordnance group alone, for example, and improvements resulted, factors other than the program (e.g., training, supervision) could not be ruled out as possible causes.

Consequently, within-group designs were considered. The first consideration was the feasibility of collecting data regularly during an extended period of time. In this situation it was feasible: data collection could be scheduled *weekly* for a year.

Given that the first requirement of a within-group design was met (i.e., the repeated measurement of behavior), the requirements of the reversal design were then considered. It was determined that it was impossible to revert to baseline conditions. The feedback component of the intervention could be discontinued, but one could not revert to a situation in which the personnel were not aware of task requirements, which was the second component of the intervention. As a result, the reversal design was rejected.

The multiple-baseline design was then considered. It was determined that on-site personnel were willing to introduce the program in a staggered fashion. Two variations of the multiple-baseline design were then examined: across behaviors and across groups. Either variation initially appeared to be reasonable. There were two work groups:

Ordnance—with six medium-sized howitzers

Motor Transport—with approximately 50 vehicles

and two behaviors:

Detecting deficiencies—accurate identification of items needing repair, adjustment, or replacement

Follow through—timely corrective action taken on the above deficiencies

Closer examination, however, led to the rejection of the multiple-baseline design across behaviors. Questions arose regarding whether the two behaviors were sufficiently independent of one another. If the program were introduced for detecting deficiencies, improvements in follow through might occur because people responsible for the first behavior might goad others into doing the second behavior. If these carryover effects occurred, then it would be difficult to assess whether the program or some other event was responsible for the changes.

The multiple-baseline design across groups was the eventual choice. All three design requirements were met:

Information could be collected regularly on two groups.

The program could be introduced in a staggered manner, first to one group and later to the other.

The groups were independent of one another.

The groups were considered independent of one another for the following reasons: First, although both groups belonged to the same organizational unit and worked in the same general area, there was no overlap in personnel assignments, and the supervisory structures were entirely separate. Second, the intervention was not readily transmissible, consisting of specific clarification procedures for the different types of equipment and specially tailored performance feedback. Thus it was thought that introducing the intervention to the first group would not affect the performance of those in the second group.

Plans were then made to implement the multiple-baseline design across groups described below.

Design Execution

Just as existing on-site conditions impacted on the choice of an experimental design, execution of the design was also influenced by environmental events.

Implementation of Data Collection. Plans were made to collect information weekly on both behaviors in both groups. However, a number of developments prevented the strict following of these plans. There were weeks when a substantial portion of the camp was on leave, the entire regiment was engaged in field exercises, personnel in one entire section had other commitments, equipment was unavailable for assessment, or no deficiencies were accurately detected, thereby precluding an assessment of follow through. As a result, baseline information was collected for a fairly lengthy period of time (e.g., 14 weeks in Ordnance) to ensure a representative picture of the initial level, and the effects continued to be monitored for 11 months.

Stepwise Introduction of Intervention. On-site personnel agreed to wait and see what happened in the first group. Once improvements were forthcoming in the first group, they were easily persuaded to allow the same program to be introduced in the second group eight weeks later.

The major problem that occurred involved the difficulty of postponing the intervention when the data or the data collection went awry at the last moment. Intervention dates were arranged in advance around other commitments. When an upward trend in detecting deficiencies became apparent in the first group immediately prior to the scheduled introduction of the program, it was not possible to delay the intervention. Similarly, various events prevented data from being collected on the second behavior the three weeks before the intervention was introduced in the second group.

The factors described, while specific to this study, are nonetheless representative of those that occur in work and other applied setting. It is hoped that this brief, behind-the-scenes discussion will alert future researchers to the many real, but surmountable problems encountered in applied settings.

COMPARING WITHIN-GROUP AND BETWEEN-GROUP DESIGNS

Within-group designs are similar to between-group designs in one important way. They allow one to draw sound conclusions about program effectiveness. The range of independent variables is also just as broad; virtually any program that can be evaluated using a between-group design can be evaluated using a within-group design. The unique features—the substantial increase in the frequency of data collection and the within-group manner of comparison—expand and limit, respectively, the types of research questions that can be appropriately addressed using within-group designs.

Similarities

Ability to Draw Causal Conclusions.   One can say with a reasonable degree of certainty—using either a between- or within-group design—that a given program produced the desired results or that a program package was responsible for the obtained improvements. The primary advantage of within-group designs is that one can establish these causal relationships without having to create a control group. In the multiple-baseline design across groups, for example, intact groups, which are so common in applied settings, can be incorporated as an integral and valuable part of the design.

Range of Independent Variables.   Virtually any independent variable that can be evaluated using a between-group design can be assessed using a within-group design: for example, the impact of psychodynamically based therapy, the effect of a behavioral feedback program, the impact of L. L. Bean's duck shoes.

Any dependent variable can be included as long as it can be measured periodically over time: for example, client behavior, worker performance, treatment outcome. Even employee attitudes were assessed in a recent study (Hall & Hursch, 1981).

Differences

Increased Frequency of Data Collection.   A hallmark of within-group designs is the collection of data *regularly* during a period of time. Instead of assessing a group once before and once after the treatment, as is typically done with control group designs, the group is evaluated regularly over time: for example, three times each week for six weeks before the treatment, and three times each week for 12 weeks after the treatment. The repeated measurements provide a descriptive record during the entire course of the experiment, permitting an examination of trends throughout time.

The fact that data are collected repeatedly over time results in several distinct benefits. First, one can determine the immediacy of treatment effects. One can watch and see exactly how long it takes before the desired goal is reached. Second, if armed with up-to-date information, one can quickly spot problems and make adjustments accordingly. Third, one can determine whether initial improvements are maintained.

Manner of Comparison. Another difference between the two types of designs concerns the way in which comparisons are made. With between-group designs, comparisons are made *between* groups; the performance of the treatment group is compared with the control group. With within-group designs, the comparisons are made, as their name suggests, *within* groups; the performance of the group is compared under different conditions. Each group essentially serves as its own control. In a reversal design, for example, one examines how well a single group performs under Baseline, Treatment, Reversal to Baseline, and Treatment conditions. In a multiple-baseline design, one compares how well people performed under Baseline and Treatment conditions; this process is then repeated for each group, behavior, or setting.

The within-group comparison allows one to assess program effectiveness without relying on a traditional control group, a distinct advantage in applied settings where control groups are difficult to arrange. On the other hand, the within-group comparison limits the types of questions that can be appropriately addressed. Comparison questions such as "Is Program B better than Program C?" are particularly difficult to answer unambiguously. As a result, between- rather than within-group designs are recommended for comparing programs. (For further discussion, refer to the following section on research questions.)

Quasi-Differences

Other differences between within- and between-group designs have been noted. These are differences of degree rather than kind, however.

Number of Subjects. One frequently cited difference involves the number of subjects. Reversal and multiple-baseline designs, sometimes subsumed under the rubric of single-case, $n = 1$, or intensive designs, are said to be different from control-group designs because individual entities comprise the entire sample. Although reversal and multiple-baseline designs can be used with only one individual, they are *not* limited to individuals or small groups. The performance of large groups of people can and has been assessed by means of within-group designs (e.g., Carlson & Hall, 1982; Ford, 1981; Kempen & Hall, 1977).

External Validity. Another frequently mentioned difference between these designs concerns the relative external validity of their results. In particular, questions have been raised about the extent to which results obtained through within-group designs can be generalized to the rest of the population. Naturally, when a small number of subjects is involved in a design, one should be cautious about drawing general conclusions. Questions of external validity, however, are not inherent in reversal and multiple-baseline designs, nor are they effectively answered by using a control group or a large number of subjects. The only way to extend external validity is to conduct a series of related experiments in which the populations, settings, response variables, *and* intervention strategies vary from study to study (Runkel & McGrath, 1972). Unless the findings are replicated with people in different organizations using different treatment strategies and response measures, it is not clear to what extent any one set of results is unique to the specific individuals, organizations, intervention strategies, or response measures involved, *regardless* of the design used.

Statistical Analysis.   Another often noted difference concerns the use of statistical analyses. To date, within-group data are rarely analyzed. Instead, the data are visually inspected, as this chapter describes. As settings become more open and the desired improvements less dramatic, however, statistical techniques are being used with increasing frequency (e.g., Komaki, Collins, & Penn, 1982; Komaki, Heinzmann, & Lawson, 1980). Consequently between-group designs are not different from within-group designs in their use of statistical analysis techniques per se.

STATISTICAL ANALYSES OF TIME-SERIES DATA

Although the use of statistical procedures has been controversial (Michael, 1974) and statistical analyses are not yet the norm, they are being used with increasing frequency and are recommended to confirm conclusions based on a visual analysis of the data (Jones, Vaught, & Weinrott, 1977).

What analysis techniques are appropriate for data from within-group designs? Time-series analysis techniques, such as the nonlinear autoregressive integrated moving averages analysis (ARIMA) or the linear autoregressive analysis, are now considered to be the appropriate statistical approach. These time-series techniques are different from the standard ways of testing statistical significance (e.g., ANOVA) because they take into consideration a unique property of data in within-group designs—the fact that the data are in the form of a series of repeated measurements collected over time on a single individual or group. Because of the repeated nature of the data, it is necessary to assess whether the data may contain serial dependencies or autocorrelations, in essence, whether the current and future values of the measured variable may be partially a function of the past values. This determination is made, because tests of significance assume that the data are not autocorrelated. If autocorrelations exist, then the data must be modeled in a manner that accounts for the autocorrelation before one can conduct and draw sound conclusions from the tests of significance.

The steps involved in using a time-series analysis procedure are as follows: One first assesses the degree of autocorrelation in the data by fitting either a nonlinear (ARIMA) or a linear (autoregressive) model to the obtained data. If the data are found to be correlated, a least-squares estimation technique is used that accounts for the autocorrelation in the data. Then, a test of significance, a specialized t-test, which also considers the autocorrelation, is calculated. The t-tests provide information concerning whether there are changes in the *mean* of the series from phase to phase, for example, and from the baseline phase to the intervention phase in a multiple-baseline design, from the baseline phase to the intervention phase, and from the intervention to the reversal phase in a reversal design. One can also determine whether there are changes in *trends* from one phase to another that lead the data to drift in a different direction. For a comprehensive discussion of the use of nonlinear models, refer to Box and Jenkins (1970) or to Glass, Willson, and Gottman (1975). The use of linear models is summarized by Kazdin (1982) and discussed at length in a readable text that accompanies a computer program by Gottman (1981).

An example of the use of a time-series analysis is contained in a recent study (Komaki, Heinzmann, & Lawson, 1980). One question addressed was whether training alone would be effective in improving on-the-job performance. Training was introduced in a staggered manner to four groups. Data were collected repeatedly; a total of 165 observations were made over 45 weeks in each of the four groups. An ARIMA model was used. Results of the specialized t-test indicated that there were no significant level or drift changes between baseline and the training-only phase. Thus the authors concluded that training alone was not sufficient for improving on-the-job performance. This example illustrates the importance of statistically analyzing the data. One might have made an erroneous conclusion if one relied on a visual analysis of the data because performance did improve slightly over that of baseline during the training phase in all four groups. The use of a statistical analysis, however, showed that the slight improvements in performance could have occurred by chance alone. As a result, it was concluded that training alone did not result in significant on-the-job improvements.

Problems of interpretation relying solely on a visual inspection of the data are well illustrated by the Komaki et al. (1980) study. Moreover, a study by Jones, Weinrott, and Vaught (1978) demonstrated that agreements between visual inferences and statistical analysis occurred only about 60 percent of the time. Particularly noteworthy was the finding that significant intervention effects were considered nonsignificant with visual inspection more often than nonsignificant effects were erroneously considered significant. Jones et al. also noted that as the autocorrelation in the data increased, agreement between visual and statistical analyses decreased; that is, the autocor-relations in the data disrupted the ability of expert judges to detect intervention effects visually. The results of this study support the argument that results from within-group experiments should be subjected to statistical analysis in addition to visual analysis.

Thus, although statistical analyses are not yet the norm for within-group designs, the appropriate techniques are available, are being used with increasing frequency, and are highly recommended.

RESEARCH QUESTIONS AND WITHIN-GROUP DESIGNS

Questions dealing with program effectiveness (Does Program B work?) can be answered, using either within- or between-group designs. Maintenance questions (How long does the improvement continue?), however, are uniquely suited to a within-group strategy, while comparison questions (Is Program B better than Program C?) are more appropriately addressed using between-group designs. Table 4.1 lists the different types of questions that are typically raised in applied settings and designates whether reasonably sound conclusions could be drawn if a within-group design were used.

Program Evaluation Questions

Within-group designs are well suited to answering program effectiveness questions, when one is concerned with the effect of a particular program rather than with comparing two or more programs.

Using Reversal Designs. Reversal designs can be, and have been regularly, used to determine whether a particular program or program package was effective relative to *no* other program. Virtually all the studies presented in the reversal section of this chapter examned whether a particular treatment was effective in facilitating performance. An ABA reversal design, for example, was used to determine the effectiveness of a token economy program on dietary compliance (Magrab & Papadopoulou, 1977). To assess whether a program package could improve production in a light manufacturing firm, an ABAB reversal design was used (Frost et al., 1981). To evaluate the effectiveness of an incentive program on tardiness, an ABABAB reversal design was used (Hermann et al., 1973).

Using Multiple-Baseline Designs. Multiple-baseline designs have also been used to determine program effectiveness. A multiple-baseline design across behaviors was employed to evaluate the effectiveness of a program package on increasing diabetic adolescents' compliance with their medical regimens (Schafer et al., 1982). The effectiveness of a time-management training program was explored using a multiple-baseline design across groups (Havel, Martin, & Koop, 1982). Similarly, a training program's impact was assessed using a muliiple-baseline design across people (Kissel, Whitman, & Reid, 1983).

Legitimately Violating the Consistent Stepwise Progression Rule. In the examples presented thus far, the *stepwise progression rule* (Barlow & Hersen, 1984) held; that is, the same variable(s) was added to or subtracted from one phase to another. Thus researchers interested in assessing the effect

Table 4.1. Research Questions Appropriately Addressed Using Within-Group Designs

Questions	Within-Group Design Appropriate?
Program Evaluation	
1. *Program:* B vs. *No* B	Yes
Is the program (B) effective relative to *no* other program?	
2. *Program package:* $P^{1,2,3}$ vs. *No* P; $P^{1,2}$ vs. *No* P	Yes
Is the program package (P) with all its components (1,2,3) effective relative to *no* package?	
Is the program package with some of its components (1,2) effective relative to *no* other package?	
Assessing Trends Over Time	
3. *Maintenance*	Yes
What happens over time? Do the improvements continue? For how long? Does performance decline gradually or precipitously?	
4. *Immediacy of Effect*	Yes
Does the treatment have an immediate, a gradual, or an accumulating effect?	
Program Comparison	
5. *Comparative:* B vs. C	No
Is Program B more effective than Program C?	
6. *Parametric:* B1 vs. B2	No
Is variation 1 of Program B more effective than variation 2 of Program B, where variations can be different levels, times, frequencies, and so on.	
7. *Component:* $P^{1,2,3}$ vs. $P^{1,2}$	No
Is the program package (P) with all its components (1,2,3) more effective than the partial package with only some of its components (1,2)?	
Other	
8. *Necessary Component*	Yes
Is one component essential for an effect to occur? Does performance change when this component is present, but return to baseline conditions when the component is absent?	
9. *Necessary and Sufficient Component:* P^1 vs. *No* P	Yes
Is one component (P^1) alone effective?	
10. *Facilitative Component:* P^1 vs. P^{1+2}	Yes
Does one component (P^2) add to the effectiveness of the other component(s)? (Note: This is not P^1 vs. $P^{1,2}$)	

of Program C used an ACA, an ACAC, or an ACACAC reversal design in which Program C and only Program C was added or subtracted in each phase. Researchers did not use an ABCA or an ABCAC reversal design to evaluate the effect of Program C.

An exception exists, however, in which one can violate the consistent stepwise progression rule, but this can be done under only one condition: *when the initial treatment is shown to be essentially the same as Baseline.* For example, if one first introduced Program B and found that it resulted in little or no change from baseline, then one would be justified in violating the rule. One could evaluate Program C using an A=BCAC reversal design in which A=B denotes that the initial program (B) was essentially equivalent to Baseline (A).

An A=BCBC reversal design was used to evaluate the effectiveness of a smoking reduction program (Jason & Liotta, 1982). As Figure 4.16 shows, when the posting of "no smoking" signs (B) was not effective in reducing cigarette smoking in a university cafeteria, the authors introduced a second

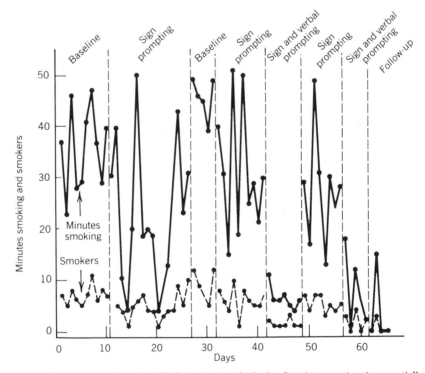

FIGURE 4.16. Example of an A=BCBC design in which the first intervention is essentially equivalent to the baseline condition. Shows number of people who were smoking and the number of minutes they smoked in a "no smoking" area of a cafeteria during a 60-day period. The first intervention was introduced after day 10, removed after day 26, and reintroduced after day 31. The second intervention was introduced after day 41, withdrawn, and then reintroduced after day 56. (From Jason & Liotta, 1982. Copyright 1982 by the Society for the Experimental Analysis of Behavior, Inc. Reprinted by permission.)

program (C) which involved personal requests to smokers to put out their cigarettes. When the second program (C) was found to be effective in reducing smoking, it was then discontinued and later reinstated, with a corresponding increase and reduction, respectively, in smoking. Because the initial program (B) was essentially the same as baseline, the investigators were justified in drawing conclusions about the efficacy of the second program (C) using an A=BCAC reversal design.

One can also legitimately violate the rule and use a multiple-baseline design when the initial treatment (B) is essentially the same as baseline. One first staggers the introduction of Program B and then introduces Program C and evaluates its efficacy. For example, to evaluate the effectiveness of a combination training and feedback program, a multiple-baseline design across groups was used in which training was first introduced in a staggered manner to group 1, group 2, and groups 3 and 4 (Komaki, Heinzmann, & Lawson, 1980). When the training did not result in significant improvements in performance over and above that of baseline, thereby indicating the equivalence of the initial treatment and baseline, feedback was added to the training and introduced in a staggered manner to each of the groups. Because performance improved during the training and feedback phase, it was concluded that the combination was effective in improving performance over and above that of baseline.

In summary, both reversal and multiple-baseline designs are appropriate for answering questions regarding program effectiveness. In general, only one treatment is introduced, in keeping with the consistent stepwise progression rule. An exception exists, however, in which more than one treatment may be introduced—when the initial treatment is demonstrated to be essentially the same as baseline. Nevertheless, the conclusions are still restricted to the effect of one and only one program.

Questions Concerning Trends Over Time

Within-group designs are uniquely suited to addressing questions concerning effects over time.

Maintenance. Questions about the long-term effects of a particular treatment are frequently raised by people in applied settings. Zohar and Fussfeld (1981), for example, were particularly interested in assessing what would happen after the termination of a successful token economy program. They had established the efficacy of the program for increasing workers' use of ear protection, using a multiple-baseline design across groups. After six weeks, they terminated the token program, but they continued to collect data periodically for 12 more weeks. As can be seen in Figure 4.17, they found that ear protection usage was maintained at the same high level that had been attained during the intervention.

Immediacy of Effects. Another advantage of within-group designs is that one can closely examine the effects of a treatment over time. One can see if a treatment has an immediate effect, as in Figure 4.5, in which a behavioral

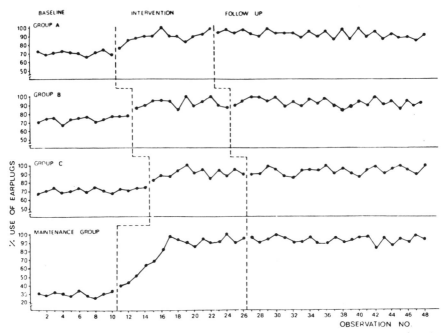

FIGURE 4.17. Example of multiple-baseline design across groups with a maintenance phase in which maintenance of performance during an extended period is assessed. After demonstrating the efficacy of a token economy program on the usage of ear protection, data were collected for 12 weeks after the termination of the program. (From Zohar & Fussfeld, 1981. Copyright 1982 by the Haworth Press, Inc. Reprinted by permission.)

safety program resulted in an immediate and dramatic change in performance (Komaki, Barwick, & Scott, 1978). Or one can find that the treatment resulted in a more gradual improvement, as in Figure 4.11 (Komaki, Waddell, & Pearce, 1977). Or one can find that progress was not being made as expected, and then attempt to rectify the problem, as in Figure 4.15 (Komaki, 1985). When it was seen that improvements were not forthcoming after four weeks with one behavior (Detected Deficiencies) for one group (Motor Transport), attempts were made to determine what the problem might be and to make the recommended changes. When the data still showed no improvements the following two weeks, efforts were renewed to ensure that supervising personnel knew what to do and carried out the recommended procedures.

In summary, the fact that information is regularly collected enables one to determine whether the changes are immediate or delayed, and whether improvements are maintained.

Comparison Questions

Within-group designs are poorly suited, however, to answering comparison questions (such as "Is Program B more effective than Program

C?" and "Is one variation of a program more effective than another variation of the same program?"). Consequently, they are not recommended when one wishes to compare the relative efficacy of one program versus another.

External Validity Limitations of Within-Group Designs. The conclusions that can be drawn about generality or *external validity* are severely limited when one uses a within-group design to compare programs. Because of the nature of within-group designs, one cannot assess what would happen if people were exposed to one and only one treatment and then compare those results with people exposed to another program.

Within-group designs, by their nature, require within-group comparisons. For example, if one uses a within-group design strategy to compare two programs, one must present both programs to the same individuals so that comparisons can be made about their performance under the two programs. The fact that the same individuals are exposed to more than one program, however, may result in what is referred to as *multiple-treatment interference*: that is, one treatment or program may interfere with the effects of another treatment and thus limit the generality of the results to individuals who experience the treatments in the same way or in the same order.

Two types of interference can occur. One can result from the *sequence effect*: the potential effect of introducing treatments in a particular order to the same individuals. The other type can result from the *contrast effect*: the effect of juxtaposing treatments in such a way that individuals can compare or contrast one treatment with the other and perhaps behave differently as a result of the juxtapositioning of the treatments. Both types of interference can be major threats to the external validity of the results.

Is it a consequential problem when the results may be limited to situations in which individuals have been exposed to more than one treatment? In a word, yes. When comparing programs, one virtually always wishes to determine which treatment *when presented by itself* is the most effective. Rarely, if ever, is one interested in generalization to the multiple-treatment situation.

Problems with Reversal and Multiple-Baseline Design Strategies. Modifications of the reversal design are sometimes used when comparing programs. Because of the problems of multiple-treatment interference, however, the conclusions that may correctly be drawn are definitely limited. Van Houten and Nau (1981) used a modification of the reversal design to compare the effects on speeding of two programs— namely, increased surveillance (B) and feedback posted on billboards (C). On two highways, they introduced the programs in a mixed order, for example, ABABACABACAB. After examining the results, they concluded that Program C was substantially more effective than Program B in reducing speeding. The problem with this conclusion is that the conditions under which these findings are likely to recur are limited to situations in which drivers are exposed to both treatments. One cannot readily rule out multiple-treatment interference, in particular, the contrast effect. Since drivers had the opportunity to experience and compare both methods of speed control, one cannot confidently assume that Program C will have the same beneficial effect when presented by itself. This external validity problem is not specific to the above study. Indeed, any investigator who compares treatments using a reversal design strategy will be subject to the same limitations on external validity. As a result, within-group design strategies are not recommended when answering comparison questions, no matter how many reversal phases are involved.

The same rationale holds for the multiple-baseline design strategy. Consider the case in which one first introduces in a staggered fashion Program B (increased surveillance), followed by Program C (feedback posted on billboards) on two different highways. Would one be justified in recommending that Program C be implemented statewide? In a word, no. Why? Because the results are ambiguous. The result may have been due to the *sequence effect*: once again, the effect of introducing different conditions in a particular order to the same group. Program C may have resulted in more improvement than Program B simply because it followed Program B. The effect of Program C may have been quite different if it had preceded Program B. Because one cannot rule out the sequence effect, one cannot confidently conclude that Program C by itself would be superior. Thus the multiple-baseline design strategy, which requires that programs be presented in a fixed sequence, is not recommended when addressing comparison questions.

One possibility that is sometimes suggested to avoid the sequence effect is counterbalancing the order of the programs: for example, Group 1 receives the treatments in the order ABAC, while Group 2 experiences the treatments in the order ACAB. Close examination of this design possibility reveals that problems still exist with multiple-treatment interference, specifically the contrast effect. People within each group can still compare and contrast the two treatments. As a result, their performance could partially be a function of the contrast effect. Furthermore, a new confound has been introduced. The unique characteristics of the people in each group are confounded with the order of the treatments. (It is assumed that people have not been randomly assigned to Groups 1 and 2, but that the groups are intact). One cannot eliminate the possibility that Program C appears to be superior only because it occurred first in Group 2, rather than Group 1. In short, although counter-balancing the order of treatments may be a partial attempt to solve the sequence effect problem, this design possibility still cannot rule out the contrast effect, and it introduces the new problem of treatment sequence being confounded with group membership. As a result, it too is not recommended when comparing programs.

Problems with Multiple-Treatment Designs. Another set of designs, multiple-treatment designs, are sometimes suggested as ways in which one can still use within-group design strategy and make comparisons between two or more treatments. However, they too suffer from multiple-treatment interference (Kazdin, 1982, pp. 194–199). As a result, they are not recommended for comparing programs.

All the variations of multiple-treatment designs have three common characteristics. First, a single behavior of one person or group is the focus of the intervention attempt. Second, two or more interventions are implemented in the same phase after baseline data have been collected. Third, while the interventions are implemented in the same phase, they "are in effect at the same time . . . they must take turns in terms of when they are applied" (Kazdin, 1982, p. 173). The manner in which the interventions "take turns" distinguishes the several types of multiple-treatment designs. The presentation of the interventions can take turns across different time periods (e.g., morning and afternoon), settings (e.g., the recreation room and the cafeteria), or program managers (e.g., the line supervisor and the department supervisor). For example, in the alternating-treatment design across times, treatments B and C are compared by presenting treatment B to a group of subjects on day one and treatment C to the same group on day two. The presentation order is typically randomized for a period of several days: that is, the treatments are not presented on alternating days in a simple BCBCBC fashion. Rather, presentation is randomized (e.g., BCCBCBBC). (For further information about these designs, refer to Barlow & Hersen, 1984; Kazdin, 1982.) Despite their elegance, these designs still suffer from multiple-treatment interference, particularly contrast effects. Because individuals experience all the treatments, it is not possible to predict the effect of a single treatment presented by itself.

In summary, within-group designs are not appropriate for answering comparison questions because of the potential effects of multiple-treatment interference. Reversal designs and multiple-treatment designs are particularly influenced by contrast effects; multiple-baseline designs are particularly influenced by sequence effects; and designs that counterbalance the order of treatments are particularly influenced by the confounding of sequence effects and the unique characteristics of the group. Therefore, between-group designs are the recommended alternative for comparison questions.

Necessary?   While problems exist in comparing two programs directly, other frequently raised questions can be unambiguously answered using reversal or multiple-baseline design strategies. For example, one can determine whether one component (P^2) of a treatment package ($P^{1,2}$) is *necessary*; that is, whether it is essential for an effect to occur. If it is essential, then the effect should occur in its presence, but not in its absence.

Therefore, to test whether a component is necessary, one examines what happens in the presence of, and in the absence of, the component. Typically, a treatment package is first assessed using a multiple-baseline design strategy and found to be effective. Then, the component considered to be essential is deleted, and later it is reintroduced. The phases of the multiple-baseline design are as follows:

1. Baseline
2. Package program ($P^{1,2}$) introduced in staggered fashion

3. Reversal (P[1])

4. Package program (P[1, 2])

The essential phases are the reversal phase, during which the component in question is either discontinued or removed, and the phase in which the component in question is reinstated. If performance declines to baseline during the reversal phase when the component is removed and then improves following the reintroduction of the component, then one can conclude that the component in question is essential for an effect to occur.

This phenomenon has been demonstrated (Komaki, 1985). First, it was demonstrated, through using a multiple-baseline design across groups, that a program consisting of specification and feedback was effective. Then the investigator introduced a Reversal phase, dropping out the component considered essential, in this case, feedback. When performance declined during the Reversal phase, the feedback was reintroduced with a subsequent improvement in performance. As a result, it was concluded that feedback was a necessary component of the treatment package without which an effect would not occur.

Necessary and Sufficient?. Another frequently raised question concerns whether a component is necessary and sufficient, that is, whether it *alone* would result in an effect.

To test whether a component is necessary and sufficient, one examines what happens when only the one component is presented. Two studies are typically conducted. In the first study, a program package is evaluated, using either a reversal or multiple-baseline design, and found to be effective. Questions are then raised regarding whether the component alone would result in the same changes. Then, a *second* study is conducted. The design is a straightforward reversal or multiple-baseline design in which the treatment is the single component (P[1]) in question. An AP[1]AP[1] reversal design could be conducted, for example. Or a multiple-baseline design could be used in which the single component is introduced.

This situation occurred regarding a combination training and feedback program. In the first study, it was found that the program resulted in substantial improvements in performance (Komaki et al., 1978). Questions were raised, however, about the importance of the feedback; perhaps training alone would be sufficient. In the second study, it was assessed whether training was necessary and sufficient (Komaki et al., 1980). As can be seen in Figure 4.18, a multiple-baseline design across groups was used and Training alone was introduced following Baseline. When it was found, using a statistical analysis appropriate for time-series data, that the training *alone* had no statistically significant effect on performance, it was concluded that training alone was not sufficient to improve performance.

Facilitative?. The within-group strategy can also be used to answer questions such as the following: Does one component *add* to the effectiveness of the other component(s)? Does one component have more of an effect than another?

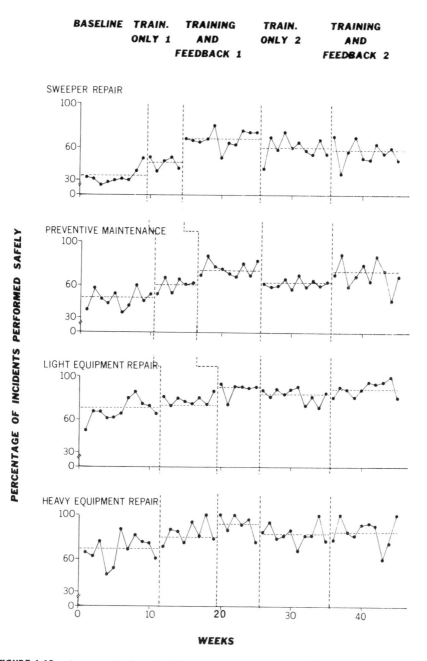

FIGURE 4.18. An example of a multiple-baseline design across groups in which an assessment is made regarding whether a single component is a necessary and sufficient condition for change. Shows percentage of incidents performed safely under five experimental conditions. (From Komaki, Heinzmann, & Lawson, 1980. Copyright 1980 by the American Psychological Association. Reprinted by permission.)

To test whether a component (P²) adds to the effectiveness of other component(s) (P¹), one first examines the effects of the other component(s) (P²). Then one assesses the effects of adding the component (P¹) in question. A multiple-baseline design is suggested with the following phases:

1. Baseline
2. Component 1 (P¹) introduced in staggered manner
3. Component 1 + 2 (P¹⁺²) introduced in staggered manner

An experiment was recently conducted addressing the following question: What was the effect of a consequence, such as feedback, *over and above* that of antecedents, such as rules and reminders (Komaki, Collins, & Penn, 1982)? A multiple-baseline design across four groups was used with the following phases:

1. Baseline
2. Antecedent Condition
3. Antecedent + Consequent Condition

When it was found that performance during the Antecedent Condition significantly improved for two of the four departments, and then during the Consequent Condition improved above and beyond that of the Baseline *and* Antecedent Condition, it was concluded that the feedback did, in fact, *add* to the effectiveness of rules and reminders despite the fact that in some groups significant changes had already occurred.

In summary, within-group designs can be and have been successfully used to address a variety of commonly raised questions in work settings. The primary exception is questions involving direct comparisons (e.g., Is program A better than program B?). One can address questions regarding program effectiveness (e.g., "Does program A work?" and "Is program A with all its components effective?"). Within-group designs are uniquely suited to addressing maintenance issues (e.g., How long does the effect last?). They can also be used to answer questions about programs involving more than one component (e.g., Is one component essential? Is one component alone effective? and Does one component add to the effectiveness of other components?).

SUMMARY

Consumers today can choose among a bewildering number of alternative programs for solving health-related problems. Determining whether a particular program produced its intended results, however, is not a simple task. Between-group or control-group designs—in which individuals are assigned by lot to one of two groups, only one of which is treated—are the standard means to measure program effectiveness. However, control groups are nearly impossible to arrange in most work settings.

Fortunately, a valuable alternative to control-group designs is available. Within-group designs are frequently used by behavioral psychologists to evaluate the effectiveness of changes introduced in applied settings. Instead of comparing a treatment group with a control group, these designs make comparisons within the group under study; that is, the group serves as its own control. With within-group designs, one is able to draw sound conclusions about effectiveness of changes without having to create a control group. The two most widely used within-group designs are the reversal and the multiple-baseline.

The reversal design includes, at least, three phases. During the initial Baseline phase, information about the dependent variable is collected regularly for a period of time prior to the implementation of any treatment. Once the baseline data stabilize, the second phase, Intervention, is introduced. During Intervention the treatment is implemented and repeated measurements of the dependent variable continue. During the third phase, Reversal, the program is removed but data collection continues. If the dependent measure improves during the Intervention phase and returns to Baseline levels during the Reversal phase, then one can say that the improvements resulted from the treatment. The Reversal phase is essential for demonstrating that the treatment was responsible for any changes in the dependent measure and for eliminating alternative explanations for the results, for example, history, maturation, and regression artifacts. The problem with the Reversal phase, however, is that it is sometimes impossible or impractical to effect.

When a Reversal phase cannot be arranged, the multiple-baseline design is a highly recommended alternative. Similar to the reversal design, the multiple-baseline design requires that data be collected regularly during Baseline and Intervention phases. As its name implies, however, it involves the collection of data across two or more baselines (consisting of people, groups, behaviors, or settings). The treatment is then introduced in a staggered fashion across these baselines. For example, in the multiple-baseline design across groups, after the baseline data are stable, the treatment is introduced with one group, while all other groups remain under baseline conditions. After a change has occurred in the dependent measure and performance has once again stabilized, the treatment is introduced to the second group. Again, after an observed change, the treatment is introduced to the next group, and so on. If performance on the dependent measure changes within each group only after the treatment is introduced for that group, and performance remains at baseline levels in the untreated groups, and this effect is replicated in the other baselines, one can conclude that the treatment is responsible for the changes. The multiple-baseline design is extremely versatile: it can be arranged across groups as just discussed, or it can be arranged across people, behaviors, or settings. In addition, it is well suited for evaluating training programs or any other treatments whose effects cannot be discontinued.

The choice of an experimental design is affected by a number of factors: the feasibility of randomly assigning people to groups, the potential for regularly collecting information on the dependent measure, the possibility of including a reversal phase, the ability to control the timing of the introduction of the treatment, and the likelihood of identifying two or more independent baselines. Although decisions about these factors are typically made behind-the-scenes, they are, nonetheless, critical to the quality of the conclusions one can draw.

Within-group and between-group designs are similar in one important way. Both types of designs can verify a cause-and-effect relationship between virtually any type of independent variable and any dependent variable that can be measured repeatedly over time. They differ, not in the conclusions that can be drawn, but in the manner in which comparisons are made. With between-group designs, comparisons are made between separate treatment and control groups. With within-group designs, the performance of a single group is compared under baseline and treatment conditions. Another difference is that data are collected repeatedly in within-group designs, resulting in a rich source of information about performance during an extended period of time. Other differences (the number of subjects, the external validity of the conclusions, and the statistical analysis of data) are differences of degree rather than kind.

Statistical analysis of the data in within-group designs is highly recommended. Statistical techniques, which account for potential autocorrelations in time-series data, have been developed and are being used with increasing frequency to assess the statistical significance of possible intervention effects.

Both within- and between-group designs can be used to answer program effectiveness questions (e.g., Is treatment B effective in improving performance?). However, the between-group design provides the most unambiguous answer to questions regarding the comparison of two or more treatments (e.g., Is treatment B better than treatment C?). Within-group designs, on the contrary, are uniquely suited to answering questions regarding the timing and the maintenance of treatment effects.

In conclusion, within-group designs can verify cause-effect relationships without using control groups and are able to document and trace shifts in behavior during extended spans of time. Such designs are especially well suited and are highly recommended for examining health and safety issues in the workplace.

REFERENCES

Adair, J. G. (1984). The Hawthorne effect: A reconsideration of the methodological artifact. *Journal of Applied Psychology, 69,* 334–345.

Anderson, D. C., Crowell, C. R., Sucec, J., Gilligan, K. D., & Wikoff, M. (1982). Behavior management of client contacts in real estate brokerage: Getting agents to sell more. *Journal of Organizational Behavior Management, 4,* 67–95.

Barlow, D. H., & Hersen, M. (1984). *Single-case experimental designs: Strategies for studying behavior change.* New York: Pergamon.

Box, G. E. P., & Jenkins, G. M. (1970). *Time-series analysis: Forecasting and control.* San Francisco: Holden-Day.

Burgio, L. D., Whitman, T. L., & Reid, D. H. (1983). A participative management approach for improving direct-care staff performance in an institutional setting. *Journal of Applied Behavior Analysis, 16,* 37–53.

Campbell, D. T., & Stanley, J. C. (1963). Experimental and quasi-experimental designs for research on teaching. In N.L. Gage (Ed.), *Handbook of research on teaching.* Chicago: Rand McNally. *Reprinted separately* as *Experimental and quasi-experimental designs for research,* Chicago: Rand McNally, 1966.

Carlson, J. G., & Hill, K. D. (1982). The effect of gaming on attendance and attitude. *Personnel Psychology, 35,* 63–73.

Cooper, M. L., Thomason, C. L., & Baer, D. M. (1970). The experimental modification of teacher attending behavior. *Journal of Applied Behavior Analysis, 3,* 153–157.

Cossairt, A., Hall, R. V., & Hopkins, B. L. (1973). The effects of experimenter's instructions, feedback, and praise on teacher praise and student attending. *Journal of Applied Behavior Analysis, 6,* 89–100.

Durand, V. M. (1983). Behavioral ecology of a staff incentive program. *Behavior Modification, 7,* 165–181.

Ford, J. E. (1981). A simple punishment procedure for controlling employee absenteeism. *Journal of Organizational Behavior Management, 3* (2), 71–79.

Frederiksen, L. W., Richter, W. T., Johnson, R. P., & Solomon, L. J. (1981). Specificity of performance feedback in a professional service delivery setting. *Journal of Organizational Behavior Management, 3* (4), 41–53.

Frederiksen, L. W., & Simon, S. J. (1978). Modifying how people smoke: Instructional control and generalization. *Journal of Applied Behavior Analysis, 11,* 431–432.

Frost, J. M., Hopkins, B. L., & Conard, R. J. (1981). An analysis of the effects of feedback and reinforcement on machine-paced production. *Journal of Organizational Behavior Management, 3* (2), 5–17.

Fulton, B. J., & Malott, R. W. (1981). The structured meeting system: A procedure for improving the completion of non-recurring tasks. *Journal of Organizational Behavior Management, 3* (4), 7–18.

Glass, G. V., Willson, V. L., & Gottman, J. M. (1975). *Design and analysis of time-series experiments.* Boulder: Colorado University Associated Press.

Goldstein, I. L. (1980). Training in work organizations. *Annual Review of Psychology, 31,* 229–72.

Gottman, J. M. (1981). *Time-series analysis: A comprehensive introduction for social scientists.* Cambridge: Cambridge University Press.

Gupton, T., & Lebow, M. C. (1971). Behavior management in a large industrial firm. *Behavior Therapy, 2,* 78–82.

Hall, B. L., & Hursch, D. E. (1981). An evaluation of the effects of a time management training program on work efficiency. *Journal of Organizational Behavior Management, 3* (4), 73–96.

Hartmann, D. P., Gottman, J. M., Jones, R. R., Gardner, W., Kazdin, A. E., & Vaught, R. (1980). Interrupted time-series analysis and its application to behavioral data. *Journal of Applied Behavior Analysis, 13,* 543–560.

Havel, F., Martin, G., & Koop, S. (1982). Field testing of a self-instructional time management manual with managerial staff in an institutional setting. *Journal of Organizational Behavior Management, 4,* 81–96.

Hermann, J. A., deMontes, A. I., Dominguez, B., Montes, F., & Hopkins, B. L. (1973). Effects of bonuses for punctuality on the tardiness of industrial workers. *Journal of Applied Behavior Analysis, 6,* 563–570.

Hutchison, J. M., Jarman, P. H., & Bailey, J. S. (1980). Public posting with a habilitation team: Effects on attendance and performance. *Behavior Modification, 4,* 57–70.

Jason, L. A., & Liotta, R. F. (1982). Reduction of cigarette smoking in a university cafeteria. *Journal of Applied Behavior Analysis, 15,* 573–577.

Jones, R. R., Vaught, R. S., & Weinrott, M. (1977). Time-series analysis in operant research. *Journal of Applied Behavior Analysis, 10,* 151–166.

Jones, R. R., Weinrott, M. R., & Vaught, R. S. (1978). Effects of serial dependency on the agreement between visual and statistical inference. *Journal of Applied Behavior Analysis, 11,* 277–283.

Kazdin, A. E. (1982). *Single case research designs: Methods for clinical and applied settings.* New York: Oxford University Press.

Kempen, R. W., & Hall, R. V. (1977). Reduction of industrial absenteeism: Results of a behavioral approach. *Journal of Organizational Behavior Management, 1,* 1–22.

Kirchner, R. E., Schnelle, J. F., Domash, M., Larson, L., Carr, A., & McNees, M. P. (1980). The applicability of a helicopter patrol procedure to diverse areas: A cost-benefit evaluation. *Journal of Applied Behavior Analysis, 13,* 143–148.

Kissel, R. C., Whitman, T. L., & Reid, D. H. (1983). An institutional staff training and self-management program for developing multiple self-care skills in severely/profoundly retarded individuals. *Journal of Applied Behavior Analysis, 16,* 395–415.

Komaki, J. L. (1982). The case for the single case: Making judicious decisions about alternatives. In L. W. Frederiksen (Ed.), *Handbook of organizational behavior management* (pp. 145–176). New York: Wiley.

Komaki, J. L. (1985) *Pinpointing appropriate target behaviors of a complex task: An empirical method.* (Manuscript submitted for publication.)

Komaki, J. L., Barwick, K. D., & Scott, L. R. (1978). A behavioral approach to occupational safety: Pinpointing and reinforcing safe performance in a food manufacturing plant. *Journal of Applied Psychology, 63,* 434–445.

Komaki, J. L., Blood, M. R., & Holder, D. (1980). Fostering friendliness in a fast food franchise. *Journal of Organizational Behavior Management, 2,* 151–164.

Komaki, J. L., Collins, R. L., & Penn, P. (1982). The role of performance antecedents and consequences in work motivation. *Journal of Applied Psychology, 67,* 334–340.

Komaki, J. L., Heinzmann, A. T., & Lawson, L. (1980). Effect of training and feedback: Component analysis of a behavioral safety program. *Journal of Applied Psychology, 65,* 261–270.

Komaki, J. L., Waddell, W. M., & Pearce, M. G. (1977). The applied behavior analysis approach and individual employees: Improving performance in two small businesses. *Organizational Behavior and Human Performance, 19,* 337–352.

Kreitner, R., & Golab, M. (1978). Increasing the rate of salesperson telephone calls with a monetary refund. *Journal of Organizational Behavior Management, 1* 192–195.

Lamal, P. A., & Benfield, A. (1978). The effect of self-monitoring on job tardiness and percentage of time spent working. *Journal of Organizational Behavior Management, 1,* 142–149.

Larson, L. D., Schnelle, J. F., Kirchner, R., Jr., Carr, A. F., Domash, M., & Risley, T. R. (1980). Reduction of police vehicle accidents through mechanically aided supervision. *Journal of Applied Behavior Analysis, 13,* 577–581.

Lichstein, K. L., & Stalgarts, T. J. (1980). Treatment of cigarette smoking in couples by reciprocal aversion. *Behavior Therapy, 11,* 104–108.

Lovett, S. B., Bosmajian, C. P., Frederiksen, L. W., & Elder, J. P. (1983). Monitoring professional service delivery: An organizational level intervention. *Behavior Therapy, 14,* 170–177.

Luthans, F., & Davis, T. R. V. (1982). An idiographic approach to organizational behavior research: The use of single case experimental designs and direct measures. *Academy of Management Review, 7,* 380–391.

Luthans, F., Paul, R., & Baker, D. (1981). An experimental analysis of the impact of contingent reinforcement on salespersons performance behavior. *Journal of Applied Psychology, 66,* 314–323.

McCarthy, M. (1978). Decreasing the incidence of "high bobbins" in a textile spinning department through a group feedback procedure. *Journal of Organizational Behavior Management, 1,* 150–154.

Magrab, P. R., & Papadoponlou, Z. L. (1977). The effect of a token economy on dietary compliance for children on hemodialysis. *Journal of Applied Behavior Analysis, 10,* 573–578.

Maher, C. A. (1981). Improving the delivery of special education and related services in public schools. *Journal of Organizational Behavior Management, 3,* 29–41.

Maher, C. A. (1982). Improving teacher instructional behavior: Evaluation of a time management training program. *Journal of Organizational Behavior Management, 4,* 27–36.

Mann, R. A. (1972). The behavior therapeutic use of contingency contracting to control an adult behavior problem: Weight control. *Journal of Applied Behavior Analysis, 5,* 99–109.

Marholin, D., II, & Gray, D. (1976). Effects of group response-cost procedures on cash shortages in a small business. *Journal of Applied Behavior Analysis, 9,* 25–30.

Michael, J. (1974). Statistical inference for individual organism research: Mixed blessing or curse? *Journal of Applied Behavior Analysis, 7,* 647–653.

Miller, L. K., & Miller, O. L. (1970). Reinforcing self-help group activities of welfare recipients. *Journal of Applied Behavior Analysis, 3,* 57–64.

Miller, L. K., & Weaver, F. H. (1972). A multiple baseline achievement test. In G. Semb (Ed.), *Behavior analysis and education—1972.* Lawrence: Support and Development Center for Follow Through, Department of Human Development, University of Kansas.

Page, T. J., Iwata, B. A., & Reid, D. H. (1982). Pyramidical training: A large scale application with institutional staff. *Journal of Applied Behavior Analysis, 15,* 335–351.

Panyan, M., Boozer, H., & Morris, N. (1970). Feedback to attendants as a reinforcer for applying operant techniques. *Journal of Applied Behavior Analysis, 3,* 1–4.

Pedalino, E., & Gamboa, V. U. (1974). Behavior modification and absenteeism: Intervention in one industrial setting. *Journal of Applied Psychology, 59,* 694–696.

Roethlisbeger, F. J., & Dickson, W. J. (1939). *Management and the worker.* Cambridge, MA: Harvard University Press.

Runkel, P. J., & McGrath, J. E. (1972). *Research on human behavior.* New York: Holt, Rinehart & Winston.

Schafer, L. C., Glasgow, R. E., & McCaul, K. D. (1982). Increasing the adherence of diabetic adolescents. *Journal of Behavior Modification, 5,* 353–362.

Stokes, T. F., & Kennedy, S. H. (1980). Reducing child uncooperative behavior during dental treatment through modeling and reinforcement. *Journal of Applied Behavior Analysis, 13,* 41–49.

Sulzer-Azaroff, B. (1978). Behavioral ecology and accident prevention. *Journal of Organizational Behavior Management, 2,* 11–44.

Sulzer-Azaroff, B., & de Santamaria, M. (1980). Industrial safety hazard reduction through performance feedback. *Journal of Applied Behavior Analysis, 13,* 287–296.

Swain, J. J., Allard, G. B., & Holborn, S. W. (1982). The good toothbrushing game: A school based dental hygiene program for increasing the toothbrushing effectiveness of children. *Journal of Applied Behavior Analysis, 15,* 171–176.

Van den Pol, R. A., Reid, D. J., & Fuqua, R. W. (1983). Peer training of safety-related skills to institutional staff: Benefits for trainers and trainees. *Journal of Applied Behavior Analysis, 16,* 139–156.

Van Houten, R., & Nau, P. A. (1981). A comparison of the effects of posted feedback and increased police surveillance on highway speeding. *Journal of Applied Behavior Analysis, 14,* 261–271.

Van Houten, R., & Sullivan, K. (1975). Effects of an audio cueing system on the rate of teacher praise. *Journal of Applied Behavior Analysis, 8,* 197–201.

Zohar, D., & Fussfeld, N. (1981). Modifying earplug wearing behavior by behavior modification techniques: An empirical evaluation. *Journal of Organizational Behavior Management, 3*(2), 41–54.

TWO

Health Concerns in the Workplace

5

Weight Control at the Workplace: The Power of Social and Behavioral Factors

KELLY D. BROWNELL

Enormous potential exists for weight-control efforts at the worksite. This potential exists from various perspectives. Weight-control programs may further the aims of public health advocates, management and labor leaders, and clinicians and researchers. I will approach the issue from these different viewpoints, to show that the combined knowledge and interests from many professionals must be considered for the health promotion movement to survive in the workplace.

Although the needs of such different audiences seldom merge into common goals, this seems to be the case with health promotion programs at the worksite. Worksite programs for weight reduction have begun to appear in the past few years, and the results from the recent reports are encouraging. These results are informative, not only for the area of weight control, but also for the general area of health promotion.

In this chapter, I will first outline the potential for worksite programs, and will then discuss the degree to which this potential has been realized. This discussion requires some specific information on the obesity field, to provide a background against which to judge the effectiveness of programs held at the worksite. The chapter will end with a discussion of key issues in program planning, design, and evaluation, including the issues of cost-effectiveness, public health implications, and dissemination practices.

THE POTENTIAL

For many reasons, the worksite may provide a valuable forum for confronting weight problems. The different reasons depend on the background and perspective of the individuals involved. The diversity in these reasons may be a major strength of the field, because benefits must be perceived by many groups for worksite programs to occur on a large scale.

The Public Health Perspective

One virtue of worksite programs is that many people can be screened and treated. In the first review of weight programs in industry, Foreyt, Scott, and Gotto (1980) cited this as the primary reason for considering the worksite as a place for treating obesity. The theory here is that so many people are employed that the "eligible" population for worksite programs is large. This potential existed when Foreyt and colleagues wrote their review and still exists now. However, several questions can be raised about this logic.

Until recently, most worksite programs for weight reduction have used the standard clinical-behavioral approach. Professional clinicians conduct small groups similar to the way they would in a clinic. The use of the worksite merely transfers the location in which professionals treat a small number of people, so the public health potential could be expanded only by training more professionals. This is not feasible and would probably not be cost-effective.

Another question relates to the ability and willingness of companies to conduct health programs. A large percentage of the work force is employed in small businesses, and it is difficult to determine whether it is possible to reach the thousands of small companies or whether these companies could or would support health promotion programs. In their survey of health programs in California businesses, Fielding and Breslow (1983) found that only 21.7 percent of companies had more than 500 employees. The likelihood of having programs was correlated with increasing size of the companies. Fielding and Breslow did not even consider companies with less than 100 employees, presumably because few programs would exist.

Recent programs have evaluated new and innovative weight-control programs that may help address these issues. For example, Brownell et al. (1984a) tested weight-loss competitions between and within industries. The typical clinical program was not used, and the programs were administered at low cost. Therefore, the aim of reaching a large number of people was achieved, the programs were done in both large and small businesses, and the low cost made it possible for the small businesses to support a program.

Another reason to view worksite programs in a favorable light is their possible effects on people who are only mildly overweight. At least one program found especially positive results for people who were less than 10 percent overweight (Brownell et al., 1984a). This degree of overweight does not constitute a clinical problem, but these people may benefit either medically (Bray, 1976) or socially and psychologically (Brownell, 1982) from weight reduction. People less than 30 percent overweight account for more than 90 percent of all overweight people (Bray, 1976; Stunkard, 1984), so the public health implications of helping this group may be important.

The benefits of worksite programs for mildly overweight people may be viewed as a prevention strategy. Some percentage of these people are destined to gain weight and become clinically obese. Early intervention with these people who would otherwise not seek assistance may help prevent these later problems.

The Business Perspective

If health programs are to survive in business, the benefits for employers and employees must be salient (Fielding, 1979, 1982). This issue will be discussed in more detail later, because it has received minimal attention in the health promotion literature. It is not correct to assume that the benefits perceived by public health and research professionals—namely, reduced health risk, or even more specifically, weight loss–are sufficient to encourage management and labor leaders to support programs that can be costly and time consuming.

It is useful to consider the viewpoints of both management and employees. Management may support a weight-reduction program to improve the appearance of the work force, to improve morale, or to show the employees that management is concerned with their welfare. This may occur quite independent of health risks. The employees may perceive a weight program as a way to help them with a personal problem and to improve their relations with co-workers and management.

These factors are typically not evaluated by professionals in the health promotion field. One reason is that health professionals are accustomed to evaluating health benefits. In addition, the study of morale, productivity, and so forth, is concentrated in the fields of management and industrial psychology, neither of which has been an impetus for health programs at the worksite. This situation is changing, however, and the business perspective is now being considered.

The Clinical Perspective

From the clinical perspective, both clinic and work settings have advantages and disadvantages. The clinical setting permits the magic of special programs administered by professionals with impressive credentials. The cost is high for these programs, a factor that may motivate participants to succeed. There are, however, several major advantages of conducting programs in work settings. One such advantage is the possibility for providing a sustained program and for following participants through the long-term. This continuity of care may provide the support and encouragement that can be so important for long-term success.

Another advantage of the work setting is that it removes the problem of overweight from the medical and psychological context inherent in clinical programs. To the extent that medical programs imply the disease model and psychological programs imply emotional disturbance, the self-blame and personal attributions will only impede progress. The work settings may be useful because a weight problem can be recast as a motivational phenomenon in which the support from the environment can be useful. Programs can be given in an educational context that may highlight the skills training aspect of most behavioral programs.

Perhaps the most important advantage of the work setting is that it provides a unique social environment in which to change health behaviors. The social bonds that develop in work settings can be powerful. It is clear that social factors play a major role in health and disease (Cobb, 1976), and extensive research reveals the importance of these factors in health promotion in general and weight control in particular (Brownell, 1982; Janis, 1983).

The social forces that exist in work settings may be the key to making health promotion programs succeed. Several aspects of this phenomenon will be discussed later. Unfortunately, these forces are not well understood and much is to be learned about intervention in this social setting. It is a positive development, however, that the worksite is now seen as more than a convenient setting in which to deliver programs.

BACKGROUND ON OBESITY AND ITS TREATMENT

This section will cover some general information on physiological and psychological factors associated with obesity, as well as an overview of the current status of treatment approaches. My purpose is not to cover these factors in detail; that has been done elsewhere (Bray, 1976; Brownell, 1982, 1984; Stunkard, 1980). Rather, I would like to show the complex interaction of factors that govern the regulation of body weight and to show why treatment of obesity is so difficult.

Descriptive Information

It is important for scientific reasons to distinguish obesity from overweight. Obesity is excessive body fat, and overweight is excessive weight. A person can be obese but not overweight (a thin person with too little muscle and too much fat) or can be overweight but not obese (a football player might weigh more than standards allow but have little body fat). Because body fat is difficult to measure accurately (Bray, 1976), most clinical and research programs use weight as the measure of outcome.

Used in a colloquial manner, overweight means slightly or moderately heavy, while obese means extremely heavy. Since extremely overweight people are likely to need more structured treatment than a worksite program can provide, it is the slightly or moderately overweight people for whom this approach is most applicable. Therefore, we call out worksite programs "weight reduction programs" rather than "obesity programs." In fact, obesity is the target because it is body fat and not body weight that is most strongly related to health risk (Bjorntorp, 1983). Since the public view is that obesity means extremely heavy, we avoid using the term. For the purposes of this chapter, both terms will be used.

Obesity is characterized by three factors, which make it one of modern society's most important public health problem. First, it has serious medical and psychological consequences. There is some dispute about whether moderate degrees of obesity are dangerous, even though it is clear that people more than 50 percent overweight are at greatly increased risk for serious health problems (Drenick et al., 1980). At the very least, obesity is a risk factor for coronary heart disease via its effect on other risk factors, such as hypertension, hyperlipidemia, and diabetes (Van Itallie, 1979). There is also mounting evidence that it is an independent risk factor (Sorlie, Gordon, & Kannel, 1980).

The health risks of obesity are far less important to most people than are the psychological and social consequences (Allon, 1979; Brownell, 1982; Stunkard, 1976). These reasons prompt people to seek treatment and are the source of the continuous torment experienced by so many overweight people.

The second notable characteristics of obesity is its prevalence. Estimates of prevalence range from 15 to 50 percent, depending on definitions for obesity (Bray, 1976). If one considers people between 10 and 20 percent overweight eligible for worksite weight programs, the percentage of people in the work force who would qualify would reach 75 percent or more. Certainly there is no health problem to remedy obesity in this regard.

The third characteristic of obesity is its resistance to treatment. It has been said that if cure from obesity were defined as reduction to ideal weight and maintenance of that weight for five years, a person is more likely to recover from almost any form of cancer than from obesity (Brownell, 1982). It is a difficult problem to remedy.

This resistance to treatment is important to consider in devising goals for worksite programs and for evaluating results. The average person who is more than 20 percent overweight cannot be expected to reach and maintain goal weight. Progress must be viewed in terms of progress toward this goal, not attainment of the goal.

Physiological Factors

I inserted a discussion of physiological factors in this chapter for one reason. Most people, including health professionals, feel that obesity is caused by psychological factors and that it is lack of willpower and motivation that inhibits a fat person from obtaining goal weight. These factors cannot be discounted, but physiological events almost certainly play a role in the maintenance of obesity, if not in its etiology (Bray, 1976; Garrow, 1978, 1981). The psychological and emotional attribution influences attitudes about overweight people, which in turn helps determine whether and how programs are established to handle the problem.

I will discuss two popular physiological theories related to obesity–the fat cell theory and the set point theory. Both are useful ways to conceptualize the regulation of body weight, and although neither has been proven, they do show the complex nature of the problem. For more detailed descriptions of these areas, please refer to other works (Bennett & Gurin, 1982; Bray, 1976; Garrow, 1978, 1981; Keesey, 1980; Stunkard, 1980).

The Fat Cell Theory.   The fat cell theory originated with the work of Hirsch and Knittle (1970), who reported that adult onset obese people had fat cells much larger (hypertrophy) than those of normal weight persons, and that childhood onset obese persons had increased fat cell number (hyperplasia). The number of fat cells can be increased by as much as fivefold in childhood onset people (Bjorntorp, 1983; Hirsch & Knittle, 1970; Sjostrom, 1980).

The fat cell theory maintains that cell number can be increased in childhood for nutritional or genetic reasons. Once cell number stabilizes sometime in adolescence, weight loss can occur only be reducing the size of the cells, not the number. The corollary of this theory is that weight loss should be difficult in childhood onset obese persons because of the excessive number of cells.

Indirect evidence for this theory was reported by Bjorntorp and colleagues (1975). They found that obese women tended to cease losing weight and drop out of treatment when cell size reached the level of normal weight people, as if a biological limit had been reached. For those patients who were hyperplastically obese, they were still overweight (by actuarial standards) even though cell size was normal (because there were too many of these cells of normal size). The ony way for such a person to attain ideal weight is to further reduce cell size, thus having lipid-depleted cells. This could create strong biological pressure to replenish the cells and to gain weight.

Several aspects of the fat cell theory have been challenged. First, clinical studies do not show that childhood onset obese people lose less weight than adult onset people (Brownell et al., 1978; Jeffrey, Wing, & Stunkard, 1978). Second, it is now known that fat cell multiplication *can* occur in adult life, perhaps when cell sizes reach an upper limit (Sjostrom, 1980). However, regardless of the validity of this particular theory, fat cells are an important part of the physiology that governs body weight.

Set Point Theory.   Several groups have proposed a body weight *set point*, which is the internal mechanism that determines whether an organism will gain or lose weight (Bennett & Gurin, 1982; Keesey, 1980). The body may have a series of regulating mechanisms to defend against deviations above or below that weight. If some people have elevated set points, they will fight strong biological pressure to reduce to society's ideal. Work by Keys et al. (1950) on starvation, by Sims and Horton (1968) on overfeeding, and by Keesey (1980) and others on animals, supports the set point theory.

The set point is intuitively appealing for both patients and professionals. It helps explain why some people have such trouble sustaining weight loss and why most people have fairly stable body weights during their adult lives. If the theory is valid, there will be a rush to find ways to reset the set point. In the meantime, it would imply that some obese people should accept their weight rather than enduring the physical and psychological torment of remaining below their set point.

As with the fat cell theory, the set point theory has not been proven. Collectively, these two theories, combined with other evidence on physiology, lead me to appreciate the possible contribution on nonpsychological factors in obesity. I endorse this notion of complexity, if not the theories used to support it. This conceptual shift leads to less blaming of the obese person and to more constructive approaches to treatment and prevention.

The Current Status of Treatment

Nearly all information about the treatment of obesity has derived from clinical studies in universities, and to a lesser extent, from sparse information on self-help and commercial groups. In clinical, self-help, and commercial groups, the populations are different from each other and from the groups most likely to be encountered in work settings. Therefore, the principles derived from this research may be applicable even though the results may not transfer perfectly to work settings. More detailed reviews of the treatment literature can be obtained from several sources (Bray, 1976; Brownell, 1982, 1984; Foreyt, Goodrick, & Gotto, 1981; Garrow, 1981; Jeffery et al., 1978; LeBow, 1981; Stunkard, 1980; Stunkard & Penick, 1979; Wilson & Brownell, 1980; Wing & Jeffery, 1979).

Behavior therapy now forms the foundation for most safe and effective weight-reduction programs. Its focus is on the permanent alteration of eating and exercise habits, attitudes and cognitive factors, and social support. The program has been systematized to some extent, and many comprehensive books and manuals describe the behavioral principles and techniques (Brownell, 1979; Mahoney & Mahoney, 1978; Stuart, 1978).

The behavioral program has been tested in dozens of studies. It is in use in the major commercial groups, such as Weight Watchers, and it is referred to frequently in the media. It is the treatment of choice for mild to moderate obesity. How well does it work?

Most behavioral programs produce average weight losses of approximately one pound per week in programs that last from 8–20 weeks. In current behavioral programs, which average 16 weeks in duration, mean losses are 20–25 pounds (Brownell & Stunkard, 1981; Craighead, Stunkard, & O'Brien, 1981). More important, the losses are well maintained at follow-ups of one year (Brownell, 1982; Foreyt et al., 1981; Jeffery et al., 1978; Stunkard & Penick, 1979; Wilson & Brownell, 1980).

It appears that behavior therapy is necessary for all levels of obesity and is sufficient for some. For people with small amounts of weight to lose (30 pounds or less) the behavioral program may be sufficient. It certainly offers the best hope for long-term success. For heavier people, the behavioral program is best used in combination with other methods for weight loss; the most promising current possibility is the very-low-calorie diet, also known as the protein sparing modified fast (Bistrian, 1978; Wadden, Stunkard, & Brownell, 1983).

The behavior therapy program is typically used in an educational context in which people are taught sensible changes in nutrition, physical activity, and other factors. Since the program itself is well established, attention has turned to ways in which the program can be used more effectively. Therein lies the search for social factors to boost treatment effectiveness.

The social context in which a program is delivered, regardless of the program itself, may be a central determinant of its success (Brownell, 1982; Janis, 1983). The search for information about this context has led to studies of the family, the schools, group process, and the worksite. This change in emphasis from education to motivation promises to advance the field far beyond where it stands currently. This is why worksite programs hold so much promise from a clinical perspective.

As a final note about treatment, exercise is now considered an essential part of treatment for obesity (Brownell & Stunkard, 1980; Epstein & Wing, 1980; Stern, 1984; Thompson et al., 1982). Exercise has many metabolic effects that extend beyond the simple calorie expenditure of the activity. Its association with long-term success at weight reduction is striking, and in studies in which exercise has been manipulated experimentally, the results support its use (Dahlkoetter, Callahan, & Linton, 1979; Stalonas, Johnson, & Christ, 1978).

The challenge with overweight people is not to convince them that exercise is beneficial, but to encourage them to do the exercise. Behavioral techniques seem to be useful in this regard (Martin & Dubbert, 1982). In addition, lessons from exercise programs in the workplace may be relevant for developing worksite weight-control programs (Haskell & Blair, 1980).

EARLY WORKSITE PROGRAMS

In 1979, the Office of Health Information of the US Public Health Service sponsored the National Conference on Health Promotion Programs in

Occupational Settings. The conference covered many aspects of health promotion and covered topics such as smoking, hypertension, stress, and weight control. The background papers for this conference were published in 1980 in *Public Health Reports* (Volume 95, March–April).

As part of this conference, Foreyt, Scott, and Gotto (1980) reviewed the literature on weight control and nutrition education programs in business and industry. So little information was available, that the authors contacted a number of companies by phone and letter to ask about weight or nutrition programs. Fewer than a dozen programs existed, and these varied widely in content and quality. There were no published evaluations of weight-control programs. These authors concluded that:

> The few weight control programs in industry are part of general exercise programs. The weight loss effort may consist of a lecture or two by the company nurse. Some firms have contracted with organizations such as the local health department or a group like the American Diabetes Association or a commercial venture such as Weight Watchers. (Foreyt et al., 1980, p. 130)

From the Clinic to the Worksite: An Imperfect Transfer

About the time of the Foreyt et al. (1980) review, the picture began to change and more systematic and thoroughly evaluated programs were initiated. These early programs came from researchers in psychology and psychiatry and from practitioners in fields such as dietetics. The programs resulted from the few successful attempts of nonbusiness professionals to "sell" programs to industry, but were typically not initiated by business leaders.

These early programs reflected the backgrounds and biases of the professionals involved. The psychologists and psychiatrists used behavior modification programs, while the nutrition professionals used dietary programs. Considering this diversity, there were remarkable similarities in the results of the programs, particularly in the rates of attrition.

The first controlled trial of weight reduction at the worksite was reported by Stunkard and Brownell (1980). The study was done with members of the United Storeworkers Union who were employed by the Gimbels retail department stores in New York City. This was the setting for the successful hypertension screening and follow-up program of Alderman and Schoenbaum (1975).

The study involved the assignment of 40 obese females to groups, which varied according to whether groups were conducted at the worksite or a medical site, whether the groups were led by professional psychologists or by union volunteers, and whether meetings were held once or three times per week. The standard behavioral program used in the clinic (Brownell, 1979) was used in the worksite, and the program was delivered in groups of 8–10 members.

The most striking findings of the study by Stunkard and Brownell (1980) were the high attrition and low weight losses. Fully 50 percent of the participants dropped out of the program, and the average weight loss for those who remained was 7.9 pounds. This compares with attrition of 15 percent and losses of approximately 20 pounds when the same program was used in the clinic (Brownell, 1982; Jeffery et al., 1978; Wilson & Brownell, 1980).

Viewed in isolation, these discouraging findings might reflect nothing more than a poor program or inadequate organization. However, a number of other reports appearing about the same time, even though from uncontrolled studies, showed similar results. Sangor and Bichanich (1977) offered a dietary instruction program for employees of a hospital in Milwaukee; 50 percent of the participants dropped out and weight losses were small. A similar program in a hospital in West Allis, Wisconsin found somewhat smaller attrition (22 percent) and greater weight losses (9.1 pounds), but no follow-up data were reported (Schumacher et al., 1979). Fisher and colleagues (1982) reported attrition of 44 percent in a program for bank employees in St. Louis.

Several findings emerge from these programs. First, programs do not transfer perfectly from the clinic to the workplace, presumably because participants in worksite programs are different from those in the clinic and because the work setting is a different social environment. Second, attrition is a major problem with these programs. The average attrition of 50 percent in the early programs is confirmed by more recent reports (Abrams & Follick, 1983; Brownell, Stunkard, & McKeon, 1984). These factors spurred the development of new approaches.

THE NEW GENERATION OF WORKSITE PROGRAMS

The new generation of worksite programs are characterized by two factors. The first is that techniques are no longer borrowed solely from clinical programs, but are being altered to suit the needs and interest of workers. The second factor is that new conceptual approaches to program delivery are being examined. One exciting development in this regard is the use of weight-loss competitions between and within industries. The gains from both factors are impressive, so that we are on the verge of having an effective program, which can be demonstrated to business leaders in good faith.

Improvements in Procedures

Two groups have done the work on the development of worksite procedures. Their emphasis has been on the educational and structural aspects of weight-control programs. One group has studied the employees of the Miriam Hospital in Providence, Rhode Island (Abrams & Follick, 1983;

Follick, Fowler, & Brown, in press). The other group evaluated programs for the employees of the Bloomingdale's and Gimbels department stories in New York (Brownell et al., 1984b; Stunkard & Brownell, 1980).

Abrams and Follick (1983) studied 133 participants who enrolled in a 10-week behavioral program, which was offered to the employees of the Miriam Hospital, a 250-bed general hospital in Providence. Most (91 percent) of the subjects were nurses, and all but 11 were female. The subjects were divided into three groups for the 10-week program; all groups received identical treatment. For the subsequent 8-week program phase, half the subjects received a structured, four-session maintenance program, while the other half served as a nonspecific group to control for contact time. Follow-ups were then done three and six months later.

The attrition rate during the initial phase of the Abrams and Follick (1983) study was 48.1 percent. Follick et al. (in press) later found that an incentive procedure could reduce the rate of attrition. The mean weight loss was approximately 9.5 pounds. This high attrition rate was similar to that reported in earlier programs, but the weight losses were greater.

The most important results of this study came during the follow-up period. The subjects who received the structured maintenance program maintained their loss significantly better than the control subjects, at both the three-month and six-month follow-ups. It is instructive, therefore, to examine in more detail what these investigators included in their program.

Abrams and Follick (1983) made several innovative additions to the standard behavioral program. The first series of procedures were derived from the principles of organizational behavioral management. Subjects were encouraged to make a public commitment by wearing bright buttons announcing their participation in the program. The average weight loss for all participants was posted in the hospital cafeteria. To foster intergroup competition, the average losses for the three initial groups were posted in the cafeteria and the winning group was announced during the weekly meetings.

During the structured maintenance program, other innovative procedures were used. Specific behavioral skills were emphasized via fading of the self-monitoring records. A structured problem-solving approach was used to teach subjects skills to handle difficult situations. Cognitive procedures were used in a "relapse prevention" effort. Finally, a buddy system was used in which participants contracted with a co-worker to achieve specific goals.

Several studies of union workers in New York City also add to our knowledge of procedures for worksite programs. The first study, by Stunkard and Brownell (1980), was described above. Brownell, Stunkard and McKeon (1984) followed with a series of studies at the same worksites. Attrition dropped during the course of the three studies, from 58 percent to 43 percent, and to 34 percent. Weight losses averaged approximately 8.5 pounds during the studies. Unlike the Abrams and Follick (1983) study, which examined program variables, Brownell et al. (1984b) examined program delivery factors.

In their first study, Brownell et al. (1984b) found that nonprofessional union volunteers who were trained to conduct groups produced weight losses equivalent to those of professional psychologists, and had even lower rates of attrition. These investigators followed with two studies to further test the feasibility of nonprofessional leaders. The results were consistent; the psychologists were no better than the union volunteers at keeping people in the program or at helping these people lose weight. Since the cost for the volunteers is much lower, the cost-effectiveness was more favorable for the lay leaders.

The Brownell et al. (1984b) studies also compared the traditional weekly meetings with a more intensive schedule of three or four meetings weekly. Attrition was lower with the intensive schedule; weight losses were equivalent for the two approaches.

Much has been learned from the work of these two research groups (Abrams & Follick, 1983; Brownell et al., 1984b). Organizational and social support principles can be used to make the behavioral program more successful in the workplace. The structured maintenance program of Abrams and Follick was successful in sustaining the weight losses achieved during the initial program. In addition, it appears that program delivery factors can influence both the cost and effectiveness of worksite programs.

A New Conceptual Approach: Weight-Loss Competitions

A new approach to weight control at the worksite was proposed recently by Brownell, Cohen, Stunkard, Felix, and Cooley (1984). The new approach involved weight-loss competitions, first between businesses and then within businesses. The work was done in Williamsport, a city of approximately 30,000 in north central Pennsylvania, under the sponsorship of the County Health Improvement Program (CHIP), a coronary primary prevention trial. The results of these competitions were striking.

The first competition in the series of Brownell et al. (1984a) took place between three banks, each of between 200 and 300 employees. The presidents of the banks agreed to a weight-loss contest and challenged each other in a formal announcement. The employees in each bank who wanted to join the program were weighed and given a weight-loss goal, which was ideal weight subtracted from actual weight, to a maximum of 20 pounds for the 12-week program. Each participant paid $5 for the program. The money for all participants was pooled and was to be awarded to the participants in the winning bank. A large board in the lobby of each bank plotted the weekly progress of all three banks.

The only "treatment" given to the participants consisted of a weekly weigh-in by a research assistant and the distribution of weekly installments of a behavioral treatment manual (Brownell, 1979). No group meetings were held. There was no contact with a professional, and there was no organized forum for the participants to discuss the program with one another. Therefore, the behavioral and educational aspect of the program was far less intensive than was used in other worksite projects.

Fewer than 5 percent of the participants dropped out. Average weight losses were 18.7 pounds for men and 11 pounds for women. A six-month follow-up was completed and the average participant had maintained 80 percent of the weight lost during the program. These results are more positive than those from other worksite programs, and the cost was less because the program required little more than the cooperation of management and employees.

Brownell et al. (1984a) conducted two other competitions, both involving competition within industries. The first was with a manufacturing firm of 225 employees (Litton Industries) in which participants were assigned randomly to teams. The second was in a manufacturing plant of 1,200 employees (Koppers Industries) in which participants within existing divisions of the company formed the teams for the competition. The $5 payment, the scoreboard, and all other aspects of the program were identical to those used for the bank program.

The results for the two competitions within industries were also favorable. Intuitively, it would seem that competition between industries would promote the most intragroup coherence and intergroup competition, yet the within-industries competitions were nearly as successful as the bank competition. The results, however, need to be tested in a controlled manner.

The competitions had particularly good effects for people who were only mildly overweight (<10 percent overweight). The average weight loss for these subjects was 7.3 pounds; their average goal was a loss of only 6.2 pounds. Nearly three-fourths (74 percent) of these participants reached their goal weights.

In addition to the weight losses, there were several notable aspects of the Brownell et al. (1984a) findings. Employees and management were given questionnaires after the program to evaluate work-related factors, such as morale, energy level, employee-management relations, absenteeism, and so forth. The results were uniformly positive. Fully 71 percent of employees and 75 percent of managers reported improvements in morale. The results were also positive for the other measures, and for those who did not report improvement, all reported no change. Each of the managers stated that they would welcome another competition in their business and that they would recommend the competition to other colleagues.

The most impressive findings of these weight loss competitions derived from a cost-effectiveness analysis done by the authors. The cost to make a 1% reduction in percentage overweight, which amounts approximately to the cost per pound lost, was calculated for the competitions and compared to figures reported in other studies. The results are shown in Figure 5.1. The figures for the university clinic and for Weight Watchers are taken from Yates (1978), and for the other two worksite programs are taken from the studies discussed above with the department store workers (Brownell et al., 1984b).

The competitions have the most favorable cost-effective ratio of any program for which cost data are available. The favorable ratio is due both to a high level of effectiveness and low cost. This argues strongly for the potential public health impact of such a program and for generalizability of the program to other work settings.

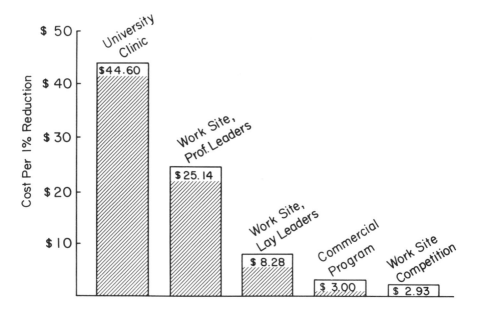

FIGURE 5.1. The cost per 1 percent reduction in percentage overweight for different weight-loss programs. Estimates were obtained from Yates (1978) for the University Clinic and for Weight Watchers; for the two worksite groups with lay and professional leaders from Brownell, Stunkard, & McKeon (1984b), and for the weight-loss competitions from Brownell (1984a) (From Brownell, Cohen, Stunkard, Felix, & Cooley, 1984a. Copyright 1984 by the American Public Health Association. Reprinted by permission.)

KEY ISSUES

Education or Motivation?

The greatest improvements in program effectiveness will come from a shift in emphasis from educational to motivational factors. It is still important to perfect program techniques, especially considering their use in the new settings of the worksite. However, the motivational variables will contribute more to the success of a program than will changes in the program itself. The weight-loss competitions are one way to exploit the natural social factors at the worksite (Brownell et al., 1984a).

A change in emphasis to motivational factors could stimulate a new line of inquiry for health promotion planners. Since the social factors that are specific to the worksite may be key determinants of motivation, these factors deserve special study. This study will necessarily involve contributions of professionals in many disciplines and will draw heavily on the experience of people in the work settings.

I will list just a few of the issues that need to be addressed. One such issue is health economics. Both the costs and the benefits of health promotion programs will be important to evaluate. Economics in general will be important because the financial stability of a business may influence the decision to support health programs. Labor relations and management policy are central to this area. Epidemiology is important to determine what risks are greatest; public health has a role in evaluating sites and avenues for intervention; and psychology can contribute to the knowledge of behavior change in individuals. This interdisciplinary work is needed, but has not yet been done.

Understanding the Needs of Business and Labor

During our work in department stores, banks, and manufacturing firms, one issue arose repeatedly. In discussing the results of a program with business leaders, our inclination was to point to attrition rates, average weight losses, reductions in cardiovascular risk, and so forth. There was only mild interest in these matters. However, there was great interest in the influence of a program on morale, productivity, absenteeism, employee–management relations, and other business factors. In many cases, the managers did not need our questionnaire data to know that morale improved or that employees perceived the programs in a positive light.

It is clear that business and health professionals view health promotion from different perspectives (Fielding, 1979, 1982). The business professionals will make the ultimate decisions about whether these programs will be adopted and maintained, so we must be sensitive to their needs. This will involve careful study of the reasons why some businesses have health programs and others do not, of the benefits perceived by business leaders, and of ways to quantify the business-related factors that may be influenced by health promotion efforts.

The perspective of labor is also crucial to the success of a program. In our work with the United Storeworkers Union in New York, the union offered health programs to employees as a special benefit because the financial climate did not permit the union to negotiate large increases in financial benefits. Organized labor must perceive some benefit from the programs and must not feel that health programs infringe on the rights of their employees (as with smoking cessation programs). What may be even more important are the perceptions of individual workers. In this area research on "quality of work life" will be important (Lawler, 1982).

Another important issue for program dissemination is that of entering industry to propose programs. Personnel managers, for example, are more responsive to health programs than are medical directors, and the personnel managers hold more powerful positions in the corporate structure. Fielding and Breslow (1983) conducted a survey of 1,000 businesses in California, and found that programs tended to be initiated by personnel departments (35.1 percent), top management (20.5 percent), safety group (18.2 percent), and then by medical departments (14.3 percent) and health benefits groups (7.4 percent).

Cost-Effectiveness

Cost-effectiveness is the final issue I will discuss in this chapter. It may be, in the final analysis, the most important aspect of the work we do. Many business leaders have initiated programs with the assumption that the benefits will outweigh the costs. Most have not called for proof on this issue, but the honeymoon will soon be over. When the tough questions are asked, health promotion professions must have the answers. For the most part, these questions have not been asked, much less answered.

Most health professionals are concerned with effectiveness and not with cost. Programs underwritten by grants or by private sources may have excessive costs even if they are effective. It is time for program managers and researchers to be concerned with costs and to report this information in descriptions of their programs.

REFERENCES

Abrams, D. B., & Follick, M. J. (1983). Behavioral weight loss intervention at the worksite: Feasibility and maintenance. *Journal of Consulting and Clinical Psychology, 51,* 226–233.

Alderman, M. H., & Schoenbaum, E. E. (1975). Detection and treatment of hypertension at the worksite. *New England Journal of Medicine, 292,* 65–68.

Allon, N. (1979). Self-perceptions of the stigma of overweight in relationship to weight losing patterns. *American Journal of Clinical Nutrition, 32,* 470–480.

Bennett, W., & Gurin, J. (1982). *The dieter's dilemma: Eating less and weighing more.* New York: Basic Books.

Bistrian, B. R. (1978). Clinical use of a protein-sparing modified fast. *Journal of the American Medical Association, 21,* 2299–2302.

Bjorntorp, P. (1983). Fat cells and obesity. In M. R. C. Greenwood (Ed.), *Obesity: Issues in contemporary nutrition.* New York: Churchill Livingstone.

Bjorntorp, P., Carlgren, G., Isaksson, B., Krotkiewski, M., Larsson, M., & Sjostrom, L. (1975).

Effect of an energy-reduced dietary regimen in relation to adipose tissue cellularity in obese women. *American Journal of Clinical Nutrition, 28,* 445–452.

Bray, G. A. (1976). *The obese patient.* Philadelphia: Saunders.

Brownell, K. D. (1979). *Behavior therapy for obesity: A treatment manual.* University of Pennsylvania, (unpublished manuscript).

Brownell, K. D. (1982). Obesity: Understanding and treating a serious, prevalent, and refractory disorder. *Journal of Consulting and Clinical Psychology, 50,* 820–840.

Brownell, K. D. (1984). The psychology and physiology of obesity: Implications for screening and treatment. *Journal of the American Dietetic Association.*

Brownell, K. D., Cohen, R. Y., Stunkard, A. J., Felix, M. R. J., & Cooley, N. B. (1984a). Weight loss competitions at the work site: Impact on weight, morale, and cost-effectiveness. *American Journal of Public Health, 74,* 1283–1285.

Brownell, K. D., Heckerman, C. L., Westlake, R. J., Hayes, S. C., & Monti, P. M. (1978). The effect of couples training and partner cooperatives in the behavioral treatment of obesity. *Behavior Research and Therapy, 16,* 323–333.

Brownell, K. D., & Stunkard, A. J. (1980). Exercise in the development and control of obesity. In A. J. Stunkard (Ed.), *Obesity.* Philadelphia: Saunders.

Brownell, K. D., & Stunkard, A. J. (1981). Couples training, pharmacotherapy, and behavior therapy in treatment of obesity. *Archives of General Psychiatry, 38,* 1223–1229.

Brownell, K. D., Stunkard, A. J., & McKeon, P. E. (1984b). *Weight reduction at the work site: A promise partially fulfilled.* (Paper submitted for publication.)

Cobb, S. (1976). Social support as a moderator of life stress. *Psychosomatic Medicine, 38,* 300–314.

Craighead, L. W., Stunkard, A. J., & O'Brien, R. (1981). Behavior therapy and pharmacotherapy for obesity. *Archives of General Psychiatry, 38,* 763–768.

Dahlkoetter, J., Callahan, E. J., & Linton, J. (1979). Obesity and the unbalanced energy equation: Exercise vs. eating habit change. *Journal of Consulting and Clinical Psychology, 47,* 898–905.

Drenick, E. J., Bale, G. S., Seltzer, F., & Johnson, D. G. (1980). Excessive mortality and causes of death in morbidly obese men. *Journal of the American Medical Association, 243,* 443–445.

Epstein, L. H., & Wing, R. R. (1980). Aerobic exercise and weight. *Addictive Behaviors, 5,* 371–388.

Fielding, J. E. (1979). Preventive medicine and the bottom line. *Journal of Occupational Medicine, 21,* 79–88.

Fielding, J. E. (1982). Effectiveness of employee health improvement programs. *Journal of Occupational Medicine, 24,* 907–916.

Fielding, J. E., & Breslow, L. (1983). Health promotion programs sponsored by California employers. *American Journal of Public Health, 73,* 538–542.

Fisher, E. B., Jr., Lowe, M. R., Levenkron, J. C., & Newman, A. (1982). Reinforcement and structural support of maintained risk reduction. In R. B. Stuart (Ed.), *Adherence, compliance and generalization behavioral medicine.* New York: Brunner/Mazel.

Follick, M. J., Fowler, J. L., & Brown, R. (in press). Attrition in work site interventions. *Journal of Consulting and Clinical Psychology.*

Foreyt, J. P., Goodrick, G. K., & Gotto, A. M. (1981). Limitations of behavioral treatment of obesity: Review and analysis. *Journal of Behavioral Medicine, 4,* 159–174.

Foreyt, J. P., Scott, L. W., & Gotto, A. M. (1980). Weight control and nutrition education in occupational settings. *Public Health Reports, 95,* 127–136.

Garrow, J. S. (1978). *Energy balance and obesity in man* (2nd ed.). Amsterdam: Elsevier.

Garrow, J. S. (1981). *Treat obesity seriously: A clinical manual.* London: Churchill Livingstone.

Haskell, W. L., & Blair, S. N. (1980). The physical activity component of health promotion in occupational settings. *Public Health Reports, 95,* 109–118.

Hirsch, J., & Knittle, J. L. (1979). Cellularity of obese and nonobese human adipose tissue. *Federation Proceedings, 29,* 1516–1521.

Janis, I. L. (1983). The role of social support in adherence to stressful decisions. *American Psychologist, 38,* 143–160.

Jeffery, R. W., Wing, R. R., & Stunkard, A. J. (1978). Behavioral treatment of obesity: State of the art in 1976. *Behavior Therapy, 6,* 189–199.

Keesey, R. E. (1980). A set point analysis of the regulation of body weight. In A. J. Stunkard (Ed.), *Obesity.* Philadelphia: Saunders.

Keys, A., Brozek, J., Henschel, A., Mickelson, O., & Taylor, H. L. (1950). *The biology of human starvation* (Vols. 1 & 2). Minneapolis: University of Minnesota Press.

Lawler, E. E., III. (1982). Strategies for improving the quality of work life. *American Psychologist, 37,* 486–493.

LeBow, M. D. (1981). *Weight control: The behavioral strategies.* New York: Wiley.

Mahoney, M. J., & Mahoney, B. K. (1976). *Permanent weight control: A total solution to the dieter's dilemma.* New York: Norton.

Martin, J. E., & Dubbert, P. M. (1982). Exercise applications and promotion in behavioral medicine: Current status and future directions. *Journal of Consulting and Clinical Psychology, 50,* 1004–1017.

Sangor, M. R., & Bichanich, P. (1977). Weight reducing program for hospital employees. *Journal of the American Dietetic Association, 71,* 535–536.

Schumacher, N., Groth, B., Kleinsek, J., & Seay, N. (1979). Successful weight control program for employees. *Journal of the American Dietetic Association, 74,* 466–467.

Sims, E. A. H., & Horton, E. S. (1968). Endocrine and metabolic adaptation to obesity and starvation. *American Journal of Clinical Nutrition, 21,* 1455–1470.

Sjostrom, L. (1980). Fat cells and body weight. In A. J. Stunkard (Ed.), *Obesity.* Philadelphia: Saunders.

Sorlie, P., Gordon, T., & Kannel, W. B. (1980). Body build and mortality: The Framingham Study. *Journal of the American Medical Association, 243,* 1828–1831.

Stalonas, P. M., Johnson, W. G., & Christ, M. (1978). Behavior modification for obesity: The evaluation of exercise, contingency management, and program adherence. *Journal of Consulting and Clinical Psychology, 2,* 225–235.

Stern, J. S. (1984). Is obesity a disease of inactivity? In A. J. Stunkard & E. Stellar (Eds.), *Eating and its disorders.* New York: Raven.

Stuart, R. B. (1978). *Act thin, stay thin.* New York: Norton.

Stunkard, A. J. (1976). *The pain of obesity.* Palo Alto, CA: Bull Publishing Co.

Stunkard, A. J. (Ed.). (1980). *Obesity.* Philadelphia: Saunders.

Stunkard, A. J. (1984) The current status of treatment of obesity in adults. In A. J. Stunkard & E. Stellar (Eds.), *Eating and its disorders.* New York: Raven.

Stunkard, A. J., & Brownell, K. D. (1980). Work site treatment for obesity. *American Journal of Psychiatry, 137,* 252–253.

Stunkard, A. J., & Penick, S. B. (1979). Behavior modification in the treatment of obesity: The problem of maintaining weight loss. *Archives of General Psychiatry, 36,* 810–816.

Thompson, J. K., Jarvie, G. J., Lahey, B. B., & Cureton, K. J. (1982). Exercise and obesity: Etiology, physiology, and intervention. *Psychological Bulletin, 91,* 55–79.

Van Itallie, T. B. (1979). Obesity: Adverse effects on health and longevity. *American Journal of Clinical Nutrition, 32,* 2723–2733.

Wadden, T. A., Stunkard, A. J., & Brownell, K. D. (1983). Very low calorie diets: Their efficacy, safety, and future. *Annals of Internal Medicine, 99,* 675–684.

Wilson, G. T., & Brownell, K. D. (1980). Behavior therapy for obesity: An evaluation of treatment outcome. *Advances in Behavior Research and Therapy, 3,* 49–86.

Wing, R. R., & Jeffery, R. W. (1979). Outpatient treatments of obesity: A comparison of methodology and clinical results. *International Journal of Obesity, 3,* 261–279.

Yates, B. T. (1978). Improving the cost-effectiveness of obesity programs: Three basic strategies for reducing cost per pound. *International Journal of Obesity, 2,* 249–266.

6

Conflicting Perspectives on Stress Reduction in Occupational Settings: A Systems Approach to Their Resolution

JEFFERSON A. SINGER, MICHAEL S. NEALE,
GARY E. SCHWARTZ, and JEANNE SCHWARTZ

This research project began with a request from NIOSH to conduct an inquiry into innovative stress management programs for blue-collar workers. We had originally intended a survey and site visit of stress management programs run by enlightened corporations and unions. Extensive phone-calling to consultants, corporate health departments, and academics, resulted in a rethinking of our intended purpose. Health (how to get it and keep it) sounded the dominant theme in all conversations and written materials. If stress received mention, and it seldom did, the reference inevitably invoked deleterious effects of stress on health.

Meanwhile, we called and wrote to labor unions, with a premonition that the phrase "stress management" would not be well received. We did not anticipate the vocal hostility that many representatives expressed, when quizzed about existing stress management programs for their members. What we heard instead amounted to impassioned explanations of collective bargaining, grievance procedures, governmental intervention, and increased worker control. The daily exercise of a union's duties, we were told, provided the best prescription for stress.

As researchers with experience in defining and isolating the experimental

This chapter was funded in part by the National Institute for Occupational Safety and Health. For a copy of the questionnaire materials and a complete list of corporations and unions contacted, please write to Jefferson A. Singer, Department of Psychology, Yale University, Box 11A, Yale Station, New Haven, Connecticut 06520.

concepts of a stressor and a stress response, we were simultaneously amused and frustrated to see how these words were being stretched. Accordingly, we shifted the focus of our original task. We assumed that by making explicit our understanding of stress, we could systematically compare our definition to the corporate and labor conceptions. This comparison would allow us to understand different efforts at stress management both in the terms of a research psychologist and in their own terms. Our task included the following questions:

1. How does an experimental concept such as stress become altered in the real world of special interests?
2. How does one's choice of a definition for stress influence the subsequent interventions applied to its amelioration?
3. What are the corporate and union definitions of stress, as determined by our questionnaires and conversations?
4. What are the corporate and union strategies for stress reduction?
5. Do these strategies flow logically from the definitions they use for stress?
6. To come full circle—what effect would the corporate and union strategies for stress reduction have on the problem of stress as originally defined by our research?

This last question carried considerable weight. It suggests a test of dual validity. First, were the experimental findings on stress actually applicable to the experience of stress in the field? Second, if we assumed they were, were stress reduction practitioners in labor and industry honestly attacking the real problems of stress, as opposed to merely improving health or raising productivity?

As our starting point for understanding stress, we used the factors of controllability, level of stimulation, and predictability, which have been well-established in the psychological literature. (Frankenhaeuser, 1979; Levine, 1980). At the same time, we asserted the importance of using a biopsychosocial perspective (Leigh & Reiser, 1980; Schwartz, 1982). Stress is usually characterized as disharmonies among the biology, psychology, and environment of a particular individual. Schwartz (1983) describes the disregulation of any one of these systems with each other as both stressful and potentially harmful to health. Loss of control, excessive demand, and lack of predictability can occur within each of these systems, in the linking of these systems, and in their interactions. A sleepy body cannot tolerate stimulation; an anxious mind cannot attend to the body's exhaustion; an impersonal organization will overlook this frazzled state of affairs. Recognizing the importance of this systems perspective, we developed an assessment tool, the Occupational Stress Evaluation Grid (OSEG), which would allow us to observe both individual and institutional conceptualizations of stressors and efforts to reduce them (see Table 6.1). The OSEG reflects a systems

Table 6.1. Occupational Stress Evaluation Grid

Levels	Stressors	Interventions	
		Formal	Informal
Sociocultural	Racism Ecological shifts Economic downturns Political changes Military crises Sexism	Elections Lobbying/political action Public education Trade associations	Grass roots organizing Petitions Demonstrations Migration Spouse employment
Organizational	Hiring policies Potential plant closings, Lay-offs, relocation, Automation, market Shifts, retraining Organizational priorities	Corporate decision Reorganization New management model Management consultant inservice/retraining	Social activities Contests Incentives Manager involvement and ties with workers Continuing education Moonlighting
Work Setting	Task (time, speed, autonomy, creativity) Supervision Co-workers Ergonomics Participation in decision making	Supervisor meetings Health/safety meetings Union grievance negotiations Employee involvement Quality circles Company intitiated redesign Inservice training	Slow downs/speed up Redefine tasks Support of other workers Sabotage, theft Quit, change jobs
Interpersonal	Spouse – divorce, separation, marital discord	Legal services Time off Counselling,	Seek social support and advice Seek legal or financial

	Death Child problems Friends Parents In-laws	Psychotherapy Insurance plans Family treatment Loans/credit unions	assistance Self-help groups Vacations/leave of absence Childcare
Psychological	Personality, Coping behavior (e.g., Type A, repression) Emotion/mood/"stress" expectations, beliefs, goals Self-inefficacy Mental illness	Employee assistance (referral/inhouse) Counselling, therapy (cognitive, behavioral, biobehavioral) Medication Supervisory training	Social support (friends, family, co-workers, church) Self-help groups/books Self-medication Recreation, leisure, Sexual activity "Sick" days
Biological	Circadian rhythm Nutrition, sleep, exercise, weight, age, sex, race (genetic) Current illness impair- ment/disabilities Drugs Pregnancy	Rescheduling Placement and screening Health education Counselling Substance abuse treatment Biobehavioral treatment Maternity leave Health promotion	Change sleep/wake habits Sleep on job Bag lunch Drugs Cosmetics Diets, exercise Medication Self-care Dietary change
Physical/ Environmental	Climate Poor air Noise Toxic chemicals Pollutants Poor lighting Architecture Radiation	Clothing and equipment Climate control Ventilation Chemical control Interior decoration Muzak Protection Medical office	Own equipment, decoration Walkman Soap operas Music Personal physician

perspective, which orders phenomena into a hierarchy that ranges from physical dimensions to sociocultural levels of analysis (Miller, 1978; Schwartz, 1982, 1983; Von Bertallanfy, 1968). In addition, it recognizes that any stressor produces a reaction that at some level serves an adaptive function. The distinction of formal and informal interventions in the OSEG allows one to gauge the amount of personal and organizational control inherent in any one adaptive reaction. We discuss the OSEG as an integrative instrument in the last section of this chapter.

After presenting our own understanding of stress, we will turn to the world of occupational stress as seen through the differing visions of labor and management. We will briefly review the methodology used to gather our data on their respective perspectives. Then each view of stress and stress reduction will be presented separately. First, we attempt to extract a definition of stress from the diverse opinions of each sector. These general definitions will be described with attention to the biopsychosocial levels of the OSEG. After the stressors are enumerated, the matching stress reduction strategies will be listed and discussed. General remarks will be made within each section. Finally, an integrative conclusion is offered.

A RESEARCH DEFINITION OF STRESS

In formulating a basic definition of stress, we drew on the psychological and psychophysiological concepts of controllability, level of stimulation, and predictability. The research of Frankenhaeuser (1979) and Levine (1980) served as our reference guides.

Frankenhaeuser and her laboratory in Stockholm have produced more than twenty years of studies on the psychological factors that precipitate catecholamine increases above baseline levels (Frankenhaeuser, 1971; Frankenhaeuser, 1979; Frankenhaeuser & Jarpe, 1963). In reviewing this work, Frankenhaeuser has highlighted the importance of moderate stimulation and limited task demand in holding down excessive catecholamine release. Laboratory research and field studies of factory workers indicate that people function best subjectively and show the least stress physiologically when work is neither too monotonous nor too stimulating. Frankenhaeuser has also demonstrated that the following stressors produced increased catecholamine secretion—task conflict (having to choose between two stimuli demands), lack of control over work pace, and overcrowding during commuting to work. These stressful dimensions closely parallel those outlined by Robert Karasek of Columbia University, who has developed a stress grid with coordinates of autonomy and job demand (Karasek, Baker, Marxer, Ahlbom, & Theorell, 1981).

A common theme linking this research is the amount of control an individual exercises over external demands. According to Frankenhaeuser (1979), "Conditions characterized by uncertainty, unpredictability, and *lack of control* usually produce a rise in adrenaline output" (p. 134).

Levine has added a second dimension of *predictability* to the theme of control as a stress inhibitor (Weinberg & Levine, 1980). Levine has shown in several studies that animals receiving warning of shocks, who also possess the ability to escape or avoid the shocks, will show lower levels of plasma corticoids than rats without control or signal warnings. His work suggests that the excessive stress reaction can be contained if individuals have enough information about potential threats to formulate plans of action to minimize their resulting harm.

Lazarus (1977) and Glass (1977) have written at length about two other individual determinants of the stress response. Lazarus's work has revealed the overwhelming importance of *cognitive appraisal*—in essence, how our learning history and current psychic resources allow us to interpret the emotional value of a stimulus. Glass's contribution to the Type A literature implicates a behavioral pattern of time urgency, aggressiveness, competitiveness, and achievement-striving in development of coronary heart disease.

METHODOLOGY

We began our investigation in a somewhat traditional manner with a file of clippings about stress interventions and several lists of employers, labor groups, and trade associations. Using telephone contacts with these individuals and groups as a starting point, we developed an extensive list of programs and people who were involved with some attempt to reduce occupational stressors and improve health in the workplace. In all, 217 organizations and individuals were contacted (see Appendix 1 for the specific breakdown). Of these, 67 were corporations and 53 labor unions; 61 organizations of this 120 sent us written material in addition to answering structured questions on the phone. After selecting particularly representative or unique programs, an initial stress management questionnaire was mailed to 63 labor and corporate representatives. A second questionnaire, focused less on programs and more on ideology, was mailed specifically to labor representatives. The overall response rate to the two questionnaires was 46 percent, with corporate participation at 57 percent and labor response at 32 percent.

Characteristics of the Two Questionnaires

The first questionnaire was organized around general concepts about stress management that had emerged from our initial phone conversations. These included assumptions that stress management was approached in diverse ways; that it was usually packaged or programmatic; and that it was most often directed by a health or mental health professional.

In addition to a lengthy list of stress management techniques (i.e., relaxation training, yoga, biofeedback, meditation, exercise, discussion groups, etc.), we sought more general information about counseling, recreation, and educational presentations. We also included queries about work redesign and worker participation. As our knowledge developed through phone calls and written materials, we introduced some major revisions into the survey. Faced with the paucity of stress management programs for blue-collar workers, we revised our target population to include nonprofessionals—clerical, technical, and service workers. We continued to expand our definition of stress management to include any active interventions that sought to improve "life at work." We tried to include questions that discriminated inhouse programs from programs purchased from consultants or health professionals.

The second questionnaire was sent only to selected labor representatives and was designed to allow a more open-ended discussion of stress reduction. Rather than a programmatic focus, it simply asked for definitions of stressors affecting their specific workplaces and strategies that had been employed to combat these stressors. The responses to this questionnaire, though few in number, tended to be rich in content and observation.

Data Analysis

The format of this investigation of stress management was decidedly qualitative. Though we tabulated some rough results, their purpose was more to structure our telephone inquiries and written materials than to generate meaningful statistics. Our selection of questionnaire recipients was nonrandom and, in fact, biased toward groups whom we believed to have particularly interesting intervention strategies.

The tables presented in this chapter are not meant to give exact depictions of the state of stress management in the corporate and union sectors. Rather, they represent the converging data we managed to cull through our diverse methods of gathering information. In this respect, their main purpose is to stimulate hypotheses and models that may be submitted to more rigorous and quantitative methods of inquiry.

RESULTS OF LABOR UNION STRESS EVALUATION

Definition of Stress

Our central question concerns the outcome of an operationalized experimental phenomenon when it makes its debut in the "real world" arena of competing social and economic interests. The parsimony the honest researcher values in describing stress may be neglected for the political or economic advantage conflicting groups might gain in crying the "stress wolf."

Through telephone contacts, questionnaires, and the extensive mailings we received, we were able to ascertain how more than fifty major labor unions and organizations across the country conceptualize stress. By focusing more specifically on how these groups chose to prevent, reduce, or cope with stressors, we discovered what they think causes stress, and at what levels (biological, psychological, or social) they believe it is most effectively battled. It surprised us to see how closely their stress complaints corroborated the basic research findings on stressors. This fact is most clearly highlighted by the choice of specific stress-reducing strategies that confront head-on the majority of experimentally defined stressors. As we shall see, the labor movement in its more progressive stance argues for worker control over stimulation, work pace, task demand, and role conflict. Similarly, it seeks to build more predictability into workplaces, by demanding in contracts advance notice of technological change, plant closure, and downgrading of work.

Finally, though for its own strategic purposes it has been slow to handle this issue, the labor movement has begun to confront injurious lifestyles and habits of its members (including diet, substance abuse, and emotional illness). In general, a positive correlation exists between the newness of a portion of the labor sector and its commitment to attacking stress on multiple levels. The most progressive organizations in stress reduction are those representing the rapidly growing clerical, service, and health care work force. It is probably no coincidence that women comprise the rank-and-file majority of these sectors; our work strongly showed women to be more outspoken and less tolerant of intrusions upon their mental and physical equilibrium.

THE STRESSORS MOST COMMONLY NAMED BY LABOR ORGANIZATIONS

Physical and Biological Stressors

These stressors are the ones most likely to be handled by the OSHA statutes and NIOSH recommendations. They range from acute and chronic life-threatening hazards to subtle and persistent hassles that may be just as lethal in creating a long-term stress problem. In our survey, the older, more established industrial trade unions named the kind of stressors one would expect to find on a shopfloor of an assembly belt. A more revealing class of physical hazards was offered by VDT and CRT operators, whose ergonomic considerations have only recently begun to receive attention. Table 6.2 is meant to be a representative selection of the biological/physical/environmental stressors we uncovered, ordered from industrial to office settings (see Table 6.2). It illustrates various physical and environmental stressors named by the participants in our survey, but is not meant to be exhaustive or to imply that our respondents were the only groups exposed to these stressors.

Table 6.2. Biological/Physical Environmental Stressors (Labor Perspective)

Stressor	Source	Specifics
Heat	Flint Glass workers and related industries, electrical workers, glass bottle blowers	Temperatures above and below 17°C to 23°C
Ventilation	Mining and chemical workers and many other industrial workers	Inadequate exhaust systems, blowers, and hoods
Noise	Machinists, automobile and aircraft assembly heavy assembly, work	Punch presses, drills, grinding
Hazardous machinery	Electrical workers	Unguarded punch presses, frayed or ungrounded wiring; absence of safety shields and hoods
Dust and toxic fumes	Painters, mining, chemical workers, rubber workers, textile workers, asbestos workers	Emphasis on personal prevention (respirators, masks, safety goggles), as opposed to structural alteration of the work site or process
Toxic solvents	Paint and chemical workers	Anilines – suspected carcinogens
Radiation	Nuclear plant workers, also a worry of office workers exposed to low level radiation from CRT	Inadequate knowledge of long-term effects
Poor ergonomic design of work	All workers contacted	Chronic back pain, carpal tunnel syndrome, pulled and strained muscles and general fatigue. Building architecture – windowless, declining industrial workplaces, overcrowding of work stations
Office work hazards	Clerical and technical workers represented by nonprofit organizations	"Sealing-in" – where toxic fumes, vapors, bacteria, and dust are recirculated by central air conditioners, toxic office substances – methanol in duplicators, solvents in stencils and correction fluids, nitropyrene in toners, VDTs – poor lighting, glare, muscle fatigue, inadequate breaktime, constant breakdown

Many of the stressors identified in Table 6.2 have direct biological consequences, such as silicosis, cancer, deafness, and tissue damage. Even without producing actual health effects, they wield the constant psychological threat of their potential injury. In this sense, biophysical stressors are particularly relevant to the experimental stressors we have identified. For example, a NIOSH-supported study by Ramsey, Burford, and Beshir (1982) reveals the behavioral influences of heat exposure beyond the physical danger of prostration: "Temperatures below and above the preferred level (approximately 17° to 23°C) have a significantly detrimental effect ($p < .01$) on worker safety related behavior." Workers exposed to extremes of temperature are more likely to use their hands instead of the appropriate tools, lose control over their work pace, neglect safety equipment, and fail to maintain their tools properly. Physical stressors, by reducing a workers' ability to control his or her environment, and by posing an unpredictable threat of injury, create circumstances in which an elevated stress response would be naturally evoked.

Psychological/Interpersonal/Work-Setting/Sociocultural Stressors

At higher levels of stress-inducing events, union representatives appear more willing to identify stress as emerging from both the workplace *and* the homelife. Economic hard times, unemployment, plant relocation, and automation have tended to blur the boundaries among job stress, family stress, marital problems, and social problems. This type of systems analysis by workers is particularly prevalent in white-collar sectors. A stress manual used by the Graphic Artists International Union divided stressors into environmental and lifestyle categories, including nonoccupational stressors such as sex, the natural environment, and bereavement.

An American Federation of Teachers publication, *Stress and Burnout in the Schools*, divides the work setting and interpersonal realms into three levels: the microsystem, the mesosystem, and the exosystem, representing the smallest work unit; the organization; and the outside social supports (a fourth, the macrosystem, encompasses the sociopolitical-economic conditions). Borrowing this conceptualization from two researchers on burnout (Carroll & White, 1981), the union identifies stress as ranging from role mismatch and overload in the microsystem to the effects of family demands and school-board decisions in the exosystem.

Bearing in mind this comprehensive definition of stressors used by the less traditional unions, we may list the variety of psychological, work-setting, and interpersonal stressors identified by our study's participants (see Table 6.3). The stressors ascend roughly from the formal work process to the less formal psychological and social influences that affect work. (See Appendix 1, for an enumeration of the organizations that provided information about exposure to stressors.)

Table 6.3. Psychological/Interpersonal/Worksetting/Sociocultural stressors (Labor Perspective)

Stressor	Source	Specifics
Work pace (too fast or too monotonous)	Electrical workers, glass manufacturers, automobile assembly, aircraft assembly, chemical workers, communication workers	Continuous motion, understimulation, speedups, short cycles of repetitive movements
Workload	Electrical workers, glass manufacturers, automobile assembly, aircraft assembly, mining, chemical workers, communication workers, teachers, correction officers	Inflexible demands, poor equipment to do requested work, underemployment, conflicting task demands, inadequate pay for work requested
Work content	All unions contacted	Lack of creativity, no decision-making power, no say over the product you produce
Shiftwork	Mining, glass manufacturers, physical plant operators, health workers	Swing shifts, lack of stable schedule
Automation	Automobile assembly, service workers, office workers	Downgrading of job, lack of preparation for technological change, lack of retraining, redundancy
Supervision	All unions contacted	Unsympathetic managers and supervisors, incompetent supervision, too much stress on supervisors, arbitrary enforcement of rules
Psychological job pressures	Firefighters, stewardesses, police, office workers, health workers, correction officers	Sense of guilt over survival while others die in fire or plane crash, grieving for victims or patients who die, inability to communicate special stressors of job, sexual harassment, trouble controlling loss of temper and physical anger, the spilling over of work stress on to marital relations and the reciprocal effect on work, general personal problems – mental illness, family deaths, substance abuse
Job security	All unions contacted	Lay-offs, plant relocations, plant closings

We should qualify this list of stressors by stating that because of the limitations on time and resources, our sample did not draw on the vast number of small, unrepresented workplaces or the largely unorganized work force of the Southwest. What our summary does show is a certain current of repeated concerns that span the work environment from the punch press to the word processor. Labor representatives seem to agree that the following stressors are most troublesome:

1. **A Lack of Control over the Work Content.**   Individuals are unable to influence the choice of product, the design of the work, and the level of effort it requires.

2. **A Lack of Control over the Work Process and Pace.**   Pacing of the work is often machine-determined, leaving the individual little opportunity to regulate the task to his or her own rhythm. The result is a tense secretary waiting for the terminal to reply or a worker in a SO^2 plant watching one warning gauge for an eight-hour shift.

3. **Unrealistic Task Demands.** Individuals are requested to handle conflicting tasks. (For example, teachers must be sensitive role models and plainclothes police officers at the same time; or, given inadequate resources, correction officers must confront unresponsive lawmakers and overcrowded prisons).

4. **A Lack of Understanding by Supervisors and Management.**   Principals are characterized as too enmeshed in their own paperwork demands to listen to the teachers' problems. Supervisors on the assembly line were described as seldom listening to workers' suggestions about how a production process could be done more efficiently or cheaper.

5. **An Inability to Keep Work Stressors and Home-life Stressors from Interacting Multiplicatively and Negatively.** Many unions are beginning to admit that all the stress at work does not arise simply from the work itself. People have personal troubles, divorce, mental illness, family death, regardless of what work they do. Though unions are not about to concede that such problems are the major stressor for their members, they have become more willing to acknowledge their own foibles or misfortunes rather than blame the company blindly for all stress.

6. **A Lack of Predictability and Security about the Job's Future.**   This factor includes questions about plant relocation, automation, general economic downturns, and skill obsolescence. Workers in the older industrial trades face the accelerating phasing-out of their livelihoods, while younger workers are unsure what skills will be appropriate in a technological workplace that innovates yearly. Unions intimately involved with assembly work must grapple with their position toward robotics and entirely automated belts. The problem of downgrading of

existing jobs is dramatized by office workers' claims that their jobs as key puncher and word processor are leading to the creation of a "mental assembly line." Successful drives to organize clerical and technical unions at large universities (most notably at Yale) suggest that a new force of female workers is demanding control over the future structure and pace of their work.

RESPONSES TO STRESSORS

Unions show remarkable consistency in tailoring their stress reduction efforts to combat the stressors they named in our survey. A statement by Lee Schore, *UAW* #805, in a pamphlet she sent us summarizes the dominant union approach to stress reduction:

> A union approach is based on the belief that hazards of job stress are related to the structural conditions at the workplace and affecting all workers there, not merely the "troubled" worker. We do not seek to help workers manage their stress better or to adjust to the conditions that are potentially dangerous to their health. Rather when we speak about stress reduction we are really talking about stress prevention—creating healthy workplaces. This calls for collective solutions and requires collective forms.

Accordingly, of the 53 unions included in our inquiries, none of them had its own in-house stress management or counseling service, though such member assistance programs have begun to emerge (e.f., *DC 37* in the public sector and the Amalgamated Clothing and Textile Workers in the private sector). A number of unions contracted psychologists, including the state *AFL-CIO* office in Hawaii, the Montgomery county police in Maryland, and a *UAW* local in New York City (which has made referrals to a psychoanalytic institute for 20 years). Municipal workers in New York City pay into an *HMO*, which gives them access to short-term counseling and social work services. Many San Francisco area local unions in California are affiliated with the Institute for Labor and Mental Health, which runs 12-week support and educative groups on occupational stress, covering topics such as the physiology of stress, organization of work, and self-blaming and anger. The Institute also offers steward training and documentation of disabling workplace stressors.

Aside from these more recent efforts to offer more individualized stress treatment, the majority of unions who provided information had not gone far beyond the day-long workshop and educational material format. There was little emphasis on lifestyle reformation or health promotion programs. A health and safety representative for an electrical workers union stated,

> I think health promotion is important but does not belong in the conflict-ridden

workplace atmosphere, but in the community. It is difficult for workers to enthusiastically embrace a program when the company offers health promotion with one hand and toxic exposure with the other hand.

BIOLOGICAL/PHYSICAL/ENVIRONMENTAL STRESS REDUCTION

Turning to the preventive rather than stress management tactics unions employ, we might start again with biological/physical/environmental stressors. Though the stressors might be located on the lower end of the OSEG, the unions' formal intervention occurs at the work setting and above. Table 6.4 illustrates the dominant responses of labor representatives regarding how they might handle physical stressors in the workplace. Their emphasis upon legislative restraints and strong contractual control in regard to physical health hazards fits well with their conceptualization of stress as a lack of control over their work environment. As many union leaders explained in our discussions, there will be little improvement in working conditions, if workers simply learn ways of improving their leisure and deep breathing. The lengthy latency period of many occupational illnesses precludes an extensive emphasis on short-term adjustment strategies.

As a case in point, labor representatives cite the problems of office workers in changing working conditions. Eighty percent of public-employed clericals are not unionized, while 90 percent of private sector clericals remain unorganized. Without an institutional advocate, these workers have little recourse if health and safety conditions are unsatisfactory. OSHA statutes primarily serve older industrial worksites; the proper safeguards for office environments have yet to be promulgated. The lack of organization and of federal guidelines forces the clerical worker to rely on the social support of fellow workers. Perhaps, for this reason, international unions, such as the *UAW, AFSCME*, and the Hotel Employees and Restaurant Employees who already represent some office worker locals, have begun to put time and money into organizing this sector. As the supportive network of the coffee-break develops into a fledgling organizing committee for a union drive, we see a powerful example of how an informal intervention strategy can gain legitimacy and structure until it is formalized by larger institutions.

RESPONSES TO PSYCHOLOGICAL/INTERPERSONAL/WORK-SETTING/SOCIOCULTURAL STRESSORS

The union response to nonphysical stressors is outlined in the second part of Table 6.4. In controlling work-setting stressors, the strongest active strategy recommended by the union spokespeople was a contractually empowered health and safety committee, composed of union and corporate members.

Table 6.4.   Labor Responses to Stressors

Stressor	Informal Response	Formal Response
	Labor Responses to Biological/Physical/Environmental Stressors	
Heat, ventilation, noise, dust, etc.	Slow downs, self-regulation, prophylactic devices and clothing, frequent breaks, postural adjustments, denial, apathy and accommodation	Federal statutes through OSHA – education, monitoring, requesting inspections, NIOSH research to substantiate claims of adverse health effects, health and safety committee with equal representation of employer and worker and contractually guaranteed powers, write health and safety clauses in to contract to formalize grievances
	Labor Responses to Psychological/Interpersonal/Work Setting/Sociocultural Stressors	
Work pace, load, content, shiftwork, supervision	Slow down/speed up, sleep on job, sabotage, redefine task, quit*	OSHA statutes, contract language assuring grievances, Health and Safety committee, quality circles, training of supervisors
Automation	Sabotage, develop new skills, hide redundancy, go into management*	Advance notice and retraining clauses in contracts, participation in corporate planning through board membership or stock ownership
Psychological job pressures	Social support, self-help, recreation, substance abuse, vacation	Referral to health agencies, leaves of absence, joint company and union alcohol and counselling programs, union sponsored therapy or counselling
Job security	Become "indispensable," find "recession-proof" line of work, try not to make waves, take lower wages and less benefits	Security clause in contract, gain voice in corporate decisions by board membership or stock ownership

*Inclusion of a response on this list does not indicate its support by any union. Rather, it is an acknowledgement that such responses do occur informally among employees.

Worker participation at this shop-floor level returns the opportunity for self-control to individual workers. They are able to effect the ergonomic changes necessary for stress reduction, including lengthening work cycles, removal of swing shifts, flextime, monitoring of production equipment, and redefinition of task demands. Such participation through Health and Safety committees and worker-approved Quality Circles allocates partial responsibility back to the worker, relieving some of the burden from the middle supervisor. This system might in turn aid the poor relationship between workers and supervisors, encouraging them to be collaborators rather than adversaries. Some unions went further to suggest that supervisors receive educational credits in human relations and personnel management courses.

With American industry in the middle of a massive retooling for automated assembly, unions face the multiple stressors of unpredictability, redundancy, downgrading, and virtual extinction of certain job sectors. While they acknowledge that jogging might keep their minds off a future fast approaching, most union representatives advocate strong contractual guarantees to protect their members against the high tech dismantling of their livelihoods. (Appendix 2 offers an example of strong contract clauses that request advance warning of technological innovation and subsequent retraining without penalty for the displaced workers.) Recent efforts by the UAW to achieve lifetime employment clauses are also geared toward maintaining some predictability and control for workers in this disorienting period of industrial modernization.

Finally, as mentioned earlier, some unions have also begun to address the more immediate injuries these diverse stressors inflict on the worker. Beside the referral programs described in brief above, the International Association of Machinists and Aerospace Workers might serve as a useful example of a progressive union approach to the psychological health of its members. The international has given a mandate to its locals to create a "Human Contract," based on the training of union counselors in 8 to 12 week courses. These union counselors learn about the community resources of various helping agencies, health providers, psychological services, and social work facilities. They learn about listening skills, state laws, and financial assistance (including disability and insurance payments). The expectation for these union counselors is that they will help tie the union membership back into the community and encourage the collective perception and solution of common stressors. A similar program has been implemented by the Association of Flight Attendants, in conjunction with their employee assistance program. Ultimately, this attention to psychological and community stressors may pay off politically for the union. By broadening its services, a union increases the vitality and allegiance of its membership. Strongly bonded unions enter collective bargaining with the distinct advantage of solidarity.

DISCUSSION OF THE LABOR PERSPECTIVE

We began this chapter by noting some of the major experimental findings on the antecedents of stress. Control, demand, and predictability proved to be crucial psychological factors in evoking the stress response. When we moved to the field and surveyed a diverse group of labor representatives, we found their discussion of specific stressors paralleled closely the experimental findings. Furthermore, the overriding concern of their formal interventions to alleviate stressors was the increase of control over work-setting and organizational factors. This emphasis on structural change, as opposed to individual change, was reflected in the relative lack of personal lifestyle, diet, and exercise interventions advocated by labor organizations. Similarly, union spokesman were adamant in rejecting a Lazarus-type coping style approach to workplace stressors.

Cognitive appraisal is an attractive strategy with its stoic belief in the power of human expectations to alter the stress value of stimuli. If the worker changes his or her perception of the job demands, the stress level will be accordingly diminished. Unions reject this perspective, arguing that workers' complaints about stress are a sign of health. Union members see themselves as entitled to jobs with creativity, autonomy, and self-determination. Their appraisal of dissatisfaction with the limitations of their existing working conditions is strong evidence of this emerging attitude in the workforce. To alter this self-affirming appraisal would be viewed as a step backward toward the older industrial notions of the "dumb worker" and expendable labor.

In the same way, unions rejected spending large resources of time and money to alter Type A behavior in their membership. Their politically based complaint focuses on a society that fosters aggressiveness, competitiveness, and time-racing behavior. They argue that their members would learn more relaxed styles if the demands and deadlines, job-loss threats and speedups, were diminished to a more acceptable level. To adjust the worker to these stressors is to make them more bearable. The more bearable they become, the less motivation the rank-and-file will have to alter the institutions that impose these stressors. For the more adversarial representatives with whom we spoke, coping is tantamount to accepting and thus only blurs the battlelines.

Ultimately, unions may have inadvertently helped to define the concept of stress even more precisely than the laboratory. The advances of the Frankenhaeuser (1971, 1979) and Levine (1980) studies demonstrated that the crucial variable in the stress response was a stimulus's effect on the individual's psychology. Still, their method of validating this assertion relied on physiological indices correlated with behavior and self-report. The most obvious interpretation of their data suggests the potential health ramifications of prolonged stress exposure. Yet to relate stress to its physical influence on years of life or potential illness is to fail to define its meaning completely. While unions recognize the material effects of stress (and have never been ones to underplay material gains at the bargaining table), they increasingly emphasize the psychological importance of the concept. Improvements in health due to stress management are not satisfactory, if the psychological imperatives of control and moderate predictability are not obtained. In recognizing autonomy and self-determination as stress-reducers in themselves, independent of their influence on physical health, the unions argue for the integrity of the psychological world aside from its bodily concomitants. They do respond, one might add, with experimentally supported conviction that, as they alter the psychological atmosphere of their members, the physical changes will follow.

RESULTS OF CORPORATE STRESS EVALUATION

Definition of Stress

In contrast to the approach of labor groups, the most salient feature of the corporate approach to stress is its tendency to avoid characterizing stressors. Of all the materials we obtained from corporate organizations, only one included a list of the stressors to which employees were exposed. This lack of formulation typifies the main theme of the corporate conceptualization of stress. Although excessive job demands and responsibilities may receive mention in discussions of stress, the primary emphasis falls on maladaptive personal styles and poor "person-environment fits" (Chesney & Chavous, 1981).

Many corporate approaches to stress disregard workplace characteristics or work organization altogether. The focus is on personal responsibility and coping. Employees are instructed to identify stressors and strains and to recognize the role of stress as a risk factor in disease. Modification of one's lifestyle is promoted, with particular attention to exercise, nutrition, and coronary-prone behaviors. Employees are encouraged to develop personal strategies that would improve their resistance to stress or their ability to cope with strain. Relaxation, cognitive restructuring, communication skills, and stress management are among the solutions proposed.

Though the term, stress, is rather general in corporate parlance, the motivation for stress management seems to be more clearly defined. Corporations see satisfied employees as healthier and more *productive* workers. The corporate world has recently experienced an explosion of consulting groups whose main message is that the quality of an employee's life and the success of a firm are one and the same goal. As we detail various corporate health promotion and stress management programs, it will become apparent that many well-intentioned corporate members take this philosophy to heart and have vastly improved the day-to-day working environment of their employees. In this manner, they have defined stress on personal, biological, and physical dimensions. Their intervention strategies fall on the same levels of the *OSEG*. What we might remember, as researchers with a particular definition of stress, is the extent to which these approaches address the full problem of stress. Are the problems of control, demand, and predictability across the multiple levels of the biopsychosocial model confronted by this person-focused approach?

The discussion that follows offers a synthesis of the information obtained from the 67 corporations we either spoke to or from whom we received materials or questionnaires. It also draws on the comments and questionnaires of the 23 consultants we contacted who specialized in health promotion or stress management programs (see Appendix 1 for more specific demographics).

Biological/Physical/Environmental Stressors

Corporations choose to define biological and physical stressors primarily in personal, as opposed to ergonomic or workplace terms. As a brochure of one "life extension" company states, "whether an individual lives or dies, possesses vitality and health, or suffers debilitating illness is largely in his or her own hands." Still, this individual orientation has not meant a lack of imagination or breadth in corporate "stress management" strategies. The variety of personal health risks corporations have begun to identify suggests an awareness of employee behaviors affecting the entire biopsychosocial system. The following list illustrates the comprehensiveness of their approach:

1. Nutritional excesses or deficits
2. Smoking
3. Alcohol and drug abuse
4. Lack of exercise and recreation
5. "Lifestyle diseases," including cancer and hypertension
6. Chronic pain

Corporations with active attention to these concerns range in size, type of product, and location. Communications, business machines, paper products, health needs, missiles, computers and software, insurance, automobiles—the leading corporations responsible for each of these products or services have developed extensive programs to define and isolate the health concerns listed above. Eighty percent of the corporations and organizations responding to our questionnaire have some form of health promotion program.

The beginning stage of a representative corporate program is a health risk appraisal. An employee schedules an appointment with the company physician and undergoes a thorough physical and lifestyle history. Blood pressure, blood chemistry, height, weight, cholesterol, and other diagnostic clues are recorded, including an extensive family history. The physician integrates this information into a profile of the employee's general health, specifying problem areas and remedial actions that might be taken. Review of the health profile leads to a referral to one or more of the health promotion programs also offered by these companies. These programs will be discussed later.

Upbeat and attractive educational packages always accompany any corporate health appraisal program. Newsletters, departmental contests, fun runs, audiovisual guides, and illustrated brochures provide incentives for the employees to "become involved" with their own health.

By the above account, it should be clear that corporations have chosen to rely on health professionals, most often physicians and occupational nurses, to define the parameters of physical and biological stressors. The health promotion programs have almost without exception arisen from intact medical departments and in this regard reflect national trends in preventive health and lifestyle management.

Psychological/Interpersonal/Work-Setting/Sociocultural Stressors

Once again, the corporate focus tends to avoid organizational or work-related definitions of stress in their discussions of nonphysical stressors. The majority of companies center their approach in the *OSEG* levels, labeled Interpersonal and Psychological. Table 6.5 lists the kinds of nonwork stressors that corporations have begun to identify as influencing the welfare of both the workers and the company. It also includes a small set of work-related stressors that a few of the companies identified. The list demonstrates an emerging awareness in the corporate world of employees as complicated human beings with continual challenges outside their work. The breadth of what the company considers within the purview of treatable stressors suggests they will leave no stone unturned to aid the employee in achieving satisfaction and productivity. It is not unusual to hear stress consultants and managers quoting the self-actualization theory of Abraham Maslow. Believing their employees have moved up the ladder from physical needs and achieved security as well, managers attest their sensitivity to the life goals and creativity of their employees.

An emphasis on nonphysical needs necessitates an integrative perspective toward the employee. One must think of him or her in family or community roles outside the workplace. Recognition of variations in personality, temperament, and mental capacities is equally important. For humanitarian or economic reasons (or both), most large corporations will no longer send the employee to the mines and forget the person's name. How they follow-up the single employee with every conceivable program of psychological and physical health promotion is the topic of the next section.

Corporate Responses to Stress

Corporate organizations vary significantly in the kind of efforts employed

Table 6.5. Psychological/Interpersonal/Worksetting/Sociocultural Stressors (Corporate Perspective)

Stressors	Potential Ramifications

Personal psychology
  Type A
  Anxiety
  Depression
  Dissatisfaction
  Apathy
	Stress-related including all the maladaptive health behaviours listed by the corporation (i.e., hypertension, substance abuse, lack of exercise, etc.); absenteeism; lessened productivity; health services overuse; poor interpersonal relations with co-workers; no "team spirit"; sabotage; general frustration and fatigue due to over/or under-exertion on job

Life Development
  First job
  Starting family
  Raising children
  Middle age
  Retirement
	Distraction from the job itself; increased burden to earn more money; fear of "dead-end" job or too slow advancement; loss of energy and acuity; fear of uselessness and death

Interpersonal relations
  Spouse difficulties
  (Divorce, separation, abuse)
  Children or parent difficulties
  (discipline problem, illness, aging parents)
  Death of relation or close friend
  Loneliness or unsatisfying relationships
	Irritability; inattention to task; aggressiveness; time conflicts; apathy; relationships with co-workers; sexual harassment

Financial and legal problems
  Debts
  Liability suits
  Tax evasion
	Overwork; absenteeism; theft

Rapid changes in society
  Economic shifts
  Technology
	Job reorganization; new skill requirements; transfers; automation; alienation

Work-setting issues
  Job demands
  Person-environment fit
  Hazards or stresses intrinsic to work
  (i.e., police or air traffic controllers)
	Dissatisfaction; ineffective employees; burnout; loss of self-esteem; chronic anxiety; influences on mood and sleep

to promote health or to reduce stress among employees. If our contacts with many of the world's largest corporations are an accurate indication, the majority either did not address employee health or stress with specific programs, or did so exclusively through health insurance packages. Representatives from many companies greeted our inquiry with pride or defensiveness and proceeded to describe their employee assistance referral system to handle "troubled employees." On the other hand, some organizations had so many programs in progress that they were reluctant to characterize their efforts in any simple way. Instead, they sent us thick envelopes filled with glossy brochures and handouts, or referred us to the administrators, clinicians, or consultants responsible for different program components. Even within the most comprehensive efforts, however, individual programs often remained isolated from one another.

If corporate responses to stress vary across organizations, they also seem to vary across classes of industry. At this point, high technology (communication and information) concerns and large-scale manufacturing companies lead the way in the diversity and depth of their efforts. Retail and service organizations are less likely to implement health promotion or stress programs for their employees. Although economics may be the most important contributor to this discrepancy, another reason may involve the relative decentralization of retail and service personnel. The reliance of corporations on in-house provision of services and education is better suited to centralized work settings.

Moreover, several large companies now exist solely to promote health or to manage stress. Although some corporations have contracted their services, the larger corporations, represented in our questionnaire replies, provide more than 90 percent of their health promotion services and more than 80 percent of their stress management services in-house. While health promotion more clearly falls under the medical aegis, stress management programs are administered by a cross-section of corporate departments. The list includes human resources, personnel, employee advisory services, medical department, educational training, employee service organization, psychological services, and office of stress management. This lack of commonality in approaches to stress is mirrored by the different sources of funding for stress management. In our questionnaire results, 35 percent of the respondents' programs were funded by a medical plan, 45 percent by the company's general funds, and 25 percent were fee-supported (some companies using mixed funding strategies).

Table 6.6 presents corporate strategies for handling the extensive health and psychological stressors, which they have characterized as troubling workers and the workplace. The first category, health promotion and education, is usually open to all employees, both professionals and nonprofessionals. There were instances, however, in a few corporations in which executives still received certain educational materials and physical screenings that were as yet not offered to the nonmanagement employees.

Wherever programs were offered, attendance and use by all employees was reported to be remarkably high. When scanning the extraordinary breadth of health services offered by the larger corporations, it is apparent that many of the approaches extend up and down the levels of the *OSEG*. In accentuating education, prevention, and lifestyle enhancement, the more progressive companies break down mechanistic models of the body and challenge the employee to think holistically about health.

Furthermore, as Table 6.6 indicates, the corporate outlook on health is both technologically and humanistically sophisticated. Combining computer health appraisals with fun runs blends gravity and amusement in a way that promises to keep employees motivated about health promotion. The company representatives we contacted often remarked that beyond the potential health benefits, the education and prevention programs elevate the morale of the company. The programs, courses, and facilities serve to remind employees that they are more important to the firm than are their time cards or production quotas. In addition, corporate studies mailed to us indicate employee satisfaction with learning techniques of self-mastery and control. These findings suggest that the health policies of corporations may have tangible psychological benefits in reducing stress on both the biological and psychological levels of the *OSEG*.

Employee assistance and substance abuse rehabilitation programs are often closely linked in corporations, with the former developing from the latter. For example, an extensive alcohol program at one company has evolved into a "pretreatment intervention program dealing in secondary prevention." Availability extends to all employees and their dependents. The emphasis is on union/management/community support to help the "troubled employee" back onto his or her feet. This type of program appears to be more common, extending into financial, preretirement, and psycho-therapeutic counseling. A second corporation with a similar spirit employs a volunteer network of "peer counselors," whose main role is in assessment and referral. *ALMACA* (Association of Labor-Management Administrators and Counselors on Alcoholism), the *EAPs* association, has intensified epidemiological efforts in recent years to identify high-risk populations and to take more appropriate preventive actions.

Of the companies in our questionnaire pool, nearly 80 percent listed stress education offerings, including literature, classes, and workshops. Neverthe-less, actual stress management interventions, such as relaxation training or meditation, were used by 50 to 60 percent of companies responding. As opposed to health education and promotion, both stress education and management programs were less likely to apply to nonprofessional staff.

While health promotion and education remain the province of the medical department, psychologists led all other health providers in directing stress management and education programs. Perhaps even more surprising, psychologists were outnumbered as directors of these programs by the "other" category—that is, health educators, exercise physiologists, training

Table 6.6. Corporate Responses to Stressors

Stressors	Program
Corporate Responses to Biological/Physical/Environmental Stressors	
Maladaptive health behaviors	*Health Education* – brochures, workshops, audiovisual tapes, newsletters, courses *Health Promotion* – smoking cessation, weight reduction clinic, exercise gyms and jogging breaks aerobic conditioning, first aid, CPR, screening (cancer, diabetes, hypertension, teeth), healthy back, safety-accident reduction, defensive driving
Substance abuse	Drug and Alcohol Rehabilitation programs Employee Assistance Programs – offering short term counselling and referral
Corporate Responses to Psychological/Interpersonal/Work Setting/Sociocultural Stressors	
Personal psychology (Type A, Anxiety, Apathy)	*Stress Education* – Courses on time management, lifestyle changes, anger/conflict management, communication and negotiating skills, values clarification *Stress Management* – Meditation training (including place for practice), relaxation training, biofeedback, group discussion, yoga, self hypnosis, development of social support networks *Employee Assistance Programs* – screening, crisis, short term, longterm counselling, referral service to community agencies
Life development	See EAP above; retirement counselling
Interpersonal relations	See EAP above; family therapy; couples therapy; support groups; volunteer employee counsellors
Financial and legal problems	See EAP above; financial counselling; legal assistance;
Change in society	See stress management and EAP above
Work-setting issues	Prescreening Transfer EAP referral Work redesign – quality circles, employee involvement, flexitime, management – employee discussions Management training Bring in outside consultant Research

and development staff, and in one case, legal staff. This finding of our questionnaire suggests again that stress is not yet the domain of any single profession. Its diffuseness of definition in the corporate world allows for many interventions with different starting points. This idea also accounts for the frequency with which stress education programs were provided by personnel from outside the organization. Diverse professionals appear eager to share in the recent boom in stress management workshops and in-service training by consultants. Accordingly, stress reduction programs have expanded their province to the realm of management training. Conflict resolution, human relations, and value clarification were mentioned as workshop topics by many of our corporate contacts. For organizations contemplating worker participation, these efforts may facilitate structural changes.

What all the stress management perspectives in our research have in common is an overarching theme of personal responsibility and personal change. The Lazarus-coping model plays a large part in the educational materials we received. Stress educators mention control but mostly in terms of teaching individuals what they can and cannot control. Most corporate material on stressors only minimally discussed enhancement of individual control through structural change.

The emergence of the *QWL* (Quality of Working Life) movement suggests that the corporate conception of stress has finally penetrated into the work organization itself. With the automobile companies the most dramatic case, a new effort toward employee involvement and joint labor-management ventures has begun. A large utilities company offers an excellent shopfloor example. Employees, both hourly and salaried, meet weekly in an accident prevention committee. Membership is rotated, and published minutes of the meetings are distributed to all employees. The committee reviews issues of lost time due to accidents and equipment problems that might lead to accidents.

The exact nature of redesign and participation efforts remain unclear. Corporations are wary of sharing too much information about these interventions, claiming "trade secret" prerogatives. Consequently, we feel less able to offer a meaningful analysis of their actual benefits to workers as stress reducers. To the extent that they actually enhance worker participation in decisions about work organization or demands, they hold great promise for reduction of stress and enhanced perception of control by employees.

GENERAL DISCUSSION OF CORPORATE APPROACH

The corporate approach to stress concentrates on the individual employee's lifestyle and health habits. It defines stress, when it does define it, as personal styles or sets of expectations that are inappropriate for the demands of the environment. The management of stress parallels this definition by specifying techniques to alter one's habits or perceptions in an effort to achieve a better person-environment fit. Comparing this perspective to our research definition of stress provides a useful contrast. In this three-dimensional model of the biopsychosocial approach, the corporate strategy essentially omits the third dimension of the social or organizational environment. Just as labor organizations played down an individual emphasis, the corporations make little mention of how work structures might bend to fit the workers. Within the two-dimensional world of biology and psychology, the corporations demonstrate continual innovation in enhancing the personal control of their employees. Employees educate and train themselves in sensible nutrition, exercise, and substance abuse. They master self-regulation techniques and potentially become less reliant on the health care system. They begin to counsel each other and gain the valuable buffer of social support. They possess the opportunity to reduce physical demands on themselves, to increase biological and psychological control, and to achieve a more moderate and predictable pace in their lifestyles.

The corporate motives for these tangible gains must necessarily be mixed. The progressive corporations that wrote or spoke to us preferred to characterize their organizations as *humane profit-making firms*. Though the research evidence on these interventions is not available, the corporations bank on the prospect of their cost-effectiveness in reduction of health service utilization, absenteeism, and accident rates. The corporate understanding of how an individual's well-being influences other lives in his or her family, community, and, most relevantly, within the corporation, demonstrates a dexterous integration of all levels of the *OSEG*. At the same time, the corporate silence with respect to the reciprocal effect of society and company on an individual's experience of stress reveals a curious inconsistency. Whether corporations simply overlook the role of organizational stressors, believe them to be inexorable, or perceive them as necessary is also a complicated and controversial question.

Related to this last observation, corporations have thus far shown less initiative in developing networks with existing community services, often favoring an in-house or consultant approach. If their integrative approaches are to be effective in the long run, they may need to establish resource exchanges with public agencies that have long histories of human service.

CONCLUSION

The concept of stress, as experimental research has defined it, often seems quite different from the ways unions and management describe and treat it. We have observed that the manner in which unions and corporations define stress along the biopsychosocial continuum dramatically influences their intervention strategies. The union perspective defines stress primarily in terms of control and predictability, thus affirming the experimental concept.

However, its viewpoint pays less attention to the influence that an individual's personal style or set of expectations may have in creating stressful situations. Conversely, corporations recognize acutely the role of individual habits and perceptions in generating stressful "person-environment fits." Missing from their approach is the acknowledgment that organizational or work-setting constraints help to create maladaptive behaviors and cognitions.

Similarly, labor-initiated stress reduction strategies flow out of their organizational orientation. They emphasize regulation, contractual restraints, access to decision-making, and increased autonomy at the worksite. Only the most progressive service sector unions (usually highly female and urban) combine these reduction tactics with a comprehensive approach to mental and physical health (i.e., through *HMO* membership). Corporations lavish attention on the physical and psychological well-being of employees in an effort to slow down ineffective and harmful behavior patterns. They take pains to increase employees' control in the interpersonal, financial, and life-development domains. Yet only recently have they initiated efforts to involve workers in workplace decisions and organizations. Both labor and management appear consistent in providing stress reduction strategies commensurate with their circumscribed definitions of stress.

What effect do corporate and union stress reductions have on the problems of stress, as originally defined by our research? The answer is less scientific than political. The major finding of our research is that unions and corporations effectively handle certain aspects of stress, and just as effectively *choose not to handle other aspects of stress.* A union that concedes that its members' personal habits are inefficient or harmful to the company, gives ground to management in the struggle for a finite pie. A company that acknowledges safety or psychological hazards has little recourse but to institute costly changes.

Organizations in society invested with certain economic interests will adopt from the stress concept the level of understanding that supports their cause. It is both expedient and logical that unions and corporations construe stress as they do. Since corporations occupy such a dominant position in society, they would be loathe to initiate large alterations in material conditions in the society. Thus they often identify stress as the unfortunate biological and psychological reaction of certain people to the pressure of work or family life. The precipitating pressures will continue to be viewed as parts of life, "What we all must face." Meanwhile, labor unions, with a vested interest in altering material conditions (though not dramatically), note the environmental and organizational pressures that lead to stressed workers.

These selective formulations of the stress concept frustrate the psychologist because they are not wrong; they are simply incomplete. To the researcher, each of the strategies described previously makes sense in considering the diverse contributions to the stress response. To the political individual caught up in ideological wars, some of these methods seem worthless or even malevolent. Used in isolation, they indeed may be. It seems potentially dangerous to encourage jogging to an individual who leaves work pent-up with rage at a brutal work schedule. The physiological benefits of running may be greatly distorted by his or her emotional state. Similarly, a contract clause ensuring advance notice of a plant closing may not help much if little preparation has been made for familial strife or mental illness that may emerge despite the warning.

As we have emphasized by repeated reference to the *OSEG*, true stress management manages the relations between the biology, psychology, and environment of the individual and the organization. Political contingencies may cause organizations to neglect one or more of these dimensions, but in so doing, they may later face ramifications not only in the neglected area but also in the areas to which they have attended. This is the essence of a systems approach. If the problems of stress involve an interconnected whole, as the *OSEG* suggests, then selective attention to one aspect is not only doing less than a complete job, but also threatens to be useless in the long run. A corporation, busy with a yoga program, may allow a discussion of health and safety standards to slide as toxic exposures accumulate. In this case, how cost-effective will the corporate investment in stress management be, as workers develop acute reactions and take sick time? Just as in a geodesic dome, any disturbance felt in one part of the workplace will eventually resonate throughout all levels of the organization.

The systems or ecological perspective embodied in the *OSEG* suggests that labor and management positions, though unique, are necessarily dependent. An ecological approach would involve interventions that recognize the importance of the relationship between organism and environment, between worker and workplace.

The Children's Health Center in Minneapolis, Minnesota, provides an example of a top-down ecological approach to stress reduction. In this case, hospital administrators made an initial decision that the well-being of child patients and their families was of primary importance. They went one step further when they empowered a mental health/ecology officer with the authority to propose and implement changes in the system that would reduce stress and improve health for the patient. Guided by a biopsychosocial model of illness, the ecology officer (a clinical psychologist) and administration effected changes in admission procedures, visiting policies, interior design, and medical procedures. Perhaps the most exciting aspect of this program was its gradual extension to staff training in interpersonal relations—with the dual purpose of increasing sensitivity to children in a hospital setting and increasing sensitivity to each other.

In agreement with the Minneapolis project, the *OSEG* emphasizes that stress reduction, simultaneously personal and organizational, is in the interest of both unions and corporations. We suggest that a comprehensive multiple-level commitment to stress reduction is the most effective answer to occupational stress. This commitment requires a resolve by corporations to introduce true democracy and autonomy into the workplace. At the same time, it also requires hard work on labor's part to improve the health and social supports of its members. Once organizations yield more control to their members, the onus will then fall on employees to look more carefully at their maladaptive personal styles. With corporate action on control and predictability and labor initiative on personal habits, the experimental concept of stress would finally be integrated with its treatment in the field. We might then predict better health and greater satisfaction for employees, which should also result in an increase in employee cost-effective behaviors.

However, until labor gains greater control over work organization and quality, it is unlikely that union representatives will direct their membership to personal change. Our hope is that a perception by both groups of the inter-relatedness of all levels of stressors could produce a collaboration. How to inculcate that view into corporate and labor perspectives remains a major challenge of a biopsychosocial approach to stress.

REFERENCES

Carroll, J. F. X., & White, W. L. *Understanding burnout: Integrating individual and environmental factors within our ecological framework.* Paper presented at the National Conference on Burnout, Philadelphia, PA, November, 1981.

Chesney, M. A., & Chavous, A. W. (1981). Stress management in industry. *SRI—International Business Intelligence Program*, (Guidelines No. 1060), pp. 1–13.

Frankenhaeuser, M. (1971). Behavior and circulating catecholamines. *Brain Research, 31,* 241–262.

Frankenhaeuser, M. (1979). Psychoneuroendocrine approaches to the study of emotion as related to stress and coping. In B.B. Howe & R.A. Dienstbier (Eds.), *Nebraska Symposium on Motivation 1978,* (pp. 123–161). University of Nebraska Press.

Frankenhaeuser, M., & Jarpe, G. (1963). Psychophysiological changes during infusions of adrenaline in various doses. *Psychopharmacologia, 4,* 424–432.

Glass, D. C. (1977). *Behavior patterns, stress, and coronary disease.* Hillsdale, NJ: Erlbaum.

Karasek, R., Baker, D., Marxer, F., Ahlbom, A., & Theorell, T. (1981). Job decision latitude, job demands, and cardiovascular disease: A prospective study of Swedish men. *American Journal of Public Health, 71,* 694–705.

Lazarus, R. S. (1977). Cognitive and coping processes in emotion. In A. Monat & R.S. Lazarus (Eds.), *Stress and coping.* New York: Columbia University.

Leigh, H., & Reiser, M. F. (1980). *The patient: Biological, psychological and social dimensions of medical practice.* New York: Plenum.

Levine, S. (1980). A psychobiological approach to stress and coping. In S. Levine & H. Ursin, *Coping and health.* New York: Plenum.

Miller, J. G. (1978). *Living Systems.* New York: McGraw-Hill.

Ramsey, J. D., Buford, C. L., & Beshir, M. Y. *Effects of heat on safe work behavior* (NIOSH Report No. 210-79-0021). Washington, DC: US. Government Printing Office, March 1982.

Schwartz, G. E. (1982). Testing the biopsychosocial model: The ultimate challenge facing behavioral medicine? *Journal of Consulting and Clinical Psychology, 50,* 1040–1053.

Schwartz, G. E. (1983). Disregulation theory and disease: Applications to the repression/ cerebral disconnection/cardiovascular disorder hypothesis. *International Review of Applied Psychology, 32,* 95–118.

Von Bertalanffy, L. (1968). *General Systems Theory*. New York: Braziller.

Weinberg, J., & Levine, S. (1980). Psychobiology of coping in animals. In S. Levine & H. Ursin (Eds.), *Coping and health*. New York: Plenum.

APPENDIX 1:   SOURCES OF INFORMATION

Respondent	Contacted by Phone	Descriptive Material Received	Questionnaire	
			Sent	Completed
Unions and labor	53	29	28	9
Corporate/business	67	32	35	20
Hospitals/health	9	6	4	2
Government	12	11	4	3
Trade/professional	14	9	2	—
Academic/research	30	19	6	—
Consultant	23	12	7	4
Occupational health	9	5	2	1
Totals	217	128	88	39

APPENDIX 2

SECTION V—TECHNOLOGICAL CHANGE[1]

1. The Employer shall advise the Union as far in advance as possible, but no less than six months, of any proposed technological change. Technological change shall be defined as any change in equipment, material, method and/or procedure occurring after the date of this Agreement, which may result in a reduction in the number of bargaining unit employees, a decrease in employment opportunities for members of the bargaining unit, or alteration of any skill or knowledge requirements for job positions within the bargaining unit. The Employer shall be responsible for providing the Union with full information regarding proposed change(s) in order for the Union to determine the potential effects on the bargaining unit. Upon request by the Union, the Employer shall promptly meet with the Union to negotiate regarding the effects of the proposed change(s) upon the work force.

[1] From Service Employees International Union (AFL-CIO), Lifelong Education and Development Project, *Stress: Contract Provision* (SEIU brochure).

2. There shall be established a Joint Committee on Technological Change, with equal representation of Union and Management, to study the impacts resulting from technological change in relation to bargaining unit employees.

3. The impact of the proposed change on worker health and safety shall be a primary consideration in these discussions. The Employer shall not introduce change which imposes conditions detrimental to health and safety, or which results in increased stress for employees performing the new work. The employees who are expected to work under changed conditions shall be consulted during the planning procedure and shall have the right to bring relevant evidence and expertise to bear on the impacts of proposed change. Ergonomic considerations shall be addressed to the satisfaction of affected employees prior to implementation of any new technology.

4. When technological change is to be introduced, (a) the most senior employees in the affected job classification shall be given preference for new and/or revised work resulting from such change; (b) work resulting from such change shall be included under a scale of wages and conditions of work agreed upon by the Joint Union-Management Committee on Technological Change (however, in no event shall the agreed-upon earnings of the revised job be less than they were prior to the technological change); (c) wage rates agreed upon shall be retroactive to the date of the introduction of new equipment or processes; (d) all of the aforementioned conditions must be agreed upon prior to commencement of the revised work.

7

Managing Hypertension in the Workplace

C. BARR TAYLOR, W. STEWART AGRAS, and GUNNAR SEVELIUS

More than 30 million members of the American work force have some degree of hypertension (blood pressure 140/90 mm Hg) and nearly two-thirds of these workers have blood pressures greater than 160/95 mm Hg (NHLBI, 1981). Hypertensives have three times the age- and sex-specific incidence of coronary heart disease, and seven times the incidence of stroke (HDFP, 1979; NHLBI, 1981; VACG, 1972). Lowering of even moderately elevated blood pressure can reduce heart disease morbidity and mortality (HDFP, 1979), and effective interventions to lower blood pressure are available. In recent years, most hypertensives have become aware of the importance of treatment, and the majority are now well controlled. For instance, Berkson et al. (1980) found continuously increasing rates of controlled hypertensives during the last decade (1971—21 percent, 1976—59 percent, 1977—73 percent).

More effective treatment of blood pressure has been one of the main factors contributing to the reduced mortality from heart disease evident in this country during the past 20 years (NHLBI, 1979). While many hypertensives are being identified and treated, there is room for improvement. First, many hypertensives are still not identified and, if identified, are not receiving and/or not taking medication. Second, if, as suggested by some recent studies (HDFP, 1979), even lower levels of blood pressure than have been traditionally accepted would lead to reduced cardiovascular mortality and morbidity, then millions more individuals would be targeted for treatment. And third, nonpharmacological approaches, such as weight or salt reduction, and supplementing or replacing the use of medication, are not widely used, despite their probable benefit for reducing blood pressure.

The workplace has been demonstrated to be an ideal site to identify the hypertensive, to monitor blood pressures, and to begin or even carry out interventions designed to lower blood pressure. Worksite based interventions to reduce blood pressure may be one of the most cost-effective methods to reduce cardiovascular morbidity and mortality. In this chapter we will review recent studies evaluating the effectiveness of worksite blood pressure reduction programs and describe the preliminary results of our studies at Lockheed, which have parceled out some of the active elements of a worksite hypertension program.

WORKSITE INTERVENTIONS

Worksite interventions have traditionally focused on (1) screening to identify hypertensives, (2) the provision of information/education about high blood pressure, and (3) referral and/or on-site treatment.

Screening

Worksites offer ideal locations for hypertension screening, since a large number of people can be evaluated rapidly and inexpensively, and individuals with initially high blood pressure levels can be reassessed, referred, and/or treated as appropriate. Table 7.1 shows the results of eight studies that have used worksite screening. Depending on the criteria for hypertension, from 5 to 29 percent of the screened individuals were found to be hypertensive, many of these representing new cases of hypertension. Perhaps more important, of the hypertensives receiving treatment, fewer than 50 percent are usually found to be in control, although, as noted above, the situation may be improving.

Screening would seem to be an innocuous procedure, but an early study of a hypertension blood pressure screening program reported an unexpected undesirable side effect. In screening more than 3,600 Canadian steelworkers, Haynes et al. (1978) identified 245 individuals with diastolic blood pressures 95 or greater (on two occasions in three months), on no medication, and with no history of hypertension—that is, previously unaware hypertensives. These 245 individuals participated in a trial to improve medication compliance and were followed for a year. Surprisingly, illness days rose from an average of 2.3/year before identification, to 8.6/year after diagnosis, in those subsequently participating in treatment (patients more than 80 percent compliant with medication taking exhibited no increase) and from 3.3/year to 8.1/year in those who remained untreated. Thus, independent of being treated, except in the highly compliant, newly diagnosed hypertensives

Table 7.1. Worksite Hypertension Screening

Author	Site	Number Screened	Percent of Work Force Screened	Hypertensive Criteria (mm Hg)	Percent Hypertensive of Those Screened
Sackett et al. (1975)	Steel mill	5,000	95	Diastolic > 95 × 2 not dx hypertensive	5
Alderman & Davis (1976)	Gimbels Bloomingdale's	8,467	68	≥ 160 and/or ≥ 95 × 3 or hypertensive dx	17
Stamler et al. (1978)	Various Chicago sites	7,151	74	Diastolic > 95 × 1, then diastolic > 90	11
Baer et al. (1981)	Mental health employees	6,785	21.8	≥ 140/90 × 3	27
Jackson (1981)	NASA	2,100	–	Systolic > 140 or diastolic > 90	28
Wasserman (1982)	NIH	5,781	43	160/95 (if over 40) × 2 140/90 (if under 40) × 2	13.6
Alderman et al. (1983)	Massachusetts Mutual Life Insurance	2,463	98	≥ 160/95 (if over 30) ≥ 150/90 (if under 30)	11
Foote et al. (1983)	Ford Motor Co.	11,196	66–83	≥ 160 or ≥ 96	13–19

experienced an increase in sick days. Haynes et al. (1978) concluded that being labeled a hypertensive inadvertently encourages people to assume a "sick" role. This important question—"Does labeling a person hypertensive increase absenteeism?"—has received surprisingly little research attention. In one study, Alderman and Davis (1976) found no increased days of absenteeism in newly labeled hypertensives. In an extension of that study, Charlson, Alderman, and Melcher (1982) identified hypertension in 273 employees of the Massachusetts Mutual Life Insurance Company. Forty-eight of these were previously unaware that they had hypertension. Before screening, the unaware subjects had significantly less absenteeism than did the aware subjects (2.5 days versus 4.5). After screening, absenteeism due to illness of the unaware subjects increased significantly (to 4.6 days/year). A comparison population of nonhypertensive subjects had an average of 3.7 days of absenteeism before screening and an average of 4.4 days after screening (a nonsignificant difference). Among newly labeled hypertensive subjects, older subjects with diastolic hypertension did not have increased absenteeism after labeling; the increased absenteeism was found only among subjects with systolic hypertension and younger hypertensive subjects. The newly labeled, high-risk patients who were actively followed and treated with medication did not experience increased absenteeism. In a subsequent report of that population, Alderman and Melcher (1983) found that in the fourth year of their on-site treatment program, absenteeism in the newly diagnosed hypertensive had returned to the preprogram level of 2.5 days/year, a figure lower than control populations.

Two studies have not shown an effect from screening (Alderman & Davis, 1976; Rudd et al., 1983). In both of them, fewer than 50 percent of the work force volunteered to be screened. Sackett et el. (1983) has hypothesized that aggressive screening might account for the differences between his studies and these two—that is, subjects volunteering for screening under non-demanding circumstances may be less likely to be affected by labeling.

It appears that labeling of newly diagnosed hypertensives can increase absenteeism, an effect ameliorated by treatment. Worksite hypertension screening should not be done casually: the diagnosis should be firm (blood pressure elevated on at least two occasions a month or two apart) before subjects are told they are hypertensive, and screening should be done only where adequate follow-up care is possible. Particular care should be taken to ensure that younger workers with elevated systolic blood pressures are given good care. If done in this way, the benefits of screening far outweigh the disadvantages.

Information/Education

The worksite also provides an ideal location for information and education about the risks of hypertension and the benefits of treatment. Information has been distributed in employee newsletters, with notices or small pamphlets attached to paychecks, through posters at screening sites, cafeterias and union halls, or through pamphlets available in the medical or nursing office. The effects of this informational material are not known, as controlled evaluations of worksite hypertensive programs have not been reported. However, studies done with other populations have demonstrated benefits from information and education (Farquhar et al., 1977).

Referral/Treatment

Identified hypertensives can be referred for treatment, with follow-up monitoring occurring at the worksite, or the care can be provided at the worksite. In recent years, more worksite based treatments have been provided, and some worksite programs have been evaluated (Table 7.2).

Demonstration Studies. In an early study, Alderman and Davis (1976) reported on the effects of a worksite treatment provided by a nurse–physician–paraprofessional team for bringing hypertensives under control. At one year, 81.4 percent of the treated patients were in control (BP <160/95 or <10% MAP), and by two years 83.6 percent of the treated patients were in control. The success of the program may have been due to a variety of factors. The treating physicians were encouraged to follow a rigid protocol, primary patient care was provided by a health team, some impediments to care (such as direct patient fees or the need to obtain drugs at pharmacies) had been removed, and on-site treatment eliminated problems of physical accessibility. The role of physical accessibility has been evaluated in studies reported in the next section.

Baer (1981) provided on-site treatment to hypertensive mental health employees. Nurses monitored patients' blood pressures, provided advice on lifestyle changes, and, in consultation with a physician, made medication change recommendations. At 13.5 months, 67 percent of these employees were in control. At the National Institutes of Health, the Office of Management of the Budget provided education, monitoring, and referral for hypertensives identified through screening. One to three years after the program began, 49 percent of subjects were in control (average diastolic <90), compared to 37 percent before the program began (Wasserman, 1982). Alderman and Melcher (1983) reported on the effects of one worksite based program, which was instituted in 1977 at the Massachusetts Mutual Life Insurance Company for its 2,495 home office employees in Springfield, MA. The program began with an educational effort based on posters, a booklet, and letter distributed with each employee's paycheck. Patients were screened during release time. Ninety-eight percent of the work force participated in screening; 14 percent were identified as hypertensives. Before the program began, only 36 percent were controlled by the authors' criteria, at the end of Year 1 63 percent were controlled and 70 percent were controlled by Year 5. There was a significant mean blood pressure reduction. Consistent with Haynes' study, newly identified hypertensives significantly increased their

Table 7.2.   Effects of Worksite Hypertension Programs

Author	Site	Definition of Hypertension	No of Subjects	Intervention	Percent in Control at Baseline	Def. of Control (mm Hg)	Percent in Control at Follow-up
Alderman & Davis (1976)	Gimbels Bloomingdale's	⩾ 160 and/or ⩾ 95 × 3 or hyper. dx	1,413	On-site o off-site	50	bp ⩽ 160/95 ⩾ ↑ 10% × MAP	81.4 at 1 year 83.6 at 2 years
Logan et al. (1979)	41 Toronto sites	⩾ 95 × 2 or > 91-94, and > 140 not dx hyper.	1. 204 2. 206	1. Off-site 2. On-site by nurses	0	< 90 if > 95 at entry or ↑ dia > 6 if bp < 95 at entry	1. 27.5 at 6 months 2. 48.5
Alderman & Davis (1980)	United Store Workers Union	⩾ 160 and/or ⩾ 95 or hypertension	1. 555 2. 815	1. On-site treatment 2. Off-site treatment Costs covered	1. 24-37 2. 29-37	< 160/95 or reduction of ⩾ ↑ 10% MAP	1. 85.2-90.7 2. 83.5-88.6
Baer et al. (1981)	Mental health employees	⩾ 140/90 × 3	232	On-site treatment information service costs mostly covered	39	diastolic ⩽ 95	67 at 13.5 months
Wasserman (1982)	NIH	> 140/90 (if under 40) × 2 > 160/95 (if over 40) × 2	338	Education, monitoring by nurse, referral	37	diastolic < 90	49
Alderman & Melcher (1983)	Massachusetts Mutual Life Insurance	> 160/95 (if over 30) > 150/90 (if under 30)	271	Education, referral BP follow-up Costs covered	30	unclear	63 at 4 years 70 at 5 years
Foote & Erfurt (1983)	Ford Motor Co.	> 160 > 96	1. 211 2. 555 3. 493 4. 210	1. Screening only 2. Referral, semiannual follow-up 3. Referral, frequent follow-up 4. On-site treatment	0	⩽ 160/95	1. 47 at 3 years 2. 62 at 3 years 3. 56 at 3 years 4. 62 at 3 years

days of absenteeism from 2.7/year in the three years before the program began to 4.4/year in the first year of the program. (By the fourth year of the program, the mean annual absenteeism had returned to 2.5 days/year). On the other hand, there were significantly fewer cases of cardiovascular morbidity and mortality than would be predicted from comparable populations.

These studies have confirmed the fact that worksite based programs combining screening and/or referral and on-site treatment can be effective in bringing previously uncontrolled hypertensives under control. The specific importance of on-site and off-site treatment has been explored in a series of other studies.

On-Site versus Off-Site Treatment. To determine the role of physical accessibility in improving hypertension care, Alderman and Davis (1980) undertook another study in which the effects of on-site and off-site care were compared. Seven worksites were identified: four Storeworkers Union on-site clinics, two Building and Service Workers (BSW) on-site clinics, and off-site BSW clinic. In the initial year, 1,360 patients were identified. By Year 3 only 38 percent of the off-site subjects were still in therapy compared to 75 percent of the on-site patients. At Year 3, 84 to 87 percent of the patients treated off site were in control compared to 81 to 91 percent of those treated on site. The authors concluded that the site of treatment is not a critical determinant of outcome, but the much higher attrition in the off-site treatment would seem to favor on-site treatment.

In a subsequent study, two general models of care were compared: one, community-based care represented by the Massachusetts Mutual Life Insurance Company, mentioned above, and a worksite clinic established in the New York metropolitan region for service and craft employees who belong to various union-management sponsored plans. The main difference between the two models was the site of delivery: in the first, care was provided in a community setting by private physicians; in the second, care was provided at the worksite by specially trained nurses under the supervision of a physician (Ruchlin et al., 1984.)

After adjusting for differences in labor-market costs between New York and Massachusetts, average annual costs per client were found to be comparable for both programs. Blood pressure control (number of hypertensives below 160/95 mm Hg) was greater in the worksite program, and the cost per mm Hg reduction in blood pressure was $8.25/mm Hg compared to $28.84 in the community, suggesting a greater cost-effectiveness for the worksite-based program.

Logan et al. (1979) found that on-site hypertensive care provided by nurses was more effective than off-site care. In this study, 210 hypertensives were identified via screening at 41 Toronto business sites. Only patients not previously diagnosed as being hypertensive nor on treatment were included in the study; thus, by definition, no patient was under control. At six months, 27.5 percent of the employees treated off site were in control compared to 48 percent of those treated on site by nurses.

A study by Foote and Erfurt (1982) supports the effectiveness of a worksite-based program. In this study, 1,469 hypertensives were randomized by worksite to screening and referral only (711), referral and a semiannual follow-up (555), referral and/or frequent follow-up (493), or on-site treatment (210). After three years, 56–63 percent of subjects in the latter three groups were in control (bp <160/95), compared to 47 percent in the referral group; 25–34 percent of the hypertensives were less than 140/95, compared to 21 percent in the screening and referral only. The group referred to community resources and given frequent follow-ups did the best, but not significantly so, compared to other groups. The authors concluded that aggressive, routine follow-up is extremely important to achieve significant gains in control of blood pressure.

These studies suggest that a worksite program can be effective in identifying and providing successful referral of hypertensive patients and that referral and/or management at the worksite can have a significant impact on achieving blood pressure control. Nevertheless, because the diagnosis of hypertension might affect newly diagnosed individuals adversely, it is important to have the following: a good diagnosis based on accurately obtained blood pressure, adequate follow-up at least once every six months, and resources adequate to manage patients at the worksite and in the community. A target of 50 percent of subjects in control would seem reasonable.

Cost-effectiveness

Data from several studies suggest that worksite programs can be cost-effective. Estimates of the cost of the screening alone vary from $0.98/person screened to $1.70/person screened (Foote & Erfurt, 1977). Since screening without adequate follow-up may increase absenteeism and have little impact on blood pressure control, although inexpensive, it may not, on balance, be cost-effective. In a detailed cost analysis of a worksite program, Foote and Erfurt (1977) estimated that the average cost-per-person involved in their hypertensive program was $6.21. They estimated that the cost-per-person per-year from cardiovascular disease exceeded $300.00. A net reduction in cardiovascular disease costs of 3 percent would more than pay for the cost of the program. Since many of the controlled hypertension outcome studies have shown more than a 3 percent reduction in cardiovascular morbidity and mortality with effective treatment, it is likely that their program more than paid for itself in terms of reduced medical costs. Alderman and Davis (1976) estimated that their worksite program added only an additional $1.20 to $1.66 to the employees' health care insurance costs. In a later study, Ruchlin et al. (1984) demonstrated that the cost per mm Hg reduction in blood pressure was significantly less in a worksite treatment program compared to the community-based program. Taken together, these data suggest that worksite hypertension programs may reduce employer and employee health-care costs.

Contributing Factors

Across populations, blood pressure is elevated in direct proportion to weight and excessive salt intake. Stress management techniques, such as relaxation, have been shown to reduce blood pressure in poorly controlled hypertensives, and exercise, directly or indirectly, by facilitating weight loss, has been shown to reduce blood pressure. The worksite is potentially an ideal site for managing those factors that contribute to high blood pressure; worksite control of weight, exercise, nutrition, and stress management are discussed in other chapters and will not be reviewed here, except as they have been used in the Lockheed program described next.

THE LOCKHEED STUDY

In 1977, we began a study at Lockheed to determine the effects of relaxation therapy in reducing blood pressure in poorly controlled hypertensive patients, using the workplace to recruit hypertensives and practice relaxation. In the course of the study, we were able to evaluate a number of factors relevant to hypertension control at the worksite.

Many studies have shown that progressive muscle relaxation and related therapies—those that invoke the "relaxation response," such as biofeedback, hypnosis, progressive muscle relaxation, and autogenic training—produce significant reduction in blood pressure, particularly with poorly controlled hypertensives (Taylor, 1980). But few studies had examined the effects of relaxation for long periods and in a large number of subjects. Also, few of the studies reporting long-term follow-up had monitored practice. We wanted to follow subjects for 30 months posttreatment, to monitor practice during the entire period, and to have numbers sufficiently large to detect differences. Lockheed was chosen as one of the sites because of its large work force (21,000 employees at the Sunnyvale plant, where the study was conducted), willingness to cooperate in long-term controlled studies, and low turnover.

Screening.   Screening was conducted at employees' work stations, using a hand-held sphygmomanometer. Ninety-nine percent of the workers asked to be screened. Subjects with diastolic blood pressures > 90 mm Hg on the average of the second and third readings were reevaluated two to three weeks later. Those with blood pressures greater than 90 mm Hg on the average of the second and third readings of two further screenings were classified as out-of-control hypertensives and randomized into the study. Of approximately 7,200 individuals screened, 442 poorly controlled hypertensives were identified on the first screen; of these, 88 refused to participate or were ineligible for the project, and 159 were no longer hypertensive at the second screen. Of the remaining 195 subjects, 65 were not hypertensive on the third screen, and 2 refused to participate, leaving 128 subjects eligible and interested in the study. These 128 individuals were then randomized equally to receive relaxation therapy or monitoring (control) only.

Individuals in the relaxation therapy group were taught relaxation exercises in a 60-minute group once each week for eight sessions spread over 10 weeks, using a series of three relaxation tapes. Each of the tapes taught subjects to learn to relax using progressively fewer tension/relaxing cycles, thus inducing more rapid relaxation. Subjects were told to listen to and practice with the tape, five or more times per week. A relaxation practice was held during each group session, problems with practice were discussed, and solutions developed. Subjects were not given work time off for practice, so sessions were held during the lunch hour, between shifts, and in the evening. During the first year, attendance at practice sessions was greater than 90 percent.

Following the 10-week initial training program, subjects met with the therapist in a group once every other month for the next 30 months. During these sessions, subjects practiced relaxation, reported on problems that they were having with practice of the procedure, and went through the relaxation session.

Health Education. All subjects participated in the general health education programs available at Lockheed. These programs, begun in 1976, were carried out by the medical department under the direction of Gunnar Sevelius, M.D. More than 15,000 hypertension pamphlets were distributed at Lockheed, and we can assume that most of the subjects in this project had received them.

Blood Pressure Follow-up. Blood pressures were obtained on all subjects every two months for the entire 30 months of the study. The medical records of subjects receiving care at Kaiser were also examined at the end of the study to determine the types and amounts of medication used. Weights were also obtained on subjects at the end of the study.

Dropouts. During two years, there were eight dropouts in the monitoring-only group (four subjects retired, three were on medical disability, and one subject was unaccounted for) and nine dropouts in the treatment group (four subjects retired, two were on medical disability, and three subjects were unaccounted for).

Results

The mean pretreatment, posttreatment, and two-year follow-up systolic and diastolic results were 144/95, 136/87, and 141/90 mm Hg for the treatment group, and 144/97, 141/92, and 137/91 for the control group. Systolic and diastolic blood pressures were significantly lower in the relaxation group posttreatment compared to the control group ($F = 8.3$, $p < .001$, $F = 9.6$, $p < .003$, for systolic and diastolic, respectively). A slopes analysis from posttreatment to two-year follow-up revealed no significant differences between the two groups, suggesting equal maintenance for the two groups. Equally important, with a criterion of diastolic less than 90 representing being in control, at posttreatment, 63 percent of the relaxation subjects were in control compared to 37 percent of the control subjects. At two-year follow-up, 56 percent of relaxation subjects remained in control, compared to 44 percent of the control subjects. The results of percentage of subjects in control were even more striking for the high-risk subjects (those in the upper third of diastolic blood pressures) (see Figure 7.1). For the relaxation group, 56 percent were in control posttreatment compated to 24 percent for the control group (chi-square $= 4.06$; $p = .04$). At one year, 63 percent of the high-risk relaxation subjects remained in control, compared to 19 percent of the control subjects (chi-square $= 7.3$; $p = .007$), and at two years, 63 percent of the high-risk subjects were in control, compared to 29 percent of the control subjects (chi-square $= 4.3$; $p = .04$). Thus, for all subjects posttreatment, and particularly for the high-risk subjects over long-term follow-up, relaxation seemed to provide an additional benefit for reducing blood pressure, separate from other factors at Lockheed, leading to blood pressure reduction.

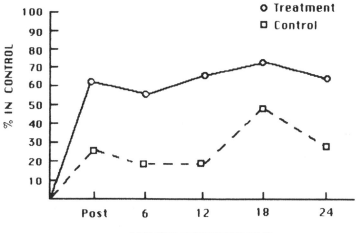

FIGURE 7.1. Shows the percentage of high-risk hypertensives in control (diastollic < 90 mm Hg) from posttreatment to 24-month follow-up.

Adherence to Practice. To determine the adherence to relaxation practice, 23 subjects in the study were given tape recorders designed to measure when a relaxation tape was used and were told to listen to the tape five or more times a week (Taylor et al., 1983). The mean frequency of practice per week as measured by the computer was 4.6; 39 percent of subjects accurately adhered to instructions. Mean frequency of practice by self-report was 5.9, with 71 percent of subjects reporting that day adhering to instructions. Thus subjects tended to overreport the frequency of their practice compared to a direct measurement of practice; practice was highest in the first week and dropped off during the next eight weeks. Nevertheless, even by the eighth week of practice, subjects, as measured by microprocessor, averaged 2.5 practice sessions a week (and 4.2 by self-report). Since subjects also had a therapist-led practice session, they were practicing on the average more than 3.5 times a week. Using the same technique, four participants had their relaxation practice monitored at one year. The average number of home practice episodes recorded by the electronic device was 6.2 per week (Agras, Schneider, & Taylor, 1984).

Generalization of Relaxation Training. In another worksite, we examined the generalization effects of the relaxation procedures used in the Lockheed study. Forty-two hypertensive subjects, randomized into the relaxation training program or no treatment, wore an ambulatory blood pressure monitoring device before and after training (Southam et al., 1982). Subjects wore the monitor during the working day, 8:00 A.M. to 5:00 P.M., and were instructed to inflate the cuff to a predetermined level every 20 minutes. This provided approximately 24 recordings on magnetic tape of the individual's blood pressure in his or her working environment on the day of measurement. The results of the blood pressure assessment at the worksite before and after eight weeks of relaxation training or no treatment are given in Table 7.3. At the worksite, the relaxation-trained group exhibited changes in systolic and diastolic pressures of −7.8 and −4.6 mm Hg respectively, while the control group's mean changes were −0.1 and +1.7 mm Hg, significantly different for both systolic ($p < .05$) and diastolic pressures ($p < .01$). At 15 month follow-up, the worksite and clinic diastolic blood pressures changes remained significantly different (Agras, Southam, & Taylor, 1983).

Weight Changes. The reductions in blood pressure were not due to weight changes; in fact, the relaxation group gained 6.0 pounds during the two-year course of the study, compared to 5.8 pounds gained in the control group (a nonsignificant difference).

Table 7.3. Mean Blood Pressure (mm Hg) at the Worksite

		Baseline	Posttreatment
Systolic			
Treatment	12	142.7 ± 10.0	133.6 ± 14.2*
Control	18	140.0 ± 10.6	141.5 ± 14.1
Diastolic			
Treatment	12	92.5 ± 9.3	86.8 ± 12.0*
Control	18	91.6 ± 7.6	93.9 ± 8.5

Source: Adapted from Agras, Southam, & Taylor (1983). (Reprinted by permission).

Note: p values are identical between groups and within groups from baseline.

*p < .01

Weight Changes. The reductions in blood pressure were not due to weight changes; in fact, the relaxation group gained 6.0 pounds during the two-year course of the study, compared to 5.8 pounds gained in the control group (a nonsignificant difference).

Medication Changes. To determine if there were different patterns of medication changes between the two groups, the records of a sample of 41 individuals receiving care at a local HMO were reviewed. During the two years of the study, for the relaxation group, 7 subjects had a net increase in medication, and 13 a net decrease or no change; for the control group there were 6 subjects who had a net increase and 15 who had a net decrease or no changes in blood pressure medication. Also of interest, in reviewing 249 medication decisions made on these 41 subjects in the two years before the study began, 60 represented more vigorous treatment than recommended by the HDFP (1979) guidelines. The same pattern continued for the two years following the study: there were 260 decisions, 65 of which represented "overtreatment." Both the relaxation and control groups appeared to receive aggressive, but not different, medication management of their blood pressure during the study, compared to two years before the study began.

Kaiser Clinic Visits.   The number of visits to Kaiser clinics before and after the relaxation program began was also examined. In the year before the program, subjects in the medical and relaxation treatment groups made an average of 0.9 visits/year for treatment of hypertension, compared to an average of 1 visit/year after the program began. The medical treatment and monitoring-only group made an average of 0.8 medical visits in the year before, and 0.7 in the year after. There were also no significant changes in visits for nonhypertension visits: 1.0 in the year before, and 0.4 in the year after for the relaxation-only group. The apparent interest in blood pressure measurement did not seem to have an effect in increasing visits for other medical problems.

Lockheed Visits.   We also examined the Lockheed clinic visits. Before randomization, the relaxation group had 0.7 visits/year, compared to 2.8/year afterwards; the monitoring-only group went from 0.4 visits/year to 2.4/year afterwards. This increased monitoring was apparent throughout the company, increasing from a mean of 1.2/year in the year before to 2.9/year in the year after. The research program and the worksite education program may have stimulated a general interest in blood pressure monitoring.

Retraining in Relaxation.   Although most subjects showed a significant effect from relaxation during the eight-week training period, about 25 percent of those who showed an initially positive response had returned to baseline by one year. To determine if intensive retraining in relaxation therapy at one year would bring these subjects back under control, 22 participants in the large-scale study were randomly allocated either to retraining in the relaxation procedures or to the routine follow-up care (Agras, Schneider, & Taylor, 1984). While retraining was associated with reductions in blood pressure, these changes were not significantly different from those occurring in subjects receiving only routine follow-up care. We concluded that retraining in relaxation procedures is no more effective in lowering blood pressure than are routine booster sessions.

Sick Days.   The rate of sick days lost per 1,000 at-risk hypertensive patients from 1978 to 1982 can be seen in Figure 7.2, provided by Dr. Sevelius. Between 1980 and 1981, there was a dramatic reduction in the rate of sick days lost because of hypertension, dropping from 52 lost days per 1,000 employees with hypertension in 1980, to 12 days in 1981, and 15 in 1982.

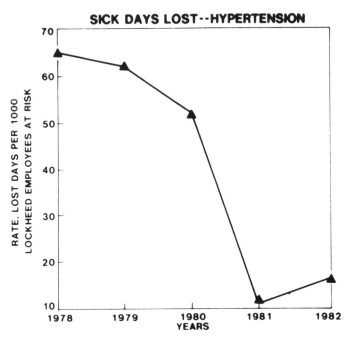

FIGURE 7.2. Shows the rate of lost days due to sickness per 1,000 Lockheed hypertensive employees from 1978 to 1982.

Implications of the Lockheed Study

At Lockheed, screening of hypertensives, a relaxation therapy program, monitoring of poorly controlled hypertension, an aggressive pamphlet-based hypertension education program, easy monitoring of blood pressure at the worksite, and aggressive medical treatment of hypertension—occurring against the background of considerable general media attention directed at the importance of managing hypertension and the proliferation of programs in the community to do so—all led to a significant reduction in blood pressure, improvement in the percentage of patients under control, and a significant reduction in illness days lost to hypertension. The study did not permit us to determine the relative effectiveness of the various aspects of the treatment program, except for the added specific benefit of relaxation, particularly evident in the upper third of poorly controlled hypertensives. Furthermore, the fact that illness days due to hypertension declined after aggressive screening and, in this case, screening of 99 percent of employees, calls into question the conclusion made by Sackett et al. (1975) that mass screening is harmful, at least if treatment is made available.

PREVALENCE OF HYPERTENSION PROGRAMS

Hypertension programs have achieved some popularity and distribution in industry, although less than would seem justified by their merit. In 1978, the Washington Business Group on Health surveyed its 160-member companies regarding their health promotion and risk reduction activities. Of 59 companies responding to the survey, 63 percent reported sponsoring hypertension control programs for their employees (Fielding & Breslow, 1983). A 1979 survey of 3,000 companies of varying size by the National Interagency Council of Smoking and Health yielded a 30 percent response rate. Of those responding, 27.1 percent sponsored high blood pressure programs. A sample of 511 organizations was randomly selected from a listing of all California employers with more than 100 employees at one or more sites. Eighty-three percent of these were called. Surprisingly, only 10 percent of the organizations offered hypertension control activities, and they were seventh in frequency among 11 types of health activities. Larger companies were more likely to offer programs than were smaller ones. Eighty-six percent of the hypertension programs were provided on site; and 90 percent were offered on a continuous basis. The majority of the programs consisted of both lectures and instruction and employed a wide assortment of methods. Screening, referral, and follow-up were offered in half the programs. For those companies not offering a hypertension program, about six percent had plans to offer one in the ensuing year.

FUTURE DIRECTIONS

Given the increasing concern of industries with the cost of health care and the relative cost-effectiveness of worksite hypertension treatment programs, both pharmacologic and nonpharmacologic, such programs should be more actively instituted by industries, particularly those with populations at high risk for hypertension, in essence, middle-aged populations. If widely applied, worksite hypertension programs would have a major impact on reducing employees and employers' health care costs.

Such worksite programs should be multifactorial, providing programs for the risk factors contributing to hypertension and other cardiovascular disease. Weight reduction, smoking cessation, exercise, nutrition, and stress reduction should be combined with identification of hypertensives and medication treatment.

Finally, more research must be done in determining how the work environment itself can be modified to reduce blood pressure. Many physical factors, such as noise levels, are known to raise blood pressure. Interpersonal factors also affect blood pressure as do other "stressors" in the environment. Some companies in America and abroad have already taken steps to alter shift cycles, time at highly stressful jobs, and to provide personal resources, such as day care, which might have an impact on reducing stress and blood pressure. Too little research has been undertaken to determine the cost-effectiveness of such programs on blood pressure reduction, but a plethora of laboratory and analogue studies would indicate a benefit from such changes in reducing blood pressure. Such environmental changes, combined with multifactorial worksite programs, could provide significant impact on the entire nation's morbidity and mortality from hypertensive disease.

REFERENCES

Agras, W. S., Schneider, J. A., & Taylor, C. B. (1984). Relaxation training in essential hypertension: A failure of retraining relaxation procedures. *Behavior Therapy, 15,* 191–196.

Agras, W. S., Southam, M. A., & Taylor, C. B. (1983). Long-term persistence of relaxation-induced blood pressure lowering during the working day. *Journal of Consulting and Clinical Psychology, 51,* 792–794.

Alderman, M. H., & Davis, T. K. (1976). Hypertension control at the worksite. *Journal of Occupational Medicine, 18,* 793–796.

Alderman, M. H., & Davis, T. K. (1980). Blood pressure control programs on and off the worksite. *Journal of Occupational Medicine, 22,* 167–170.

Alderman, M. H., & Melcher, L. A. (1983). Occupationally sponsored, community-provided hypertension control. *Journal of Occupational Medicine, 25,* 465–470.

Alderman, M. H., & Schoenbaum, E. E. (1975). Detection and treatment of hypertension at the work site. *New England Journal of Medicine, 293,* 65–68.

Baer, L., Parchment, Y., & Kneeshaw, M. (1981) Hypertension in health care providers: Effectiveness of worksite treatment programs in a state mental health agency. *American Journal of Public Health, 71,* 1261–1263.

Berkson, D.M., Brown, M.C., Stanton, H., Masterson, J., Shireman, L., Ausbrook, D. K., Mikes, D., Whipple, I. T., & Murkel, H. (1980). Changing trends in hypertension detection and control: The Chicago experience. *American Journal of Public Health, 70,* 389–393.

Charlson, M. E., Alderman, M., & Melcher, L. (1982). Absenteeism and labelling in hypertensive subjects. *American Journal of Medicine, 73,* 165–170.

Drazen, M., Nevid, J. S., Pace, N., & O'Brien, R. N. (1982). Worksite-based behavioral treatment of mild hypertension. *Journal of Occupational Medicine, 24,* 511–514.

Farquhar, J. W., Maccoby, N., Wood, P. D., Alexander, J. K., Breitrose, H., Brown, B. W. Jr., Haskell, W. L., McAlister, A. L., Meyer, A. J., Nash, J. D., & Stern, M. P. (1977). Community education of cardiovascular health. *Lancet, 1,* 1192–1195.

Fielding, J. E., & Breslow, L. (1983). Worksite hypertension programs: Results of a survey of 424 California employers. *Public Health Reports, 98,* 127–32.

Foote, A., & Erfurt, J. C. (1977). Controlling hypertension. A cost-effective model. *Preventive Medicine, 6,* 319–343.

Foote, A., & Erfurt, J. C. (1983). Hypertension control at the worksite. *New England Journal of Medicine, 308,* 809–13.

Haynes, R. B., Sackett, D. L., Gibson, E. S., Taylor, D. W., Hackett, B. C., & Roberts, R. S. (1976). Improvements on medical compliance in uncontrolled hypertensives. *Lancet, 1,* 1265–1268.

Haynes, R. B., Sackett, D. L., Taylor, D. W., Gibson, E. S., & Johnson, A. L. (1978). Increased absenteeism from work after detection and labelling of hypertensive patients. *New England Journal of Medicine, 299,* 741–744.

HDFP Cooperative Group. (1979). Five-year findings of the Hypertension Detection and Follow-up Program 1. Reduction in mortality of persons with high blood pressure, including mild hypertension. *Journal of the American Medical Association, 242,* 2562–2577.

Jackson, A. S., Squires, W. G., Buxton, V. P., White, R., Bergtholdt, C. P., Heil, W. R., & Beard, E. F. (1981). Evaluation of the NASA/JSC Blood Pressure Screening Clinic. *Journal of Occupational Medicine, 23,* 1175–8.

NHLBI. (1979). *Proceedings of the Conference on the Decline in Coronary Heart Disease Mortality* (DHEW Publication No. NIH-79-1610). Bethesda, MD: U.S. Government Printing Office.

NHLBI. (1981). *Cardiovascular primer for the workplace* (NIH Publication No. 81-2210). Bethesda, MD: U.S. Government Printing Office.

Ruchlin, H. S., Melcher, L. A., & Alderman, M. H. (1984). Work-related hypertension care programs. *Journal of Occupational Medicine, 26,* 45–49.

Rudd, P., Price, M. G., Graham, L. E., Fortmann, S. P., Beilstein, B., Bachetti, P., & Tarbell, S. V. (1983). Absenteeism and psychosocial dysfunction after hypertensive screening. *Clinical Research, 31,* 303A.

Sackett, D. L., Haynes, R. B., Gibson, E. S., Hackett, B., Taylor, D. W., Roberts, R. S., & Johnson, A. L. (1975). Randomized clinical trial of strategies for improving medication compliance in primary hypertension, *Lancet, 1,* 1205–1207.

Sackett, D. L., Macdonald, L., Haynes, R. B., & Taylor, D. W. (1983). Labeling of hypertension patients. *New England Journal of Medicine, 309,* 1253.

Southam, M. A., Agras, W. S., Taylor, C. B., & Kraemer, H. C. (1982). Relaxation training: Blood pressure lowering during the working day. *Archives of General Psychiatry, 39,* 715–717.

Stamler, R., Goesch, F. C., & Stamler, J. (1978). A hypertension control program based in the workplace. Report on the Chicago Center. *Journal of Occupational Medicine, 20,* 618–625.

Taylor, C. B. (1980). Behavioral approaches to essential hypertension. In J. M. Ferguson & C. B. Taylor (Eds.), *Comprehensive handbook of behavioral medicine* (Vol. 1), pp. 55–88. Jamaica: SP Medical and Scientific Books.

Taylor, C. B., Agras, W. S., Schneider, J. A., & Allen, R. A. (1983). Adherence to instructions to relaxation exercises. *Journal of Consulting and Clinical Psychology, 51,* 940–941.

Taylor, C. B., & Fortmann, S. P. (1983). Essential hypertension. *Psychosomatics, 24,* 433–448.

Veterans Administration Cooperative Study Group on Hypertensive Agents. (1972). Effects of treatment on morbidity and mortality in hypertension. III. Influence of age, diastolic pressure, and prior cardiovascular disease: Further analysis of side effects. *Circulation, 45,* 991–1004.

Wasserman, B. P. (1982). The employee high blood pressure program of the National Institutes of Health. *Public Health Reports, 97,* 122–126.

8

Practical Indices of Compliance in Cardiovascular Risk Reduction Programs

PATRICIA M. BROWNSTEIN and J. ALAN HERD

RISK FACTORS FOR CORONARY HEART DISEASE

The risk factor hypothesis is based on the demonstrated association of coronary heart disease with high blood pressure, cigarette smoking, and high total cholesterol concentration in blood. This association was demonstrated in the Framingham study(Dawber, Moore & Mann, 1957). Since that time, many epidemiological studies have confirmed these associations and identified other associated characteristics including diabetes mellitus (Stamler & Epstein, 1972), obesity (Armstrong et al., 1951), physical inactivity (Sanne & Wilhelmsen, 1971), and the Type A behavior pattern (Jenkins, Rosenman, & Zyzanski, 1974). Comparisons between the prevalence of risk factors and incidence of complications from coronary heart disease have shown remarkable concordance in all the epidemiological studies in the industrialized countries.

Results of the Pooling Project Research (Pooling Project Research Group, 1978) in this country showed the relationship of blood pressure, concentrations of total cholesterol in blood, cigarette smoking, relative body weight, and electrocardiographic abnormalities with the incidence of coronary heart disease. When men in these studies were classified according to the three major risk factors into quintiles of expected risk, there was close agreement between predicted and observed risk with substantial differences between occurrence of coronary heart disease in the lowest and highest quintiles of predicted risks. As shown in Table 8.1, the second, third, fourth and fifth quintiles of predicted risk had progressively higher occurrences compared to the first quintile. The realization that risk factors can be altered has suggested that interventions aimed at blood pressure, cigarette smoking, and total cholesterol concentrations in blood might reduce the incidence of complications from coronary heart disease.

Table 8.1. Serum Cholesterol, Blood Pressure, and Cigarette Use for 6,875 White Men Aged 40–59, Pool 5, Pooling Project

Risk Quintile	Total Cholesterol (mg/dl)	Diastolic BP (mm Hg)	Cigarettes Smoked* (No./Day)
I	206	79	22
II	222	83	30
III	234	85	35
IV	247	88	40
V	269	96	45
All	236	86	37

Source: Poolin Project Research Group (1978). (Copyright 1978 by the American Heart Association, Inc. Reprinted by permission.)

*Smokers only.

The possibility that reducing risk factors might reduce the incidence of complications from coronary heart disease is supported by the recent decline in mortality from coronary heart disease. During the years 1963 to 1976, the age-adjusted death rate for all causes in the United States declined by more than 25 percent. Major cardiovascular diseases were the cause for more than 50 percent of the age-adjusted death rate in 1950 and accounted for just over 45 percent in 1976 (Rosenberg & Klebba, 1979) (Figure 8.1). By 1976, coronary heart disease accounted for two-thirds of all deaths from cardiovascular disease and one-third of deaths from all causes. During the years from 1968 to 1976, the death rate for coronary heart disease declined by 20.7 percent. Death rates from cerebrovascular disease declined by 27.9 percent and for hypertension by 45.5 percent. Thus the age-adjusted death rate, which had been generally increasing between 1950 and 1963, turned downward during the period 1964 to 1967 and has been declining ever since.

During the same period of time, the prevalence of cardiovascular risk factors paralleled the incidence of complications from coronary heart disease. As shown in Table 8.2, there was a decrease in the major risk factors for middle-aged white American men that paralleled the reduction in mortality from all causes and mortality from coronary heart disease. The mean values for total cholesterol concentration in blood for men averaging 50 years of age decreased from 235 to 220 mg/dl, hypertension was successfully controlled in 50 percent of hypertensives, and the prevalence of cigarette smoking was

FIGURE 8.1.   Age-adjusted death rates for ischemic heart disease, by color and sex: United States, 1950–1976. (From Rosenberg & Klebba, 1979.)

reduced from 55 percent to 40 percent. The intensity of these risk factors and extent of change might be expected to affect mortality rates. Comparisons of mortality rates between different levels of risk factors, corresponding to the change in risk factors observed during the 20-year interval, would be about 19 percent for mortality from coronary heart disease and 14 percent for mortality from all causes.

Stamler and his associates (Byington et al., 1979) have given estimates of the percentage contributed by each of the risk factor changes to the overall expected changes in mortality from coronary heart disease (Table 8.3). A change in population mean for total cholesterol concentration in blood from

Table 8.2.   Expected Decrease in CHD and All Causes Mortality for Middle-Aged White American Men

Variable	Pooling Project (1950s)	Population Studies (mid-1970s)	Change (%)
Age – (years)	50	50	
Total cholesterol (mg/dl)	235	220	– 6.4
Diastolic BP > 95mm Hg (%)	20	11	
Cigarette smoking (%)	55	40	
Expected 8.6 year mortality/1000 coronary heart disease	50.8	41.0	– 19.3
All causes	120.9	103.7	–14.2

Source: Byington et al., (1979)

235 to 220 mg/dl would account for about 28 percent of the total reduction in coronary heart disease mortality, better control of high blood pressure would account for 24 percent of the total fall in coronary heart disease mortality, and a decrease in the prevalence of cigarette smoking would account for 48 percent of the observed decline in mortality from coronary heart disease. However, the association of risk factors with mortality from coronary heart disease is not proof of a cause-and-effect relationship. Prospective studies in controlled clinical trials with interventions directed toward cardiovascular risk factors have been necessary to prove that reducing risk factors reduces prevalence of morbidity and incidence of mortality from coronary heart disease.

Table 8.3.   Contribution of Changes in Major Risk Factors Expected Decrease in CHD and All Causes Mortality for Middle Aged White American Men

Variable	Change in Variable 1950s to Mid-1970s	Change in CHD Mortality		Change in All Causes Mortality	
Total cholesterol (mg/dl)	235 –220	–2.6	28%	–0.7	4%
Diastolic BP (mm Hg)	86.0 – 84.3	–2.2	24%	–4.9	30%
Cigarette smoking %	55% – 40%	–4.5	48%	–10.5	65%

Source: Byington et al., (1979).

CARDIOVASCULAR RISK FACTOR REDUCTION PROGRAMS

Hypertension was the first cardiovascular risk factor shown to have a causal relation to mortality from cardiovascular disease. The efficacy of treatment with antihypertensive drugs has been demonstrated in clinical trials on treatment of hypertension (HDFP Cooperative Group, 1979; VA Cooperative Study Group on Antihypertensive Agents, 1967, 1970). Treatment of high blood pressure, which reduced blood pressure effectively, also reduced morbidity and mortality associated with high blood pressure. The Veterans Administration Cooperative Study on Antihypertensive Agents first showed the benefits of treatment for patients with levels of diastolic blood pressure greater than 115 mm Hg (VA Cooperative Group Study on Antihypertensive Agents, 1967).

The Hypertension Detection and Follow-Up Program (HDFP) demonstrated the effectiveness of reducing lower levels of blood pressure in reducing mortality from cardiovascular disease including deaths from myocardial infarction (HDFP Cooperative Group, 1979). This randomized, controlled trial involved 10,940 people with high blood pressure and compared the effects on five-year mortality of a systematic antihypertensive treatment program in comparison with referral to usual medical treatment in the community. Of those in the intensively treated group, 77.8 percent continued to receive medication, compared to 58.3 percent in those receiving usual treatment in the community. Control of blood pressure was achieved in 64.9 percent of those in the intensively treated group, compared to 43.6 percent in the group receiving usual medical treatment. Five-year mortality from all causes was 17 percent lower for the group receiving intensive treatment, compared to the group receiving usual treatment. Results of this study indicated that the systematic effective management of hypertension reduced mortality for people with hypertension, including those with low levels of hypertension.

Cigarette smoking has also been shown to have a causal relation to prevalence of morbidity and incidence of mortality from coronary heart disease. Many epidemiological studies support the conclusion that cigarette smokers who reduce the amount smoked or stop entirely have greater life expectancy (Kuller et al., 1982). Evidence from controlled clinical trials has been more difficult to obtain. Anti-smoking advice has been shown to improve rates of mortality from coronary heart disease but results did not reach statistical significance (Kuller et al., 1982). More convincing evidence concerning benefits from stopping cigarette smoking was obtained from the Multiple Risk Factor Intervention Trial (MRFIT Research Group, 1982).

The Multiple Risk Factor Intervention Trial (MRFIT) was a randomized primary prevention clinical trial designed to test the effect of a multi-factor intervention program on mortality from coronary heart disease. In this trial, 12,866 men, age 35 to 57 years, with high risk for coronary heart disease were assigned to a special intervention program or referred to their usual sources of health care in the community. During a follow-up period of seven years, risk factor levels declined in both groups but to a greater degree in those in the special intervention program. Although total mortality rates were not different in the two groups, a clear-cut relationship between smoking cessation and mortality was observed. In both the special intervention and the usual care groups, those who quit smoking had significantly lower rates of coronary heart disease than those who continued to smoke (Table 8.4).

Table 8.4.   Number of CHD and Total Deaths for Cigarette Smokers (Mortality Rate: No./1000) Smoking Status at Entry and at 12 Months

	No. of Subjects		CHD Deaths		Total Deaths	
	Quit	Did Not Quit	Quit	Did Not Quit	Quit	Did Not Quit
Special treatment 1 – 29/day	454	806	6 (13.2)	19 (23.6)	13 (28.6)	34 (42.2)
30 + /day	537	2036	5 (9.3)	39 (19.2)	16 (29.8)	99 (48.6)
Usual treatment 1 – 29/day	159	1021	3 (18.9)	23 (22.5)	8 (50.3)	44 (43.1)
30 + /day	215	2435	1 (4.7)	47 (19.3)	7 (32.6)	101 (41.5)

Source: Multiple Risk Factor Intervention Trial Research Group (1982). (Copyright 1982 by the American Medical Association. Reprinted by permission.)

Coronary heart disease mortality subsequent to the first year of follow-up diverged sharply, depending on smoking status at the 12-month visit. The rate in quitters who had smoked at least thirty cigarettes per day was approximately half that of those who continued to smoke. Unfortunately, this type of analysis does not preserve the randomized control design and must be interpreted cautiously. It does not provide conclusive evidence that stopping cigarette smoking is causally related to improvement in mortality from coronary heart disease.

High levels of total cholesterol concentrations in blood have also been treated in an effort to demonstrate a cause-and-effect relationship with coronary heart disease. Epidemiological studies have shown that risk for coronary heart disease is proportional to total cholesterol and low-density lipoprotein cholesterol concentrations in blood (Gordon et al., 1977). Several clinical trials of diets and drugs in lowering total and low-density lipoprotein cholesterol have been conducted but results have not been conclusive. Recently, the Lipid Research Clinic's Coronary Primary Prevention Trial (LRC-CPPT) (LRC Program, 1984) demonstrated that reduction in total cholesterol and low-density lipoprotein cholesterol in blood was accompanied by a reduction in incidence of coronary heart disease. This was a randomized double-blind study in 3,806 middle-aged men with high levels of total cholesterol and low-density lipoprotein cholesterol concentrations in blood. Those in the active treatment group received cholestyramine resin to increase the excretion of bile acids into the intestinal tract. Those receiving cholestyramine had reductions in total cholesterol concentration in blood of 13.4 percent and reductions in low-density lipoprotein cholesterol concentration of 20.3 percent. These were 8.5 percent and 12.6 percent greater reductions than observed in the group receiving placebo. During the study, which lasted and average of 7.4 years for all subjects, those receiving cholestyramine had a 19 percent reduction in risk for myocardial infarction or death from coronary heart disease. These results indicated that reducing total cholesterol by lowering low-density lipoprotein cholesterol concentrations diminished the incidence of coronary heart disease morbidity and mortality in men with high levels of total cholesterol and low-density lipoprotein cholesterol concentrations in blood.

The combination of cigarette smoking and dietary intake of fat as risk factors also has been tested for effects on incidence of complications from coronary heart disease (Hjermann et al., 1981). A report from the Oslo Study Group describes results in 1,232 men at high risk for cardiovascular disease, selected from a population of 16,202 men, aged 40 to 49 years. The subjects were selected according to total concentration of cholesterol in blood, dietary intake of fat, and cigarette smoking. All had normal levels of blood pressure and none had clinical evidence of coronary heart disease at the time of enrollment. Men in the intervention group were advised to lower their blood lipids by eating less saturated fats and to stop smoking cigarettes. During the trial, they were examined every six months. Those in the control group were not given any specific advice and were examined every twelve months. Toward the end of the trial, serum levels of thiocyanate were analyzed as a measure of validity for self-reports of cigarette smoking. Levels of total cholesterol in blood were approximately 13 percent lower in the intervention group than in the control group. In addition, cigarette smoking was decreased 45 percent more in the intervention group than in the control group. At the end of the five-year period of observation, the incidence of myocardial infarction and sudden death was 47 percent lower in the intervention group than in the control subjects. From these results, it was concluded that advice to change eating habits and to stop smoking significantly reduced the incidence of myocardial infarction and sudden death (Table 8.5).

The evidence from controlled clinical trials indicates that interventions that reduce cardiovascular risk factors have a good effect in reducing incidence of complications from coronary heart disease. The general strategy has been to find high-risk subjects and introduce measures to reduce severity of their risk factors. The possibility that reducing risk factors might reduce the prevalence of coronary heart disease has led to recommendations for reducing risk factors in all people, even youngsters and adults with low risk for cardiovascular disease. The effectiveness of such measures is influenced not only by severity of risk factors but also the extent to which people can comply with risk-reducing measures.

According to substantial evidence, compliance with measures that reduce risk factors influences their effectiveness. Each of the clinical trials already

Table 8.5.   Cardiovascular Events in Men Receiving Advice for Lowering Blood Lipid Levels and Quitting Smoking

Cardiovascular Event	Intervention Group (n=604) No. of Events (Rate: No./1000)	Control Group (n=628) No. of Events (Rate: No./1000)
Sudden cardiac death	3 (5)	11 (18)
Sudden unexplained death	0 (–)	1 (2)
Combined sudden death	3 (5)	12* (19)
Fatal MI	3 (5)	2 (3)
Fatal MI and sudden death	6 (10)	14 (22)
Non-fatal MI	13 (22)	22 (35)
Total coronary events	19 (31)	36* (57)

Source: Hjermann et al. (1981). (Copyright 1981 by Little, Brown Publishing Company. Reprinted by permission.)
* $p < 0.05$

described has presented evidence concerning effects of compliance on cardiovascular risk factors and, ultimately, mortality from cardiovascular disease. The HDFP demonstrated how control of blood pressure was related to attendance rates at clinics and reliability in taking medication as prescribed (Shulman et al., 1982). The MRFIT demonstrated that subjects who were successful in stopping cigarette smoking had reduced mortality from coronary heart disease (MRFIT Research Group, 1982). The LRC-CPPT demonstrated that subjects who consumed all the medication prescribed had reduction in death from coronary heart disease as well as reductions in total cholesterol and low-density lipoprotein cholesterol concentrations in blood (LRC Program, 1984). The Oslo Study Group demonstrated that those subjects with the best response in lowering total cholesterol levels in blood followed the dietary advice more closely than did those who had a poor response to dietary measures. In each of these clinical trials, there was almost a twofold difference in risk-factor reduction and mortality from coronary heart disease according to compliance with measures for reducing cardiovascular risk factors.

The HDFP developed measures of attendance and compliance with medication regimens for comparison with success realized in lowering blood pressure to goal levels. As shown in Table 8.6, attendance rates were

Table 8.6.   Number of Subjects in H.D.F.P. Continuing to Return for Clinical Revisits According to Month of Follow-ups by Race and Sex

Time (months)	White Males (%)	White Females (%)	Black Males (%)	Black Females (%)	Total (%)
0	1892 (100)	1185 (100)	1065 (100)	1343 (100)	5485 (100)
6	1703 (90)	1031 (87)	916 (86)	1168 (87)	4827 (88)
12	1665 (88)	1007 (85)	895 (84)	1141 (85)	4717 (86)

Source: Shulman et al., (1982). (Copyright 1982 by Elsevier Scientific Publishing Company. Reprinted by permission.)

proportional to level of diastolic blood pressure. One-year attendance rates ranged from a low of 84 percent for those who began with diastolic blood pressures between 90 and 104 mm Hg to a high of 90 percent for those with diastolic blood pressures greater than 115 mm Hg. Attendance records also showed that white males had the highest rates of active participation and black males had the lowest (Table 8.6). An Adherence Index was developed to measure the proportion of medication that was prescribed but not used. Counts of medication used by subjects in each clinic correlated with the percentage of subjects in those clinics who had levels of blood pressure within the goal range. These data indicated that subjects with poor attendance at their clinic appointments and a high percentage of unused medication, as well as those who failed to return unused medication at the time of obtaining refills, were likely to have diastolic blood pressures above the goal range.

Practical indices of compliance were demonstrated dramatically in the HDFP (Goldman et al., 1982). The greatest differences in successful control of blood pressure were apparent between subjects who remained active in treatment compared to those who dropped out of treatment as well as compliance rates with amounts of medication used and attendance at scheduled clinic visits. The largest differences in dropout and compliance rates were seen between different clinics. As shown in Table 8.7, the greatest contrasts occurred between clinics 2 and 3. Clinic 3 had the highest number of dropouts but few who failed to keep appointments. Clinic 2 had the largest percentage who failed to keep appointments but the lowest number of dropouts. Apparently, a vigorous effort to contact people who failed to keep appointments kept them in the study but they had poor rates of compliance with visits and taking medication. These data indicated the potential for using

**Table 8.7.   Drug and Appointment Compliance for Subjects in H.D.F.P.
Percentage of Visits Not in Compliance***

	No. of Subjects	Pill Count (% Not in Compliance)	Appointment Changed (% of Visits)	No Show (% of Visits)
Clinic 1	288	24.3*	27.3	10.3**
Clinic 2	333	6.5*	31.2	13.2**
Clinic 3	115	0.3*	29.8	0.04**
Clinic 4	276	18.6*	26.9	3.2**
Total	1012	14.1	28.8	8.1

Source: Goldman et al., (1982). (Copyright 1982 by Elsevier Scientific Publishing Company. Reprinted by permission.)

*Includes all appointments changed in advance of scheduled visits and appointments rescheduled because a subject did not show.

** $p < 0.05$

visit and pill compliance to identify participants at risk for preventable lapses from the program. These observations underscore the impact of different strategies and vigorous efforts to maintain compliance in success of treatment.

Measures of compliance in the MRFIT were both simple and practical. Cigarette smokers were asked to report the number of cigarettes they smoked each day. For men who reported smoking at the time they entered the study, those in the special intervention group had stated quit rates at twelve months that were 43 percent, compared to 14 percent for those receiving usual treatment in the community. At 72 months, those in the special intervention group reported quit rates of 50 percent, while those receiving usual treatment reported 29 percent. Measures of thiocyanate in serum provided validity checks for these reports. The thiocyanate-adjusted quit rates at 72 months were 46 percent for those in the special intervention group, and 29 percent for those receiving the usual treatment. These data showed no differences between self-reports and thiocyanate-adjusted quit rates for those receiving usual treatment, and less than 10 percent incorrect self-reports for those in the special intervention group. Self-reports were also used to determine dietary patterns. A review of dietary patterns was obtained using 24-hour dietary recalls at 1, 2, 3, 5, and 6 year follow-up visits. Correlations between dietary patterns and reductions in total cholesterol concentrations in blood were not reported.

Differences among the 22 centers in the MRFIT included relative differences in thiocyanate-adjusted quit rates for cigarette smoking and relative reductions in total cholesterol levels in blood. At 72 months, the differences between groups receiving special intervention and those receiving usual treatment in thiocyanate-adjusted quit rates in all 22 centers ranged from 5 to 24 percent. Differences for reductions in total cholesterol levels in blood between subjects receiving special intervention and those receiving usual treatment in 22 centers varied from -1.6 to 10.4 mg/dl at 72 months. Here again, substantial differences in compliance were observed in different clinical centers. Apparently, personnel in some clinics developed an appreciation for practical indices of compliance and an aptitude for assisting subjects in the special intervention group to reduce cardiovascular risk factors.

Measurements of the number of packets of cholestyramine provided the most practical index of compliance for subjects in the LRC-CPPT (LRC Program, 1984). The mean daily packet count was assumed to represent the amount of cholestyramine consumed by each subject. As shown in Table 8.8,

Table 8.8.   Association of Post-Treatment Variables with Incidence of Coronary Heart Disease in L.R.C. – CPPT

Variable	Cholestyramine (n=155)		Placebo (n=187)	
	Regression Coefficient	Z Score*	Regression Coefficient	Z Score*
Mean Daily Packet Count	-0.069	-1.79	-0.02	-0.53
Total Cholesterol	2.63	3.47	0.67	0.71
LDL Cholesterol	1.92	3.37	0.70	0.93
HDL Cholesterol	-1.15	-1.81	-0.25	-0.44

Source: Lipid Research Clinics Program (1984). (Copyright 1984 by the American Medical Association. Reprinted by permission.)
*Z SCORE is the ratio of the regression coefficient and its se one-sided threshold for statistical significance ($p < .05$) is $Z > 1.65$.

there was a close relationship between percent reductions in total cholesterol and low-density lipoprotein cholesterol concentrations in blood with the mean daily packet count. Also, the mean daily packet count had as strong a relation to incidence of coronary heart disease as did changes in high-density lipoprotein cholesterol concentrations in blood. Unfortunately, no data were presented concerning the relation between reductions in body weight and reductions in cardiovascular risk factors or incidence of complications from coronary heart disease.

Indices of compliance in the Oslo Study were taken from a diet score applied to a standardized questionnaire (Hjermann et al., 1981). Two subgroups were formed from subjects in the special treatment group to determine the impact of dietary patterns on total cholesterol concentrations in blood. The quintile with the most pronounced reduction in cholesterol concentration was compared with the quintile having the lowest response

(23.5 and 0.5 percent). This analysis showed a significant relation between change in dietary pattern and reduction in serum cholesterol. Those subjects in the upper response quintile adhered more closely to instructions concerning intake of fat (Borgen et al., 1980). In addition, those subjects in the special treatment group who had the greatest reduction in diastolic blood pressure as an independent response were those whose weight fell approx-. imately 3 kilograms. They also had the best response in lowering total cholesterol concentrations in blood. Of those in the special treatment group, a decrease in body weight occurred in 70 percent of subjects, whereas only 40 percent in the control group showed a decrease in body weight. Apparently, self-reports of dietary pattern and measurements of body weight provided good indices of compliance.

RISK REDUCTION PROGRAMS IN THE WORKPLACE

Community awareness of risk factors in relation to coronary heart disease has increased considerably during the past decade. Although the idea of personal responsibility for health has become popular, experience indicates that resources dedicated to reduction of cardiovascular risk factors should be organized for maximum cost-effectiveness. The logical setting for organizing preventive medicine programs is the workplace. Indeed, many employers already sponsor health promotion programs. The nature and extent of activities sponsored by 424 California employers have been surveyed by Fielding and Bresow (1983). They reported the most frequent activities as accident prevention (by 64.9 percent) and CPR training (by 52.8 percent). Other programs included alcohol and drug abuse (18.6 percent), mental health counseling (18.4 percent), stress management (13.0 percent), physical fitness (11.6 percent), hypertension screening (10.1 percent), and smoking cessation (8.3 percent). The likelihood of having health promotion activities increased with company size.

Most programs directed toward reduction of cardiovascular risk factors include a physical fitness component. One of the first programs was initiated in the National Aeronautics and Space Administration (Durbeck et al., 1972). It was designed to assess the feasibility of such a program and to identify the factors that influenced effectiveness of the program. Of the 998 federal employees eligible to enroll, 237 actually entered the exercise program. All participants had a self-administered medical and dietary history, blood samples drawn for hematology and lipid analyses, anthropometric measure-ments, physical examination, resting electrocardiogram, and exercise tolerance tests. In addition, data concerning health attitudes, habits, and practices were obtained by personal interview and self-administered questionnaires. All participants entered an exercise program. They were allowed to choose between an exercise program in the exercise testing laboratory, a jogging program in an indoor exercise facility, or an individual unsupervised program. Approximately 12 months after initial evaluation, participants were reassessed. Measures of health habits, cardiovascular risk factors, dietary patterns, physical activity, smoking patterns, and effects of the program on work, health, and health habits and behavior were assessed. As shown in Table 8.9, improvements in cardiovascular risk factors were proportional to the number of days exercised each week. In addition, a consistent

Table 8.9.   Physiological Effects of Regular Exercise According to Number of Days Exercises/Week in Three Exercise Settings for Subjects in a N.A.S.A. Exercise Program

	Correlation		
Variable	Exercise Laboratory (n=149)	Jogging Program (n=46)	Individual Program (n=42)
Total Cholesterol	-0.12	0	0
Triglycerides	-0.27**	0	0
Triceps Skinfold	-0.46**	0	0.18
Body Weight	-0.37**	-0.25*	-0.07
Systolic BP	-0.34**	-0.21	0
Diastolic BP	-0.17*	0	0
HR/ETT – 12 Min	-0.43**	-0.48**	0
HR/ETT – 15 Min	-0.47**	-0.4**	0

Source: Durbeck et al., (1972). (Copyright 1972 by Tork Medical Group. Reprinted by permission.)

*p < 0.05
**p < 0.01

relation was found between subjectively reported effects of the program on work, health habits, and behavior with improvement in cardiovascular function during exercise tolerance testing. Indices of compliance most predictive of improvement in physiological measures were attendance rates in the supervised exercise program and the subjectively reported effects of the program on work, health habits, and behavior. Individuals perceived an appropriate amount of exercise for their own programs and readily achieved those goals.

The success of HDFP in controlling high blood pressure has led to the application of hypertension control programs in the workplace. The objective of these programs has been to identify employees with high levels of blood pressure and assist them in obtaining treatment (Alderman & Davis, 1976). The hypertension control program conducted by Alderman and his associates was reported to increase the percentage of patients with hypertension who obtained normal blood pressure through treatment at the workplace. Of 8,467 employees screened at eight worksites, 16.7 percent were found to be hypertensive. Of those, 85.7 percent were aware of their condition at the time of screening, and 66.1 percent indicated they were receiving treatment. However, only 29.1 percent of treated hypertensives had normal blood pressures when screened. The outcome of the worksite treatment program was successful reduction of blood pressure to normotensive levels in more than 80 percent of those who entered the program.

Concern has been expressed concerning the effects of labeling employees as hypertensive. Both reductions (Alderman & Davis, 1976; Mann, 1977) and increases (Haynes et al., 1978) in absences from work have been reported. The largest group of working subjects examined was in the HDFP (Polk et al., 1984). For those unaware of their hypertension at screening, disability days increased during the first year for those referred for usual treatment with no increase in disability days for those in the intensive treatment group. Subjects who were aware of hypertension but untreated at the time they entered HDFP had no change in disability days during the first year. Apparently detection and treatment were associated with increased absenteeism among newly diagnosed hypertensives, while management of previously treated hypertensives in intensive treatment clinics was associated with reduced absenteeism. The results of this study suggest that the potentially negative consequences of hypertension controlled programs can be overcome. Data concerning absences from work were obtained from questionnaires administered at the first screening visit and approximately one year after the HDFP began.

A worksite program with a multifactorial approach to reducing cardio-vascular risk factors has been performed in Belgium (Kornitzer et al., 1983). In this program, 19,300 males, aged 40–59 years, were screened for cardio-vascular risk factors. Those who were in the uppermost quintile of cardio-vascular risk received individual counseling from a physician twice a year. Health education was also delivered to all employees in each worksite. After two years, the employees with high risks for cardiovascular disease had an average reduction of 20 percent in severity of cardiovascular risk factors. A control group of employees with similar intensities of cardiovascular risk factors had an increase of 12.5 percent during the same period of time. Favorable changes in total cholesterol concentration in blood and thio-cyanate measurements in serum samples were used to infer changes in dietary patterns and cigarette smoking.

The effects of an employee fitness program on absenteeism have also been studied (Cox et al., 1981; Shephard et al., 1981). A total of 1,125 men and women were recruited from a test company and a closely matched control company. Evaluations covered three months prior to and six months after introduction of the fitness program, with 534 subjects participating consistently in the study. Approximately 20 percent of the test company employees participated in the exercise classes. Gains in maximal oxygen uptake and reduction in percent body fat were greatest in adherents to the exercise program. Employee turnover in a 10-month period was substantially lower in program adherents than in the remainder of company employees, and absenteeism of high adherents was reduced by 22 percent relative to other employees. In addition, self-reports and supervisor evaluations were obtained. From these data, health-care use and absences from work for illness were lower in employees with a high rate of attendance in the exercise program. In this program, the indices of adherence to the exercise program were attendance records in exercise classes and self-reports of perceived progress.

Information on a nutrition program at the worksite has been reported (Bruno et al., 1983). Employees at the New York Telephone Company corporate headquarters were recruited from a group of 290 potential participants who had increased cardiovascular risk factors. Of these, 145 expressed interest in participating and were enrolled in a nutrition treatment program or a control group. The nutrition treatment consisted of an eight-week program of food behavior change techniques combined with nutrition education, physical activity planning, and self-management skills. After the eight-week program was completed, the treatment group showed reductions in total cholesterol concentrations in blood of 6.4 percent, compared with control subjects. The magnitude of each participant's reduction in total cholesterol concentrations in blood was proportional to the total cholesterol concentration at the beginning of the program and to the reduction in percentage of ideal body weight that occurred during the program. Table 8.10 shows a correlation analysis for body weight and total cholesterol concentrations in blood.

The most significant contributions to decreases in total cholesterol concentrations in blood were associated with the relative decrease in percentage ideal body weight and total cholesterol concentrations in blood at the start of treatment. Although weight reduction was not a primary goal of the nutrition program, the 2.4 percent reduction in percentage ideal body weight was significant for individuals in the treatment group. The amount of weight lost was not dependent on the number of treatment sessions attended or weight at the start of the treatment program. However, weight loss was accompanied

Table 8.10. Associations between Body Weight, Reductions in Body Weight, Total Cholesterol, and Reductions in Total Cholesterol Following Cholesterol Reduction Program

Variable Pair	r	n	p
Initial % IBW* with % IBW	0.02	47	0.439
Initial % IBW with % TC**	-0.01	49	0.478
Initial TC with % TC	-0.36	49	0.005
% IBW with % TC	-0.54	45	<0.001
Weight with TC	-0.52	45	<0.001

Source: Bruno et al., (1983). (Copyright 1983 by the American Health Foundation. Reprinted by permission.)

*IBW = ideal body weight (16)

**TC = total cholesterol (mg/dl)

by a reduction of total cholesterol concentration in blood, and the more pounds lost the greater was the percentage reduction in total cholesterol concentration. Attendance was not critical to an individual's success in the program since the number of treatment sessions attended was not directly correlated to the magnitude of change in total cholesterol concentration in blood. No information was provided concerning amount of exercise undertaken by any one of the participants.

INDICES OF COMPLIANCE

Several methodological problems that arise in evaluating effects of treatment on cardiovascular risk factors include evaluating the extent of health-related behavior, physiological responses to behavioral influences, and compliance with the treatment regimen recommended for improving cardiovascular risk factors. These methodological problems impose severe limitations on the accuracy of assessing response to treatments designed to reduce cardiovascular risk factors.

Smoking cessation has received the most attention concerning methodological problems (McFall, 1978). Describing smoking behavior usually consists of counting the number of cigarettes consumed each day. However, the number of puffs taken, the amount of smoke inhaled, and the type of cigarette smoked all influence the physiological effects of smoking. Self-report methods are most commonly used to count the number of cigarettes smoked each day. However, smokers may bias their reports, and they may smoke less when they are asked to monitor their own smoking behavior than when reports are not required. Reports from collaborators may be obtained, observations may be made of smokers during normal daily activities, or their pattern of smoking may be studied in the laboratory. Ultimately, correlates of smoking behavior usually are used to assess amounts of cigarette smoking. Correlates most often measured are carbon monoxide levels in samples of expired air and measurements of concentration of thiocyanate in blood or saliva (McFall, 1978). According to results from the MRFIT, those subjects provided reports of smoking behavior that usually were accurate (MRFIT Research Group, 1982).

Compliance with a nutrition program designed to reduce cardiovascular risk factors is more difficult to evaluate. Effects from reduced intake of dietary fat on total cholesterol concentrations in blood appears slowly, blood samples are inconvenient to obtain, and concentrations of total cholesterol in blood are expensive to measure. Additional complexities in assessment of compliance are added by the individual differences in metabolic responses to dietary interventions (Brown et al., 1962). Some individuals may follow dietary recommendations exactly but show little effect on their total cholesterol concentrations in blood. Others may comply poorly but still show favorable effects during an early phase of losing weight with regression when weight stabilizes at a new low level. Consequently, measurements of total cholesterol concentrations in blood cannot be used to measure compliance to a fat-controlled diet designed to reduce cardiovascular risk factors.

A variety of measurements used to assess compliance with a fat-controlled diet include self-reports (Young & Trulson, 1960) and biochemical techniques (Fleishman et al., 1967). Self-reports of food intake have been obtained through recall of previous food intake for one to three days by self-administered questionnaires or interviews (Abramson, 1963), and through records of food intake for one to seven days completed at all meals as the meals are eaten (Young et al., 1952). Hyman et al. (1982) reported that reductions in total cholesterol concentrations in blood had a correlation coefficient of .2 with intakes of saturated fat, and .24 with intakes of cholesterol measured, using three-day food records. Correlation coefficients obtained using a Survey Research Interview were even lower. They also reported that correlations between laboratory values and food records were higher with younger patients than with older patients, suggesting that food records may be particularly misleading when monitoring compliance of older subjects. The measures of compliance using food records were also more accurate in the more educated patients. Still greater accuracy was obtained in patients who perceived a high degree of success in complying. These authors did not report the relation between change in body weight and change in dietary patterns (Hyman et al., 1982). The results reported by Bruno et al. (1983) and results from the Oslo Study (Hjerman et al., 1981) suggest that reductions in body weight are a good predictor of success in reducing total cholesterol concentrations in blood through reducing intake of saturated fat.

Evaluating compliance with an exercise program also presents many methodological problems, such as measuring amount of exercise, assessing physiological responses to exercise, and determining compliance with the recommended exercise program. Dishman and his colleagues (Dishman, 1982; Dishman & Gettman, 1980; Dishman & Ickes, 1981; Dishman et al., 1980) have measured the relationships between compliance with exercise programs and various psychological and physiological variables. They constructed the Self-Motivation Inventory from an initial pool of 60 test items concerning an individual's tendency to persevere or to be self-motivated. These items were administered in a 5-point format to undergraduates in introductory psychology classes. Subjects were approximately 400 men and women who ranged in age from 17 to 27 years. Analyses revealed 11 factors that collectively accounted for 40.5 percent of the total variance. Ultimately, the Self-Motivation Inventory consisted of 19 positively keyed items and 21 items that were negatively keyed. Validations of this Inventory were obtained from results of tests administered to subjects in exercise programs. The Self-Motivation Inventory proved to be highly related to compliance with exercise programs (Dishman et al., 1980).

Psychological and physiological influences on compliance with exercise have been reported by Dishman and Gettman (1980). An exercise program involving 66 men was conducted during a period of twenty weeks. Subjects included individuals in apparently good health as well as individuals who were known to have coronary heart disease. All were engaged in both cardiovascular and muscular endurance training. Prior to participation, measurements were made of body weight, percent body fat, and exercise tolerance to exercise on a treadmill under a Balke protocol. In addition, each subject completed a battery of psychological tests, including the Self-Motivation Inventory. Data regarding compliance with the exercise program were obtained from daily exercise records maintained during the 20-week period. Those who left the program before conclusion of the 20-week period were classified as dropouts, and those who continued in the program for the entire 20-week period were classified as adherers. Results indicated that percent body fat, self-motivation, and body weight discriminated between eventual adherers and dropouts. When combined within a prediction model, these variables accurately classified actual adherers and dropouts in approximately 80 percent of all cases and accounted for nearly 50 percent of the variance in adherence behavior. Results of this study suggested that assessment of self-motivation and body composition enhance prediction for dropout-prone participants. Results from other investigators (Massie & Shephard, 1971) have characterized dropouts as heavier, with a greater percentage of body fat, stronger, and more commonly cigarette smokers than those who comply with the exercise program. Unfortunately, there are no studies indicating relation between reduction in body weight and compliance with an exercise program.

Exercise training has been shown to produce weight loss and reduction in percent body fat (Bjorntorp, 1978; Brownell & Stunkard, 1980; Dalkoetter et al., 1979; Stalonas et al., 1978; Thompson et al., 1982). Although it is difficult to be sure of any cause-and-effect relationship between compliance and weight loss in an exercise program, it is possible that weight reduction may be an index of compliance to exercise.

PRACTICAL APPROACH TO COMPLIANCE

Practical experience indicates that subjects in a cardiovascular risk reduction program are most likely to achieve the goals they say they are likely to achieve. Their goals usually are stated in practical terms, such as number of cigarettes smoked each day, amount of weight to be lost, amount of exercise to be done, and amount of medication to be taken. A regimen that exceeds these goals sets the stage for poor compliance. Consequently, an individual's perception of need for improvement in health habits and expressed intention to achieve improvement are good predictors of compliance. Also, measuring performances in units used to set individual goals has the practical advantage of improving communication between participants and counselors. Finally, practical experience shows that most people will report health habits accurately unless the counselor insists on goals that exceed those accepted by the participant.

Ultimately, a combination of explicit goals accepted by participants and counselors with self-reports in simple terms, validated by physiological measures understandable to the participant, constitutes the most practical approach to compliance. A combination of reports concerning cigarette smoking, fat intake, amount of exercise, amount of medication used, and measurements of weight are practical indices of compliance. In addition, these measures are remarkably effective in predicting reduction in elevated total cholesterol concentrations in blood and reductions in elevated levels of blood pressure.

REFERENCES

Abramson, J. H., Slome, C., & Kosovsky, C. (1963). Food frequency interview as an epidemiological tool. *American Journal of Public Health, 53*, 1093–1101.

Alderman, M. H., & Davis, T. K. (1976). Hypertension control at the work site. *Journal of Occupational Medicine, 18*, 793–796.

Armstrong, D. B., Dublin, L. I., Wheatley, G. M., & Marks, H. H. (1951). Obesity and its relation to health and disease. *Journal of the American Medical Association, 147*, 1007–1014.

Bjorntorp, P. (1978). Exercise and obesity. *Psychiatric Clinics of North America, 1*, 691–696.

Brown, H. B., Meredith, A. P., & Page, I. H. (1962). Serum cholesterol reduction in patients: Response, adherence, and rebound measured by a quantitative test. *American Journal of Medicine, 33*, 753–762.

Brownell, K. D., & Stunkard, M. D. (1980). Physical activity in the development and control of obesity. In A. J. Stunkard, (Ed.), *Obesity*. Philadelphia: Saunders.

Bruno, R., Arnold, C., Jacobson, L., Winick, M., & Wynder, E. (1983). Randomized controlled trial of a nonpharmacologic cholesterol reduction program at the worksite. *Preventive Medicine, 12*, 523–532.

Byington, R., Dyer, A. R., Garside, D., Liu, K., Moss, D., Stamler, J., & Tsong, Y. (1979). Recent trends of major coronary risk factors and CHD mortality in the United States and other industrialized countries. In R.J. Havlik, & M. Feinleib (Eds.), *Proceedings of the Conference on the Decline in Coronary Heart Disease Mortality* (NIH Publicatin No. 79–1610, 340–380). Bethesda MD U.S. Department of Health, Education, and Welfare.

Cox, M., Shephard, R. J., & Corey, P. (1981). Influence of an employee fitness programme upon fitness, productivity, and absenteeism. *Ergonomics, 24*, 795–806.

Dahlkoetter, J., Callahan, E. J., & Linton, J. (1979). Obesity and the unbalanced energy equation: Exercise versus eating habit change. *Journal of Consulting and Clinical Psychology, 47*, 898–905.

Dawber, T. R., Moore, F. E., & Mann, G. V. (1957). Coronary heart disease in the Framingham Study. *American Journal of Public Health, 47*(Suppl.), 4–24.

Dishman, R. K. (1982). Compliance/adherence in health-related exercise. *Health Psychology, 1*, 237–267.

Dishman, R. K., & Gettman, L. R. (1980). Psychobiologic influences on exercise adherence. *Journal of Sport Psychology, 2*, 295–310.

Dishman, R. K., & Ickes, W. (1981). Self-motivation and adherence to therapeutic exercise. *Journal of Behavioral Medicine, 4*, 421–438.

Dishman, R. K., Ickes, W., & Morgan, W. P. (1980). Self-motivation and adherence to habitual physical activity. *Journal of Applied Social Psychology, 10,* 115–132.

Durbeck, D. C., Heinzelmann, F., Schacter, J., Haskell, W. L., Payne, G. H., Moxley, R. T. III, Nemiroff, M., Limoncelli, D. D., Arnoldi, L. B., and Fox, M. S. III. (1972). The National Aeronautics and Space Administration–U.S. Public Health Service Health Evaluation and Enhancement Program, summary of results. *American Journal of Cardiology, 30,* 784–790.

Fielding, F. E., & Breslow, L. (1983). Health promotion programs sponsored by California employers. *American Journal of Public Health, 73,* 538–542.

Fleischman, A., Hayton, T., & Bierenbaum, M. L. (1967). Objective biochemical determination of dietary adherence in the young coronary male. *American Journal of Clinical Nutrition, 20,* 333–337.

Goldman, A. I., Holcomb, R., Perry, H. M. Jr., Schnaper, H. W., Fitz, A. E., & Frohlich, E. D. (1982). Can dropout and other noncompliance be minimized in a clinical trial: Report from the Veterans Administration–NHLBI Cooperative Study on Antihypertensive Therapy: Mild Hypertension. *Controlled Clinical Trials, 3,* 75–89.

Gordon, T., Castelli, W. P., Hjortland, M. C., & Kannel, W. B. (1977). The prediction of coronary heart disease by high-density and other lipoproteins: An historical perspective. In B. Rifkind & R. Levy, (Eds.), *Hyperlipidemia—Diagnosis and therapy,* pp. 71–78. New York: Grune & Stratton.

Haynes, R. B., Sackett, D. L., Taylor, D. W., Gibson, E. S., & Johnson, A. L. (1978). Increased absenteeism from work after detection and labelling of hypertensive patients. *New England Journal of Medicine, 299,* 741–744.

Hjermann, I. (1980). Smoking and diet intervention in healthy coronary high risk men. Methods and 5-year follow-up of risk factors in a randomized trial. *Journal of the Oslo City Hospitals, 30,* 3–17.

Hjermann, I., Holme, I., Leren, P., & Byre, K. V. (1981). Effect of diet and smoking on the incidence of coronary heart disease. *Lancet, II,* 1303–1310.

Hyman, M. D., Insull, W., Palmer, R. H., O'Brien, J., Gordon, L., & Levine, B. (1982). Assessing methods for measuring compliance with a fat-controlled diet. *American Journal of Public Health, 72,* 152–160.

Hypertension Detection and Follow-Up Program Cooperative Group. (1979). Five-year findings of the hypertension detection and follow-up program. *Journal of the American Medical Association, 242,* 2562–2571.

Jenkins, C. D., Rosenman, R. H., & Zyzanski, S. J. (1974). Prediction of clinical coronary heart disease by a test for the coronary-prone behavior pattern. *New England Journal of Medicine, 290,* 1271–1275.

Kornitzer, M., Dramaix, M., Thilly, C., DeBacker, G., Kittel, F., Graffar, M., & Vuylsteek, K. (1983). Belgian heart disease prevention project: Incidence and mortality results. *Lancet, I,* 1066–1070.

Kuller, L., Meilahn, E., Townsend, M., & Weinberg, G. (1982). Control of cigarette smoking from a medical perspective. *Annual Review of Public Health, 3,* 153–178.

Lipid Research Clinics Program. (1984). The lipid research clinics coronary primary prevention trial results. II. The relationship of reduction in incidence of coronary heart disease to cholesterol lowering. *Journal of the American Medical Association, 251,* 365–374.

McFall, R. M. (1978). Smoking-cessation research. *Journal of Consulting and Clinical Psychology, 46,* 703–712.

Mann, A. H. (1977). The psychological effect of a screening program and clinical trial for hypertension upon the participants. *Psychological Medicine, 7,* 431–438.

Massie, J. F., & Shephard, R. J. (1971). Physiological and psychological effects of training—A comparison of individual and gymnasium programs, with a characterization of the exercise "dropout". *Medicine and Science in Sports, 3,* 110–117.

Multiple Risk Factor Intervention Trial Research Group. (1982). Multiple risk factor intervention trial. *Journal of the American Medical Association, 248*, 1465–1477.

Polk, B. F., Harlan, L. C., Cooper, S. P., Stromer, M., Ignatius, J., Mull, H., & Blaszkowski, T. P. (1984). Disability days associated with detection and treatment in a hypertension control program. *American Journal of Epidemiology, 119*, 44–53.

Pooling Project Research Group. (1978). Relationship of blood pressure, serum cholesterol, smoking habit, relative weight and ECG abnormalities to incidence of major coronary events: Final report of the Pooling Project. *Journal of Chronic Diseases, 31*, 201–306.

Rosenberg, H. M., & Klebba, A. J. (1979). Trends in cardiovascular mortality with a focus on ischemic heart disease: United States, 1950–1976. In J.T. Havlik, & M. Feinleib (Eds.), *Proceedings of the Conference on the Decline in Coronary Heart Disease Mortality* (NIH Publication No. 79–1610, 11–41). Bethesda, MD: DHEW, Government Printing Office.

Sanne, H. M., & Wilhelmsen, L. (1971). Physical activity as prevention and therapy in coronary heart disease. *Scandinavian Journal of Rehabilitation Medicine, 3*, 47–56.

Shephard, R. J., Cox, M. & Corey, P. (1981). Fitness program participation: Its effect on worker performance. *Journal of Occupational Medicine, 23*, 359–363.

Shulman, N., Cutter, G., Daugherty, R., Sexton, M., Pauk, G., Taylor, M. J., & Tyler, M. (1982). Correlates of attendance and compliance in the hypertension detection and follow-up program. *Controlled Clinical Trials, 3*, 13–27.

Stalonas, P. M., Johnson, W.G., & Christ, M. (1978). Behavior modification for obesity: The evaluation of exercise, contingency management and program adherence. *Journal of Consulting and Clinical Psychology, 46*, 463–469.

Stamler, J., & Epstein, F. H. (1972). Coronary heart disease: Risk factors as guides to preventive action. *Preventive Medicine, 1*, 27–48.

Thompson, J. K., Farvie, G. J., & Lahey, B. B. (1982). Exercise and obesity: Etiology, physiology, and intervention. *Psychological Bulletin, 91*, 55–78.

Veterans Administration Cooperative Study Group on Antihypertensive Agents. (1967). Results in patients with diastolic blood pressures averaging 115–129 mm Hg. *Journal of the American Medical Association, 202*, 1028–1034.

Veterans Administration Cooperative Study Group on Antihypertensive Agents. (1970). Results in patients with diastolic blood pressures averaging 90–114 mm Hg. *Journal of the American Medical Association, 213*, 1143–1152.

Young, C. M., & Trulson, M. F. (1960). Methodology for dietary studies in epidemiological surveys II. Strengths and weaknesses of existing methods. *American Journal of Public Health, 50*, 803–813.

9

Smoking
Modification in the Worksite

ROBERT C. KLESGES and RUSSELL E. GLASGOW

The health consequences and medical costs associated with cigarette smoking are well established. There are now more than 30,000 studies linking cigarette smoking to increased morbidity and mortality from cardio-vascular diseases, various forms of cancer, and chronic obstructive lung disease (US Public Health Service, 1979, 1982). Of particular relevance to this chapter is accumulating evidence on the synergistic relationship between smoking and occupational exposure to hazardous substances (Ellis, 1978). Given the consistent demonstrations of dose-dependent relationships between smoking and disease, evidence of reductions in health risks following smoking cessation, and experimental studies documenting carcinogenic effects of tobacco smoke in animals, few people question the causal nature of the relationship between smoking and illness.

While data concerning the economic costs to employers from cigarette smoking employees are less firmly established, most estimates range from $200–$500 excess costs annually per employee who smokes (Kristein, 1982; Luce & Schweitzer, 1978). These costs derive primarily from increased rates of absenteeism, higher medical costs, and decreased worker efficiency and

---

This chapter was supported by a National Institute of Health Grant #HL30615. The authors would like to thank H. Kit O'Neill and Kelly D. Brownell for their helpful comments on an earlier draft of this paper.

productivity. These data, as well as concern for the welfare of their workers, have led a large number of employers to initiate worksite smoking modification programs. A survey by the National Interagency Council on Smoking and Health (1980) revealed that 15 percent of American businesses have established some type of antismoking program for employees and that from 1 to 3 percent of employers have offered incentives to help their employees give up smoking.

Despite the economic and medical importance of the smoking problem and the high prevalence of worksite smoking programs, until recently there have been few evaluations of occupational smoking-cessation programs. As Fielding (1982) described, "controlled experimental data on the efficacy of such worksite (smoking reduction) programs is virtually nonexistent" (p. 909). Worksite smoking modification is one of the few areas in which there appear to be more reviews (e.g., Abrams et al., 1984; Chesney & Feuerstein, 1981; Danaher, 1980; Fielding 1982; Fisher, 1982; Orleans & Shipley, 1982) than experimental evaluations!

Considering this state of affairs, we must ask what purpose yet another review can serve. A focused, critical review of worksite smoking programs can be beneficial to health professionals for two reasons. First, none of the existing reviews has systematically evaluated reports of occupational smoking programs using a set of specified criteria (e.g., participation and attrition rates, long-term abstinence rates). By consistently judging investigations against a standard set of criteria, we can identify strengths and weaknesses in existing studies and note areas in which additional research is needed (e.g., studying different promotional approaches to increasing participation rates). Thus we will present a set of questions that can serve as guidelines for future research (e.g., What are the attrition rates and cost-effectiveness figures associated with different approaches?).

A second reason for reviewing worksite smoking reduction programs is that several studies not discussed in earlier reviews have recently been reported (e.g., Glasgow, Klesges, Godding, Vasey, & O'Neill, 1984; Nepps, 1984; Schlegel, Manske, & Shannon, 1983; Scott, Denier, & Prue, 1983; Stachnick & Stoffelmayr, 1983). In general, these studies have been better controlled and provided more complete information than did earlier investigations.

Worksite programs have varied markedly in their content and their effectiveness (Orleans & Shipley, 1982). Interventions have ranged from health education campaigns about the risks of smoking (Cooper, 1978) to worksite smoking restrictions (Rosen & Lichtenstein, 1977) to physician-assisted programs (Cathcart, 1977). Some companies have sponsored self-help quit smoking campaigns (Schwartz & Dubitzsky, 1967), while others have offered incentive programs for not smoking (Rosen & Lichtenstein, 1977) or granted time off work to attend structured treatment programs (Klesges, Vasey, & Glasgow, 1984). Group leaders have ranged from ex-smoker co-workers to volunteers from health organizations to outside consultants from university or commercial programs (Orleans & Shipley, 1982). Worksite programs have also varied considerably in the intensity of the interventions offered. Treatment dimensions along which programs differ include the following: (1) number of sessions, (2) degree of structure, (3) whether the employer offers a single series of sessions or an ongoing program, (4) whether the program focuses solely on smoking or concerns broader wellness or lifestyle issues, and (5) whether the goal of the program is abstinence or controlled smoking.

Most reports of occupational smoking reduction programs are uncontrolled and difficult to evaluate. Many studies present no data on outcome: of those evaluations that do, the great majority do not include even quasi-experimental control conditions. Until recently, the few experimental comparisons that had been reported consisted of treatment versus no treatment designs, did not include any long-term follow-up information, and did not attempt to validate self-reports of quitting (Orleans & Shipley, 1982). Little was known about the types of smokers participating in these programs, the effect of treatment variations on success, variables that moderate treatment outcome, or the cost-effectiveness of such services. Given the poor quality of these early investigations and the presence of several excellent reviews of these studies (e.g., Chesney & Feuerstein, 1981; Danaher, 1980; Fielding, 1982; Orleans & Shipley, 1982), we will focus primarily on recent controlled investigations.

For the reasons cited previously, we will first briefly discuss the major findings from uncontrolled, pioneering studies and mention current worksite programs that may provide suggestions for future research, but have not yet

Table 9.1. Criteria for Describing and Evaluating Worksite Smoking Modification Programs

Program Content	Corresponding Outcome Variables
Promotional and recruiting activities	Participation rate among employees
Requirements and screening criteria for participants	Attrition Rate
Change techniques employed, treatment personnel and mode of implementation, intensity of program	Initial success rate according to self-report and objective measures
Characteristics of sample and settings	Factors affecting treatment outcome
Maintenance procedures employed	Long-term success rate
Other measures collected	Data on generalized effects, such as employee morale, absenteeism cost-effectiveness, or use of other substances

presented any data. We will then systematically evaluate controlled investigations that we were able to identify by March 1984. To collect studies for this chapter, we examined the contents of the last three years of the *Bibliography on Smoking and Health*, reviewed several journals that frequently publish smoking modification articles (e.g., *Addictive Behaviors, Journal of Consulting and Clinical Psychology*), and wrote letters to active researchers in the field requesting information on worksite programs.

In reviewing these studies, we will describe program characteristics according to the criteria listed on the left-hand side of Table 9.1. Results will be evaluated according to the criteria listed on the right-hand side of Table 9.1.

We feel that describing each study in terms of program content and corresponding outcome variables is important so that studies can be systematically compared and contrasted. Furthermore, although most reviews present abstinence rates, we have found that few describe information, such as the percentage of subjects participating, characteristics of participants, factors moderating outcome, or generalization effects.

EARLY WORKSITE PROGRAMS AND UNCONTROLLED STUDIES

An early study by Kanzler, Zeidenberg, and Jaffe (1976) reported on the effects of offering a nine-week proprietary stop-smoking program (Smokenders) to staff members of the New York Psychiatric Institute. Although this investigation is uncontrolled, it is probably the most frequently cited worksite smoking modification study. It merits consideration because of its discussion of recruitment procedures and participation rates, and because its findings are representative of many worksite programs.

Administration of a pretest questionnaire to all employees revealed that 48 percent of the 204 staff members who were smokers stated that they would attend a stop smoking clinic and 18 percent reported that they would be willing to pay for such a clinic. Despite this initial response and bulletin board announcements, and notices in elevators, and reminders enclosed in paychecks, slightly less than 10 percent of the smoking staff attended free informational meetings about the program. Only nine employees (seven women and two men) actually enrolled in the program when they learned what it involved and that there was a $55 fee, and two of these participants dropped out after one session. Participants apparently received a standard Smokenders program. It is interesting to note that a plan to offer free follow-up group meetings to prevent relapse failed because of lack of interest. Kanzler et al. (1976) reported that participants initially successful at achieving abstinence showed little interest in attending such sessions and that those who had not achieved cessation felt defeated and therefore did not participate.

It is difficult to evaluate treatment success because the results section combined data for these 9 employees with those of 21 other participants from the general community. Overall, 67 percent of the participants reported abstinence at the conclusion of the program and 60 percent of these individuals (40 percent of the initial participants) reported abstinence at a one-year follow-up.

The Kanzler et al. (1976) study produced several findings representative of other worksite programs. First, self-reported success rates were fairly high. It appears that individuals who participate in worksite programs are usually successful in modifying their smoking habits. This study also identified several potential limitations to worksite programs. First, few (4 percent) of the eligible smokers participated in the program. As the authors themselves note, "even if a given approach boasts a 90 percent success rate, it would have little impact on smoking as a public health problem if only 5 percent of smokers will agree to participate" (Kanzler et al., 1976, p. 670). Second, there were far more women smokers participating than men (78 percent were women). This program, as well as many other worksite interventions we will review, was not successful in recruiting male smokers—who are at higher risk for cardiovascular disease and also tend to smoke more cigarettes per day and higher tar and nicotine brands than do female smokers. Thirdly, questions were raised about the efficacy of booster sessions. While many health professionals may automatically assume the usefulness of booster sessions for enhancing maintenance, such sessions do not appeal to the majority of smokers.

In what appears to be the first report of an employee incentive program for stopping smoking, Rosen and Lichtenstein (1977) reported an uncontrolled study on the results of a financial contingency program conducted at a small ambulance company (31 employees). Smokers were invited to participate voluntarily in a bonus program for not smoking at work that was developed by the owner of the company. Twelve of the 16 smoking employees (75 percent) elected to participate in the program, which involved a $5.00 per month bonus for not smoking at work—for all employees regardless of their smoking status at pretest. At Christmas time, the owner also matched the total amount of bonuses received during the program. No other intervention techniques were employed and no stop smoking meetings were held. This study reveals the potential power of incentives to modify smoking behavior: at posttest, 7 of the 12 smokers (58 percent of the participants) reported no longer smoking at work. At a one-year follow-up, one-third of the smokers were still reporting abstinence at work. While this study could have been improved by including control conditions, measures of smoking status during nonwork hours and biochemical verification of smoking status, such incentive procedures are deserving of replication. Developers of worksite programs have been rather reluctant to study the effects of incentive systems. This is unfortunate since such systems appear to be relatively inexpensive and, unlike some of the complex interventions to be discussed later, they should be relatively easy to implement in different settings with little or no training of staff members.

Stachnik and Stoffelmayr (1983) recently reported on the outcome of three separate offerings of a seven-month long worksite program involving commitment and contracting, social support, 20 gradually faded group meetings, information on the dangers of cigarette smoking, and various sizable financial incentives, including no smoking contests between teams of participants and an ongoing lottery. Recruitment and promotional activities involved the use of posters and word-of-mouth advertising in three organizations, all of which employed white-collar workers and had less than 100 employees. From 47 to 70 percent of smokers enrolled in the program, and the results were encouraging and quite similar across settings. From 80 to 91 percent of participants reported abstinence after six months of participation in the program. Incentives appeared to be an important part of this program, but unfortunately the study did not experimentally evaluate the impact of incentives versus other program components or collect objective measures of smoking status. It would also seem particularly important that biochemical verification procedures be employed when incentive systems are being used.

Multiple Risk Factor Reduction Programs

Following the pioneering work of Meyer and Henderson (1974), a number of worksites have offered smoking cessation programs as part of employee wellness or lifestyle modification programs. Such programs typically focus on achieving modifications in several risk factors, such as elevated cholesterol levels, low levels of physical activity, obesity, and hypertension, as well as cigarette smoking (Naditch, 1984). Often programs will also include components on stress management, modification of Type A behavior, and/or reducing the use of various substances, such as alcohol and caffeine. Almost all programs include an initial health screening to identify risk factors, but from this point there is considerable divergence in approach.

Some programs focus solely on high-risk participants (e.g., Meyer & Henderson, 1974; Ware & Block, 1982), while others invite all employees to participate regardless of risk status (e.g., Naditch, 1984). There is also considerable variability in how smoking programs are implemented, with some programs holding separate meetings for smokers, and others including information on smoking modification as part of their general wellness program. Empirical reports on smoking reductions resulting from lifestyle modification programs are just beginning to appear and are often difficult to interpret due to difficulties in following subjects, lack of description of inter- vention components, and idiosyncratic ways of reporting results.

The STAYWELL program of the Control Data Corporation (Naditch, 1984) is one of the larger and more well-known lifestyle programs. An initial report on this project stated that 12 months after enrolling in a smoking reduction program, 30 percent of participants reported having quit smoking (Naditch, 1984). One unique aspect of the STAYWELL program is the development of computer-assisted behavior change programs. Computerized interventions offer several potential advantages, such as extensive individualization of programs based on smoker characteristics and branching statements dependent on degree of initial adherence and behavior change. Hopefully, the program developers will provide data on the efficacy of these computerized change programs versus more traditional face-to-face interventions.

The concept of providing smoking modification services as part of a more general lifestyle program is appealing. Many smokers, particularly women, appear to be concerned about potential weight gain as a result of smoking cessation and such programs would address those concerns. It makes good sense to include smoking as one of the main targets of wellness programs since studies have reported that heavy smokers are less involved than non-smokers in physical activity (Cathcart, 1977), more likely to drop out of exercise programs once they join (Oldridge, 1978), and are also more likely to be hypertensive (Cathcart, 1977). Unfortunately, most of the larger worksite wellness programs containing smoking cessation components are uncontrolled and have experienced difficulties in following subjects (e.g., Ware & Block, 1982). Additional investigations of the efficacy of multiple-risk factor versus single-component smoking cessation programs and studies of the best methods for delivering the smoking modification component of wellness programs are indicated.

CONTROLLED INVESTIGATIONS

A review of the recent worksite treatment literature revealed six controlled evaluations. These investigations are summarized in Table 9.2.

Scott, Denier, & Prue (1983) reported on a program offered to 26 nurses at a VA hospital. Participants were randomly assigned to either a treatment program lasting for six months or to a wait-list control. One hundred percent of the nurses who were personally contacted and asked to participate did so. Treatment, which consisted of brand fading (gradual reductions in the tar and nicotine content of cigarette brand) plus abstinence training, was conducted on the different nursing units to increase convenience. Intervention consisted of a manual and brief, *daily* contacts to assess expired breath carbon monoxide (CO) levels. Thus participants received immediate, public feed-back on their CO levels and instructions on how to change brands to reduce tar and nicotine content sequentially. These contacts continued on a daily basis for three months posttreatment, and were then gradually faded out to "unannounced days."

There was no attrition during the treatment or maintenance phase for nurses "who have continued to work at the VA." Of the 75 percent of participants who attempted abstinence, 75 percent were abstinent at one-month follow-up, 42 percent at six months, and 33 percent at nine months. This compared to a zero-percent abstinence rate in the control group.

The Scott et al. (1983) program was an intensive, but relatively low-cost intervention that is impressive both in terms of the zero-percent attrition rate and the outcomes observed. In addition, 100 percent of the participants asked

Table 9.2. Summary of Controlled Investigations of Worksite Interventions

Reference	Setting and n	Design	% of Smokers Participating	Attrition	Initial Success Rate	Long-term Success Rate
Scott et al. (1983)	Nurses n = 26	Immediate Tx vs. wait list control	NR[a] (100% participation of those who were asked)	0% of those who continued to work at VA	Of 75% who attempted abstinence, 75% were abstinent vs. 0% of controls	42% abstinence at 6 months, 33% at 9 months vs. 0% of controls
Schlegel et al. (1983)	Canadian military personnel n = 243 from 28 different bases	Treatment delivery (full, minimal contact, self-help manual) + nicotine gum prescription (yes/no design)	NR[a]	NR[a]	Of those with abstinence goal, 49%–72% abstinent in full Tx, 30%–41% abstinent in minimal contact, and 8%–18% in self-help. Nicotine gum facilitated self-help but decreased effect of full Tx	Of those with abstinence goal, 27%–40% in full Tx, 19%–38% in minimal contact, and 10%–13% in self-help abstinent at 12 months; nicotine gum facilitated self-help but decreased effect of full Tx
Glasgow et al. (1984)	Telephone company n = 36	Abrupt reduction vs. gradual reduction, controlled smoking vs. gradual reduction plus feedback	NR[a]	8%	21% overall abstinence; mean of 28% reduction in CO among nonabstinent Ss	33% abstinence in gradual condition; 0% in other conditions

Malott et al. (1984)	Telephone company and medical clinic $n = 24$	Controlled smoking vs. controlled smoking plus partner support	NR[a]	0%	17% abstinence; mean reduction of 49% in CO among nonabstinent Ss; No between-groups differences	21% abstinence at 6 months. No between-groups differences
Klesges et al. (1984)	Four banking institutions ($n = 91$) plus one savings and loan ($n = 16$)	Controlled smoking plus competition vs. standard controlled smoking (quasi-experimental)	87% in competition condition; 53% in standard controlled smoking condition	9%	23% abstinence; mean reduction in CO of 51% among nonabstinent Ss	NA[b]
Nepps (1984)	Johnson & Johnson, $n = 36$	Minimal contact vs. group program previously run (quasi-experimental)	NR[a]	67%	22% abstinent; judged "equally successful" to the group program	14% abstinence at 6 mos.; judged equal to previous group treatment

[a] NR = not reported.
[b] NR = not yet available.

to participate did so. However, it is unclear whether participants were screened for interest, or what role the therapist's personal relationship with participants may have played.

In a complex study, Schlegel, Manske, & Shannon (1983) investigated the contribution of nicotine gum to their multicomponent behavioral treatment program. They treated 243 armed forces personnel (65 percent male) in 28 bases across Canada and assigned bases to one of six conditions in a 3 (Treatment delivery) $\times$ 2 (Nicotine gum prescription) design. Schlegel et al. (1983) investigated the effects of differing amounts of therapist contact (17 sessions versus 4 sessions versus zero treatment sessions). This variable was crossed with the presence or absence of a prescription for 2 mg nicotine chewing gum. Treatment was conducted by nursing personnel or selection officers from each of the bases, who were trained in a 3½ day workshop.

Subjects were recruited via posters and news releases in base newspapers. All subjects received a 160-page workbook outlining exercises to be completed during a six-month period. Program components included detailed self-monitoring, development of alternatives to smoking, imaginal and in-vivo rehearsal of coping strategies in high-risk situations, aversive smoking, prescriptions for increased physical activity, relaxation training, and generalized problem-solving techniques. Subjects were allowed to self-select goals of abstinence or reduced smoking. Also collected were a variety of predictor and process measures, including intentions and attitudes toward smoking, self-efficacy levels, locus of control, degree of nicotine dependence, self-motivation, life change, job satisfaction, and adherence to treatment recommendations. Subjects were assessed 2, 6, and 12 months after targeted cessation dates.

Information was not reported on the percentage of smokers participating or attrition rates. Extrapolating from base rates of smoking among similar populations, however, participation must have been low given that on average fewer than 10 smokers per base participated in the program. Schlegel et al. (1983) found that 40 percent of their participants reported both reading the manual and completing the associated exercises. Although analyses of these data are not reported, examination of tabled data suggests that adherence was greater among full-treatment than self-help subjects and that adherence was associated with abstinence only among subjects in the full-treatment conditions.

Of the 71 percent of subjects selecting abstinence as their goal, from 49 to 72 percent in the full-treatment conditions reported abstinence at the initial follow-up, compared to 8–18 percent in the self-help condition. Self-reported abstinence rates at one-year follow-up ranged from 27 to 40 percent in the full-treatment conditions, compared to 10 to 13 percent in the self-help conditions. Abstinence rates in conditions with four group meetings fell in between those reported for the full-treatment and self-help conditions. One interesting finding was that nicotine gum appeared to increase success rates under self-help conditions (10 percent higher abstinence with gum), but to *decrease* success rates under full-treatment conditions (12 percent lower abstinence).

Although this study suffers from a number of methodological weaknesses, such as the absence of biochemical verification of self-reported cessation, differential percentages of subjects selecting abstinence goals in different conditions, failure to report interaction terms from their analyses of variance, and lack of information on subject attrition, Schlegel et al. (1983) have explored several new areas relevant to worksite smoking-cessation programs. To our knowledge, this is the first worksite study to present data from a predominantly male, blue-collar population. In terms of implementation issues, they experimentally investigated the effects of amount of therapist contact and trained base employees to administer treatment programs. Finally, although the authors only footnoted results pertaining to differences among bases, these analyses apparently revealed differential patterns of success across bases. Additional studies, such as the Schlegel et al. (1983) investigation, with sufficient sample sizes to conduct analyses to identify characteristics of worksite settings associated with treatment outcome would be extremely valuable.

Our research group has conducted three controlled worksite smoking reduction studies. We have conducted these programs with several goals in mind. First, we wanted to test the applicability of our controlled smoking program developed in a clinic setting (Glasgow, Klesges, Godding, & Gegelman, 1983) both for individuals wanting to quit smoking and for those wanting to reduce their smoking levels but not stop smoking entirely. We felt that such a program would attract a greater number of smokers than a program that was solely abstinence oriented. A second goal was to verify smoking status by measures such as carbon monoxide levels, saliva thiocyanate, and laboratory analyses of smoked cigarette butts. A final goal was to manipulate program components systematically in an attempt to enhance treatment outcome.

The first worksite program (Glasgow et al., 1984) randomly assigned 36 employees (25 women, 11 men) of a telephone company to one of three controlled smoking conditions: Abrupt Reduction, Gradual Reduction, and Gradual Reduction plus Feedback. Recruitment included having an employee organization (the Telephone Pioneers) promote the program, announcements in in-house newsletters, and posters describing the program. No data are available on the percent of smokers participating, but based on the size of the company ($n = 600$), an estimated 18 percent of the smokers participated.

Participants met in small groups of three to six smokers for seven treatment sessions during an eight-week period. Treatment focused on sequentially making reductions in nicotine content of brand smoked, percent of each cigarette smoked, and number of cigarettes per day. Subjects were encouraged to set goals of at least 50 percent reductions in each target behavior. After the fifth session, participants had the choice of whether to stop smoking entirely.

Subjects in the Abrupt condition focused on making 50 percent reductions in each of the above target behaviors (see Glasgow et al., 1983, for specifics of the standard controlled smoking program). Subjects in the Gradual condition received the same program except that more gradual weekly reduction goals (25 percent of baseline smoking levels) were employed for the first week of intervention on each target behavior. Thus, while the Abrupt group attempted a 50 percent reduction for two weeks, the Gradual group attempted a 25 percent reduction one week, followed by another 25 percent reduction the following week. Subjects in the Gradual plus Feedback condition received the same program as subjects in the Gradual condition except that, at each session, subjects were presented with individualized graphs of their daily nicotine intake calculated from self-monitoring records.

The last treatment session for all conditions focused on maintenance issues. This session included a discussion of situations that have been found to be associated with smoking relapse (Marlatt & Gordon, 1980; Shiffman, 1982); it also stressed the importance of not falling victim to the "goal violation effect"—a modification of Marlatt and Gordon's (1980) abstinence violation effect adapted to apply to controlled smoking. Finally, subjects were provided with a supply of self-monitoring booklets so that they could continue to monitor their smoking at least periodically and instructed in a series of steps to follow should they find themselves increasing their smoking. A "booster" follow-up session was held two months after the end of the program to further underscore these maintenance issues.

Ninety-two percent (33 of 36) of the subjects completed the program. Nonabstinent subjects in all three conditions achieved significant reductions in each of the three targeted behaviors with no differences between conditions. Across treatment conditions, the largest reductions were in nicotine content ($M = 50$ percent reduction), followed by percent of each cigarette smoked ($M = 34$ percent reduction), and number of cigarettes per day ($M = 28$ percent reduction). Relatively consistent improvements were also seen on carbon monoxide levels ($M = 28$ percent reduction).

At the six month follow-up, 33 percent of the subjects in the Gradual condition were not smoking (confirmed by CO readings of < 10 ppm), while none of the subjects in the Abrupt or the Gradual plus Feedback conditions were abstinent. Among nonabstinent subjects, significant relapse was observed only on the variable of percentage of each cigarette smoked.

Thus the results of our first worksite investigation (Glasgow et al., 1984) indicated that our controlled smoking program could be successfully conducted in a worksite setting. Furthermore, gradual reduction goals were more successful in producing long-term abstinence than were abrupt goals. The magnitude of change observed in this study was comparable to results from our earlier clinic-based studies. However, in retrospect we did not feel that we had used many of the potential advantages of conducting smoking programs in the worksite. Specifically, we had not systematically used either social support or incentive systems available in the worksite environment.

Therefore, our next study was designed to address the incremental effects of structured coworker support on outcome (Malott, Glasgow, O'Neill, & Klesges, 1984). We randomly assigned 24 smokers (20 females and 4 males) to either Basic Controlled Smoking or to Controlled Smoking plus Coworker Support conditions. The program was conducted at a local telephone company and a medical clinic. Recruitment efforts involved notices in employee newsletters, brochures distributed to supervisors, and posters in the lunchrooms of these organizations. No precise estimate of the percentage of smokers participating was obtained. However, based on the size of the worksites, it is estimated that only 7 percent of smokers participated.

Subjects in the Basic Controlled Smoking (CS) condition received the gradual reduction program outlined above. Subjects in the CS plus Coworker Support Condition underwent the same treatment procedures as subjects in the basic CS condition, but also selected a "co-worker buddy" or partner from among the smokers in the group to which they were assigned. Subjects in this condition received a 30-page Partner's Controlled Smoking Manual, exchanged feedback on a partner interaction questionnaire to assess the type of support behaviors each smoker would prefer his or her partner to provide, and were encouraged to discuss their progress with each other at least once per day. Both the CS and the CS plus Coworker Support conditions received the maintenance intervention described in the Glasgow et al. (1984) study discussed previously.

All subjects completed treatment and completed posttest, and 23 of 24 (95 percent) participated in the six-month follow-up assessment. Seventeen percent of the smokers were abstinent at posttest and all had maintained their abstinence at the six-month follow-up. In addition, one nonabstinent smoker had quit between posttest and six-month follow-up for an overall abstinence rate of 21 percent (confirmed by CO analyses). There were no differences between conditions on abstinence rates. Consistent with our first worksite program, nonabstinent subjects in both conditions were able to reduce their nicotine content significantly ($M = 52$ percent reduction), number of cigarettes smoked ($M = 38$ percent reduction), and percent of the cigarette smoked ($M = 22$ percent reduction). These changes were confirmed by an average pre-to-posttest reduction of 49 percent in carbon monoxide levels among nonabstinent subjects. Somewhat greater relapse was observed among nonabstinent subjects than in our previous study.

Although participants were usually successful in making significant modifications in their smoking behavior, there were no differences between the CS and the CS plus Coworker Partner Support conditions. The lack of between-group differences did not appear to be due to implementation problems since partners did report frequent interactions. Perhaps high levels of co-worker support already existed in these worksites or perhaps diffusion effects were operative, with subjects in the CS condition adopting peer support techniques intended for CS plus Coworker support subjects. One other important result was noted in this study. A questionnaire designed to assess the type of partner interactions experienced (modified from that developed by Mermelstein, Lichtenstein, & McIntyre, 1983) indicated that the *lack* of negative interactions with both co-workers and spouses, but not the presence of positive interactions, was significantly related to treatment outcome. Based on these finding, we are currently offering a worksite program that emphasizes decreasing the occurrence of nonsupportive interactions as well as increasing the amount of social support received from spouses/significant others.

Overall, while we are pleased with the behavior change and maintenance results of nonabstinent subjects in our first two worksite smoking interventions, we were disappointed with both the participation rates and the percentage of subjects achieving complete abstinence in these studies. Given the success of recent weight-loss competitions (Brownell, Cohen, Stunkard, Felix, & Cooley, 1984), we decided to offer a worksite smoking competition in an attempt to increase participation rates and enhance treatment effects.

We were able to take advantage of an existing competitive environment that existed between four local banking institutions. Four banks (all within one square mile of each other) participated in the "Smoking Contest: A Healthy Competition." The four bank presidents formally challenged each other at a press conference to see which institution could produce the greatest reductions in smoking among employees, and each bank contributed $150.00 plus $10.00 per participant. Recruitment included brochures given to employees, posters advertising the program, and informational/promotional talks at bank-wide staff meetings. In addition, unlike our earlier studies, all participants were given time off work to participate in the program. Prizes to benefit *both smokers and nonsmokers* were awarded to (1) the bank with the highest participation rate, (2) the bank with the greatest success in reducing carbon monoxide levels at posttest, and (3) the bank with the greatest abstinence rate at the six-month follow-up. The "grand prize," to be awarded to the bank with the highest cessation rate at follow-up, would be a catered meal served to all employees of the winning bank by executives of the losing banks. Individual prizes for successful participants would also be presented.

All participants in this study (Klesges et al., 1984) received our gradual-paced CS program, with the exception that there was a greater emphasis on cessation than in our previous studies. While participants were able to choose either abstinence or controlled smoking goals, the major incentives were directed at cessation. Throughout the program, participants were encouraged to wear buttons stating, "I'm in the healthy competition," to increase social support. In addition, a "Smoking Barometer" that provided employees with weekly feedback on how their bank was doing compared to the other three banks was placed in the lobby or lounge of each worksite.

While this program was being conducted, we concurrently offered our standard gradual CS program without the competition component to the employees of a local savings and loan institution, allowing for a quasi-experimental comparison of the effects of competition on outcome. The sites in which the CS plus Competition and the basic CS programs were offered were highly similar in terms of type of worksite, type of employees, and geographical location. Similar to the bank competition subjects, savings and loan participants also received various promotional brochures, encouragement from top management, and time off work for participation.

Participation rates were exceptionally high. Overall, 88 percent of all bank employees who were smokers entered the competition program with no significant differences in participation between the four banks. At one bank, all but one smoker participated. A lower, but still encouraging 53 percent participation rate was found at the savings and loan. The participation rate in the CS plus Competition condition was significantly higher ($X^2 = 14.84, p <$ .005) than in the basic CS condition. Ninety-one percent (97 of 107) of subjects who began treatment completed the program. Attrition rates did not significantly differ between the CS plus Competition (10 percent, $n = 9$) and the basic CS (6 percent, $n = 1$) conditions. Due to the differential initial participation rates and the similarity in attrition rates, a significantly higher percentage of smokers in the CS plus Competition condition completed treatment ($X^2 = 8.30, p < .005$).

Initial analyses revealed no significant differences between the CS plus Competition and the basic CS condition on demographics (e.g., sex, age), smoking history, or any of the dependent variables. At posttest, preliminary results suggest that subjects in both conditions achieved significant improvements on all targeted behaviors with no differences between conditions. Across treatment conditions, 23 percent of participating smokers (22 percent in the CS plus Competition condition; 31 percent in the basic CS condition) had quit smoking by the end of the program (confirmed by CO readings of $<$ 10 ppm). Among nonabstinent subjects, an overall reduction of 47 percent was achieved in CO levels. Collapsing across conditions (there were no between-group differences), the largest behavioral reductions among non-abstainers were in nicotine content ($M = 58$ percent reduction), followed by number of cigarettes smoked ($M = 40$ percent reduction), and percent of each cigarette smoked ($M = 39$ percent reduction). A six-month and one-year follow-up will be conducted, but these data are not yet available.

Thus the preliminary results suggest that the addition of a competition–incentive component significantly enhanced participation rates in a worksite smoking control program. This finding is encouraging since most worksite smoking programs do not attract a large number of participants. The lack of posttest differences on smoking status can be viewed as encouraging since smokers who otherwise would not have participated (and who may be less motivated to change) are included in the CS plus competition condition. Future research on incentive and competition-based programs should be conducted to determine if the present results are generalizable and to identify effective components of this procedure.

We feel that four conclusions can be drawn from our worksite interventions. First, it is clear that clinic-based smoking reduction programs can be successfully adapted to a variety of worksite environments. Second, time off work and the involvement of top management in promoting the program may improve participation rates. An illustrative example can be drawn from our competition study. The bank with the highest participation rate had a president who smoked and participated in the program. Third, minor variations in the type of program offered (e.g., feedback, formal co-worker support) do not appear to affect outcome. Fourth, it appears that the absence of negative social interactions, rather than (or possibly in addition to) the presence of positive interactions, is associated with treatment success.

Nepps (1984) conducted a quasi-experimental study of a self-administered program for 36 employees of the Johnson & Johnson Corporation participating in its *Live for Life* wellness program. Recruitment included posters, desk drops, and company newsletter articles. The percentage of smokers participating was not reported. Each smoker was given the first of nine written self-help modules, which incorporated an aversive smoking procedure (smoke holding) and self-control techniques for quitting. Participants then worked on their own and received additional modules at the rate of approximately one per week. A heavy emphasis was placed on maintenance and relapse prevention issues.

Adherence to treatment recommendations was a serious problem, as is common in self-help behavior change programs (Glasgow & Rosen, 1978). Of the 36 participants who received the first module, 17 (47 percent) did not return to receive another module. Only 17 percent completed all nine modules. Twenty-two percent of participants reported abstinence at the end of the program, while 14 percent were abstinent at the six-month follow-up. Compared to earlier therapist-administered group cessation programs offered at Johnson & Johnson, Nepps (1984) concluded that this minimal contact program was "equally effective" *if* the 24 dropouts are excluded from consideration. As the author acknowledges, however, there was a substantially higher dropout rate in this minimal-contact program (67 percent attrition) than in the previously reported (Wilbur, 1982) group treatment program (20 percent attrition). While the Nepps (1984) study yielded abstinence rates comparable to other self-help smoking-cessation programs (see Glasgow, 1984), the percentage of smokers participating was probably quite small and attrition was high.

To summarize, as can be seen in Table 9.2, some progress has been made in the worksite smoking modification area. Several recent programs (e.g., Glasgow et al., 1984; Scott et al., 1983) are characterized by methodological improvements (e.g., control conditions, biochemical verification of smoking status) and also by providing options of abstinence versus reduced smoking. In addition, some studies (e.g., Klesges et al., 1984; Schlegel et al., 1983) have reported a sizable number of subjects participating in the worksite programs.

Finally, rather high abstinence rates have been reported, at least in therapist-administered cessation-based programs (e.g., Schlegel et al., 1983; Scott et al., 1983).

It appears that incentives play an important role in improving participation, and perhaps, outcome rates.[1] Still, few studies have investigated dissemination issues, collected systematic cost-effective data, or even reported participation rates. It is, however, encouraging to see that some investigators (Klesges et al., 1984; Schlegel, 1983) are beginning to collect process measures such as job satisfaction and job stress that are potentially associated with outcome. We will conclude this chapter with some clinical recommendations for conducting worksite smoking programs as well as suggestions for future research.

CLINICAL RECOMMENDATIONS FOR IMPLEMENTING WORKSITE SMOKING PROGRAMS

Initial Contact and Recruitment

In our experience, the initial contact with a worksite has proven critical to the success of the project. We have had the best success initially meeting with the president of the organization. While this individual typically does not coordinate the program, support "from the top" appears to be important in terms of program recruitment and implementation. During these initial meetings, we recommend stressing health-related and economic benefits to both employees and to the employer (e.g., reduced absenteeism, higher productivity—see Luce & Schweitzer, 1978).

After securing permission to offer a program, we have found it helpful to conduct a short (i.e., one-page) worksite-wide needs assessment. Such a survey can be used to determine (1) the number and characteristics of smokers in the worksite, (2) the number of smokers potentially interested in participating, and (3) preferences as to both the *types* of programs that might be offered (e.g., self-help versus small group; abstinence-based versus

---

[1]Additional studies (Rand, Stitzer, Bigelow, & Mead, 1984; Stitzer & Bigelow, 1982, 1983) have investigated the effects of contingent reinforcement for carbon monoxide reduction to demonstrate temporary smoking behavior changes in hired participants. Rand, Stitzer, Bigelow, & Mead (1984) reported that 5 of 18 subjects (28 percent) were successful at quitting smoking for a three-week period and were paid up to $200 if they were successful in achieving CO levels of less than 11 ppm. Stitzer and Bigelow (1983) found that 11 smokers were successful at reducing their CO levels by 38 percent during one week of contingent reinforcement, and by 35 percent during one week of post-intervention baseline in which smokers no longer received money for CO reductions. Finally, Stitzer and Bigelow (1983), in a study of 60 hospital employees, found that increased levels of monetary reward contingent on CO reduction (e.g., $1 vs. $10) resulted in significantly lower CO levels during a two-week period.

controlled smoking) as well as the most convenient times to schedule meetings.

During the recruitment phase, information about the program should come from various sources (e.g., posters, memos, and brochures). Advertising consultants feel that it helps to have multiple exposure to a "product" (in this case, a program) to convince participants to take action regarding the product. It is helpful if at least one memo or announcement comes from top management. At this stage, human resource or personnel directors can be extremely useful in suggesting the best ways to promote the program in their particular setting. Involving the local media may also serve to increase the credibility of the program as well as to provide "no cost" advertising for the program (and the worksite).

Finally, we have found offering a "no obligation" orientation meeting helpful in giving participants detailed information about the program. Offering such a meeting prior to the program can attract a number of smokers who otherwise might not attend and can also serve to set appropriate expectations and correct possible misconceptions about the program (e.g., "I have to quit at the first session.").

Program Variables

While worksite interventions offer a number of distinct advantages, they also present a number of logistical problems that must be overcome to exploit these advantages. Programs must be maximally convenient to attract smokers. We have found higher participation rates in worksites that offer *time off work*. If management is not willing to grant time off work, it may at least be possible to negotiate that employee and employer time be shared (e.g., ½ hour of work time, ½ hour during lunch hour or after work).

Programs should be offered at little cost to participants. For motivational reasons, it has been assumed that smokers should pay something to participate in cessation programs. The cost should not, however, be so large that it decreases participation rates. We have found $10 to $20 per participant (with half returned to the participant contingent on completion of six-month follow-up) to be a reasonable fee. Here again, employers are often willing to share program costs.

In addition to being convenient, programs should be maximally *attractive* to participants. For example, allowing smokers to choose the type of program (e.g., nicotine fading versus aversive smoking), the modality of intervention (e.g., self-help manual versus a small group treatment), and/or treatment goals (e.g., abstinence versus controlled smoking) may be helpful in attracting and retaining participants. In particular, we have found that not requiring a firm commitment to total abstinence at the outset of a program seems to enhance participation. Incentives, prizes, and/or awards may enhance success rates. In addition, competition between rival worksites or departments can provide additional motivation as well as foster co-worker social support for stopping smoking. While we have investigated between-site competition (i.e., one bank versus another), within-site competitions (e.g., two competing departments, or first floor versus second floor) should also be feasible.

Attempts should also be made to maximise support from nonsmoking co-workers. For example, in our competition study (Klesges et al., 1984) prizes were awarded that benefited both smokers and nonsmokers (such as a microwave oven or television for the employee lounge). Otherwise, nonsmokers may feel that the smokers are receiving preferential treatment (time off work for participation, prizes, etc.). Finally, public feedback on progress may serve to increase the magnitude of behavior change. For example, feedback regarding weekly progress of groups can be posted in lunchrooms or lounges. Periodic progress reports to department supervisors at worksite-wide meetings might also be helpful. To avoid stigmatizing particular individuals, we recommend providing feedback on *group* rather than individual progress.

We have confronted a number of problems in conducting worksite programs that other investigators can perhaps avoid or at least anticipate. Group composition is one such sensitive issue. For example, a high-ranking executive assigned to a group otherwise comprised of clerical workers was extremely uncomfortable, and his presence seemed to almost eliminate group discussion. One must be sensitive to "negativism" that can occur in groups; the focus of groups must be kept positive. One tactic is to use humor to illustrate how co-workers can mutually reinforce complaining. A positive perspective is particularly important when conducting competition or incentive interventions in which certain individuals or groups must lose. To counter discouragement that can result, we have shared results of the Brownell et al. (1984) study in which the company with the worst posttest results was best at follow-up.

RECOMMENDATIONS FOR FUTURE RESEARCH

In the previous section, we provided a number of clinical guidelines for the implementation of worksite smoking modification programs. Given the status of the field, only a few of these guidelines are empirically based; the rest are generalizations drawn from our clinical experience. Well-meaning suggestions such as these have, unfortunately, been uncritically accepted and almost institutionalized in other areas of smoking research, and only later revealed to be nothing more than untested myths and erroneous assumptions (Johnson, 1982). To prevent this from happening with the above clinical suggestions, research is needed either to support empirically or to refute these recommendations. Thus "more research is needed," but we do *not* need still more uncontrolled studies that do not further our understanding of how or why worksite smoking modification programs work. The pioneering studies conducted to date have been valuable, but the field must now move ahead to more sophisticated designs and questions.

Much research must be conducted on *ways to increase participation and follow-through rates* in worksite programs. From both public health and cost-effectiveness viewpoints, smoking control programs in occupational settings will fail to affect significantly the smoking problem if they cannot attract and retain a sizable percentage of smoking employees. Several procedures, which can potentially increase participation rates, seem particularly well suited to worksite programs and worthy of investigation. The provision of time off work for participation and/or financial incentives for successful participants might be expected to enhance participation. As in the weight-reduction area (see Abrams et al., 1984), investigators have recently turned toward competition procedures. Investigations of the effects of competition on participation (as well as success) rates and of the types of competition that are most effective (e.g., between organizations versus between departments within an organization) are both indicated. The last issue is of methodological interest as well, since it would allow investigation of possible contamination or diffusion effects occurring within worksites.

In terms of increasing adherence to program recommendations and decreasing attrition rates, many factors could be manipulated. For example, the use of computerized instructions has great potential. Interactive computer programs can be quite engaging for users, particularly if they include personalization of information and feedback on progress in the form of high-quality visual displays. Studies of ways to enhance group cohesion and how to construct intervention groups optimally (e.g., Should good friends be assigned to the same or separate groups? Should supervisors attend sessions with their supervisees?) would also be relevant, since Yalom (1975) has indicated that rates of attendance at meetings are strongly associated with level of group cohesion.

A second set of questions for future research concerns ways to *enhance the success* of worksite smoking modification programs. In searching for such methods, it makes sense to look first at factors that are uniquely related to the worksite environment. As mentioned previously, competition among organizations and/or financial incentives from employers have the potential to improve treatment outcome as well as to increase participation. Another potential advantage of conducting programs with co-workers versus strangers who have no interaction with each other outside of treatment sessions is the naturally occurring contact and social support available throughout the work-week. Initial attempts to investigate how social support operates and to enhance supportive behaviors have not been particularly encouraging, but this area is certainly worthy of further investigation. It is also important to collect manipulation check and process measures to answer the questions of how and why particular program components achieve their effects.

Additional ways to enhance the effectiveness of smoking modification programs have been discussed in other sections of this chapter. They include (1) lifestyle or multiple-risk factor reduction programs as opposed to single modality (smoking-cessation only) programs and (2) feedback on carbon monoxide levels or other physiological measures of health risks associated with smoking. This last point raises an important caveat regarding the effectiveness of smoking control programs: it is important to verify objectively that participants have actually achieved the changes in smoking patterns that they report. Options range from carbon monoxide to saliva thiocyanate to cotinine or nicotine assays (Haley, Axelrad, & Tilton, 1983) to mechanical verification of changes in smoking topography (Henningfield & Griffiths, 1979). Each of these measures has its own advantages and limitations and future studies would optimally combine two or more of these procedures. Objective verification of smoking status would seem to be particularly important in programs offering large financial incentives, involving competition between rival institutions, attempting to maximize social pressure/support to quit, and/or focusing on controlled smoking instead of abstinence.

A final possibility for improving success rates is inclusion of a physician stop-smoking message. Recent research has found that even brief recommendations by one's physician can enhance cessation rates (see reviews by: Fisher, 1982; Lichtenstein & Danaher, 1978). Many larger companies employ full-time physicians, and it seems that such a message could be particularly powerful when combined with physiological feedback on health risk, referral to on-site smoking-cessation programs, and/or prescriptions for nicotine gum.

Our last category of recommendations for future research involves *study of subject, therapist, and setting factors that affect treatment outcome.* We need additional study of the enrollment patterns and success rates of men versus women, white-collar versus blue-collar workers, and heavy versus light smokers. We also know little about the characteristics of successful program leaders. For example, worksite programs seem a natural place to investigate the efficacy of ex-smoker co-workers versus outside smoking cessation experts versus company physicians as group leaders. A related issue concerns procedures for training facilitators and disseminating programs once effective treatments have been established. The program delivered to participants may often not be the same as that designed by program developers. Thus measures of adherence to treatment protocol on the part of *both* group leaders and participants is strongly encouraged.

In terms of setting factors possibly affecting outcome, future investigations must collect more systematic information on such variables as absenteeism, job stress, and job satisfaction, which may be related to participation and success rates. Our clinical experience suggests that other factors, such as the active involvement of key labor and management figures in a program and the establishment of no smoking areas or "no smoking on the job" rules, may also enhance treatment outcome. In summary, some promising beginnings have been made in worksite smoking modification programs. There appear to be several potential advantages to conducting worksite programs, although such variables as greater participation and enhanced social support do not necessarily occur naturally. However, the area holds great potential, and it is hoped that future studies will be better designed so that they can adequately answer some of the above questions concerning ways to improve the impact of these programs.

REFERENCES

Abrams, D. B., Elder, J. P., Carleton, R. A., & Artz, L. M. (1984). A comprehensive framework for conceptualizing and planning organizational health promotion programs. In M. Cataldo & T. Coates (Eds.), *Behavioral Medicine in Industry*. New York: Wiley.

Brownell, K.D., Cohen, R. Y., Stunkard, A. J., Felix, M. R. J., & Cooley, N. B. (1984). Weight loss competitions at the worksite: Impact on weight, morale, and cost-effectiveness. *American Journal of Public Health, 74,* 1283–1285.

Cathcart, L. M. (1977). A four-year study of executive health risk. *Journal of Occupational Medicine, 19,* 354–357.

Chesney, M. A., & Feuerstein, M. (1981). Behavioral medicine in the occupational setting. In J. McNamara (Ed.), *Behavioral Medicine*. Kalamazoo, MI: Behaviordelia.

Cooper, W. A. (1978, July–August). Johns-Mansville says no smoking. *American Lung Association Bulletin*, pp. 7–10.

Danaher, B. G. (1980). Smoking cessation programs in occupational settings. *Public Health Reports, 95 (2),* 149–157.

Ellis, B. H. (1978). How to reach and convince asbestos workers to give up smoking. In J. Schwartz (Ed.), *Progress in smoking cessation: International conference on smoking cessation*. New York: American Cancer Society.

Fagerstrom, K. O. (1978). Measuring degree of physical dependence to tobacco smoking with reference to individualization of treatment. *Addictive Behaviors, 3,* 235–241.

Fielding, J. E. (1982). Effectiveness of employee health improvement programs. *Journal of Occupational Medicine, 24 (11),* 907–916.

Fisher, E. B., Jr. (1982). Cessation of smoking: Selfmotivated quitting. *The health consequences of smoking: Cancer. A Report of the Surgeon General.* DHSS Publication No. (PHS) 82-50179. Washington, DC: U.S. Government Printing Office.

Glasgow, R. E. (1984). Smoking modification. In T. Creer & K. Holroyd (Eds.), *Self-management in behavioral medicine*. New York: Academic Press.

Glasgow, R. E., Klesges, R. C., Godding, P. R., & Gegelman, K. (1983). Controlled smoking, with or without carbon monoxide feedback, as an alternative for chronic smokers. *Behavior Therapy, 14,* 386–397.

Glasgow, R. E., Klesges, R. C., Godding, P. R., Vasey, M. W. & O'Neill, H. K. (1984). Evaluation of a worksite controlled smoking program. *Journal of Consulting and Clinical Psychology, 52,* 137–138.

Glasgow, R. E., & Rosen, G. M. (1978). Behavioral bibliotherapy: A review of self-help behavior therapy manuals. *Psychological Bulletin, 85,* 1–23.

Haley, N. J., Axelrad, C. M., & Tilton, K. A. (1983). Validation of self-reported smoking behavior: Biochemical analyses of cotinine and thiocyanate. *American Journal of Public Health, 73,* 1204–1207.

Henningfield, J. E., & Griffiths, R. R. (1979). A preparation for the experimental analysis of cigarette smoking. *Behavior Research Methods and Instrumentation, 11,* 538–544.

Johnson, C. A. (1982). Untested and erroneous assumptions underlying antismoking programs. In T. J. Coates, A. Petersen, & C. Perry (Eds.), *Promoting adolescent health: A dialog on research and practice.* New York: Academic Press.

Kanzler, M., Zeidenberg, P., & Jaffe, J. H. (1976). Response of medical personnel to an onsite smoking cessation program. *Journal of Clinical Psychology, 32,* 670–674.

Klesges, R. C., Vasey, M. W., & Glasgow, R. E. (1984). *Evaluation of a worksite smoking competition program.* (Unpublished manuscript).

Kristein, M. (1982). *The economics of health promotion at a worksite.* Mimeo available from The American Health Foundation, New York.

Lichtenstein, E., & Danaher, B. G. (1978). What can the physician do to assist the patient to stop smoking? In R. E. Brashear & M. L. Rhodes (Eds.), *Chronic obstructive lung disease: Clinical treatment and management.* St. Lewis: Mosby.

Luce, B. R., & Schweitzer, S. O. (1978). Smoking and alcohol abuse: A comparison of their economic consequences. *New England Journal of Medicine, 298,* 569–571.

Malott, J. M., Glasgow, R. E., O'Neill, H. K., & Klesges, R. C. (1984). The role of co-worker social support in a worksite smoking control program. *Journal of Applied Behavior Analysis, 17,* 485–495.

Marlatt, G. A., & Gordon, J. R. (1980). Determinants of relapse: Implications for the maintenance of behavior change. In P. O. Davidson & S. M. Davidson (Eds.), *Behavioral medicine: Changing health lifestyles.* New York: Brunner/Mazel.

Mermelstein, R., Lichtenstein, E., & McIntyre, K. (1983). Partner support and relapse in smoking-cessation programs. *Journal of Consulting and Clinical Psychology, 51,* 465–466.

Meyer, A. J., & Henderson, J. B. (1974). Multiple risk factor reduction in the prevention of cardio-vascular disease. *Preventive Medicine, 3,* 225–236.

Naditch, M. P. (1984). The STAYWELL program. In J. D. Matarazzo, S. M. Weiss, J. A. Herd, N. E. Miller, & S. M. Weiss (Eds.), *Behavioral health: A handbook of health enhancement and disease prevention.* New York: Wiley.

National Interagency Council on Smoking and Health. (1980). *Smoking and the workplace: National Interagency Council on Smoking and Health Business Survey.* (Unpublished manuscript).

Nepps, M. M. (1984). A minimal contact smoking cessation program at the worksite. *Addictive Behaviors, 9,* 291–294.

Oldridge, N. B. (1978, May). *Identification of early noncompliance with cardiac rehabilitation program physical therapy.* Paper presented at the annual meeting of American College of Sports Medicine, Washington, DC.

Orleans, C. S., & Shipley, R. H. (1982). Worksite smoking cessation initiatives: Review and recommendations. *Addictive Behaviors, 7,* 1–16.

Rand, C., Stitzer, M., Bigelow, G., & Mead, A. (1984). *Contingent reinforcement for smoking abstinence.* Poster presented at the American Psychological Association, Toronto, Canada.

Rosen, G. M., & Lichtenstein, E. (1977). An employee incentive program to reduce cigarette smoking. *Journal of Consulting Psychology, 45,* 957.

Schlegel, R. P., Manske, S. R., & Shannon, M. E. (1983). *Butt out!: Evaluation of the Canadian Armed Forces smoking cessation program.* Paper presented at the Fifth World Conference on Smoking and Health, Winnipeg, Canada.

Schwartz, J. L., & Dubitzsky, M. (1967). Expressed willingness of smokers to try ten smoking withdrawal methods. *Public Health Reports, 82,* 855–861.

Scott, R. R., Denier, C. A., & Prue, D. M. (1983). *Worksite smoking intervention with health professionals.* Paper presented at the Association for the Advancement of Behavior Therapy Annual Convention, Washington, DC.

Shiffman, S. M. (1982). Relapse following smoking cessation: A situational analysis. *Journal of Consulting and Clinical Psychology, 50,* 71–86.

Stachnick, T., & Stoffelmayr, B. (1983). Worksite smoking cessation programs: A potential for national impact. *American Journal of Public Health, 73*(12), 1395–1396.

Stitzer, M. L., & Bigelow, G. E. (1982). Contingent reinforcement for reduced carbon monoxide levels in cigarette smokers. *Addictive Behaviors, 7,* 403–412.

Stitzer, M. L., & Bigelow, G. E. (1983). Contingent payment for carbon monoxide reduction: Effects of pay amount. *Behavior Therapy, 14,* 647–656.

U.S. Public Health Service. (1979). *Smoking and Health: A Report of the Surgeon General.* DHEW Publication No. (PHS) 82-50179. Washington, DC: U.S. Government Printing Office

U.S. Public Health Service. (1982). *The health consequences of smoking: Cancer: A report of the Surgeon General* (DHHS Publication No. PHS 82-50179). Washington, DC: U.S. Government Printing Office.

Ware, B. G., & Block, D. L. (1982). Cardiovascular risk intervention at the worksite: The Ford Motor Company program. *International Journal of Mental Health, II*(3), 68–75.

Wilbur, C. S. (1982). Unpublished manuscript.

Yalom, I. D. (1975). *The Theory and Practice of Group Psychotherapy* (2nd ed.). New York: Basic Books.

10

Dental Health Promotion in the Workplace

WILLIAM A. AYER, SUSAN SEFFRIN, GERALDINE WIRTHMAN, DEBORAH DEATRICK, and DANA DAVIS

OVERVIEW OF DENTAL DISEASE

In 1982, dental diseases resulted in approximately 19.5 billion dollars in expenditures for treatment (Health Care Financing Review, 1983). Dental caries is the most common disease in the United States and constitutes the main reason for tooth loss before the age of 35. By that time, the average person has lost five teeth and has eleven more that have carious lesions. After the age of 35, the main cause of tooth loss is periodontal disease. In addition to these two categories of disease, oral cancers account for some 3–4 percent of all cancers; in addition, craniofacial anomalies and traumatic injuries account for a significant, although undocumented, amount of dental treatment.

In general a major amount of dental disease is probably preventable. The National Caries Program has recorded substantial decreases in the amount of dental caries among children (on the order of 50 percent), which is probably due to the ingestion of fluoridated water supplies. Glass and his co-workers (Glass & Fleisch, 1981) have also documented the decline in caries prevalence in nonfluoridated communities and have suggested that fluoridated toothpastes, fluoride mouthrinses, and changes in dietary

This project was supported by the American Fund for Dental Health and the American Dental Association.

practices are responsible for the decline. Although dental caries is probably most efficiently and effectively prevented through the fluoridation of community water supplies, periodontal disease appears to be preventable through the regular and judicious practice of brushing and flossing. Most oral cancers seem to be the result of cigarette smoking and require the same abstinence considerations as they do with other lung diseases attributable to smoking. Many traumatic facial injuries are the result of not using seat belts during driving or of not using protective head and mouth gear when one is engaging in certain sports and recreational activities (Liss, Evenson, Lowey & Ayer, 1982).

Although the number has risen steadily, only about 50 percent of the American population visits a dentist each year. It has been reported that the average person visits the dentist 1.6 times per year. However, this figure is somewhat misleading since those who go to the dentist actually go slightly more than 3 times per year. Also Newman and Anderson (1972) have reported that 10 percent of the population account for 75 percent of the funds expended for dental care and that some 3 percent account for more than one-half of these expenditures. Ayer (1982) has noted that, on the basis of these data, it appears that in the United States only a small number of people regularly and consistently visit the dentist.

Barriers to Seeking Dental Treatment

Many studies have examined the reasons people give for not going to a dentist and the findings have been relatively consistent. These reasons commonly include the following: fear of pain, no perceived need to go, too expensive. Fear of pain accounts for about 5–16 percent of the population (Ayer & Corah, 1984) not visiting the dentist. No perceived need to go is the reason given by about 63 percent of those who have not been to a dentist in a year or more (ADA, 1979). Some 23 percent also reported that they do not go because of the expensive nature of dental treatment. A report by Avnet and Nikias (1967), however, suggests that financial barriers may play a less significant role. In their study of a population entitled to dental service at little or no cost through a dental health insurance program, they observed that when care was available it was not used. Other reasons frequently reported for not going to the dentist include the following: having to take time off from work (8 percent); transportation problems (3 percent); inability to receive a suitable appointment (2 percent). Although financial barriers continue to be cited as a reason for not going for dental treatment, the preponderance of the data would, in fact, suggest that financial barriers have minimal significance regarding the seeking of dental care.

Factors that predict the use of dental services include educational level, socioeconomic status, geographic location, recency of last visit, and whether the individual received treatment as a child. Mother's use behavior also tends to predict children's use behavior. (See Richards, 1971, for utilization of dental services.)

Workloss Due to Acute Dental Conditions

The 1979 data from the National Center for Health Statistics (Table 10.1) estimated that approximately 6.1 million days of workloss or 0.06 days per employed adult occurred because of acute dental conditions. Bailit and his colleagues (Bailit, Beazoglou, Hoffman, Reisine, & Strumwasser, 1983) analyzed secondary sources of data and concluded that the annual workloss was closer to 32.93 million workdays or approximately 2.72 hours per employee per year, and that the cost of this time in lost wages exceeds three billion dollars. These investigators were quick to caution that their estimates were *minimal* estimates and did not include costs associated with reduced productivity. Bailit and his colleagues have estimated that workloss due to dental conditions represents about 5 percent of all health-related workloss.

Certain groups that appear to lose more time than other groups include the following: blue-collar workers; people who were divorced, separated, or widowed; lower income groups; and those without a regular dentist. Time was also lost because of the need to take children to the dentist.

Despite the fact that dentally related workloss and decreased productivity cost industry a considerable amount of money, many management and labor representatives do not consider it a serious problem, primarily because of the high visibility of other medical problems.

Table 10.1.   Dental Disability – Incidence of Acute Dental Conditions, Restricted Activity, Bed Disability, and Workloss Days

Dental Disability	MCHS 1979
Incidence of Acute Dental Conditions	6,898,000
Per 100 persons per year	3.2
Male	3,644,000
Per 100 males per year	3.5
Female	3,253,000
Per 100 females per year	2.9
Restricted Activity Days	31,517,000
Per 100 persons per year	14.6
Bed days	12,017,000
Per 100 persons per year	5.6
Workloss days	6,116,240
Per 100 persons per year	5.8

Source: From Bailit, Beazoglou, Hoffman, Reisine, and Strumwasser (1983), based on data from the National Center for Health Statistics (NCHS).

OVERVIEW OF DENTAL CARE IN INDUSTRIAL SETTINGS

The first system of dental health care for workers was initiated in 1887, when Great Western Railway of England added a dental department to its already existing medical department (Steward & McLenaghan, 1957). Workers and their dependents were eligible for dental treatment. In Canada, the Barber Match Company organized the first dental program in 1890 in North America. It was followed by the Metropolitan Life Insurance Company of New York in 1915, and the Union Health Center of New York in 1918. The Olivetti Company of Italy established dental services for its employees in 1936 (Balma & Semeraro, 1964). According to Nyyssonen, Rajala, and Paunio (1983), by 1984, 1 percent of Finland's industries employed a full-time dentist and 5.5 percent employed part-time dentists to care for their workers.

Services in these programs have ranged from visual screening examinations and referrals to comprehensive treatment. Programs may also include or be limited to preventive dental health education.

Steward and McLenaghan (1957) in their review reported that by 1920 there were more than 400 known U.S. industries with some type of dental program. This number began declining in the mid- to late-1920s with a revival of interest occurring during World War II. By 1952, approximately 7.5 percent of the industries indicated they had an employee dental service.

More recently, the situation appears to have changed. In 1980, the Council on Dental Care Programs of the American Dental Association identified seven corporately owned and operated dental care facilities within the United States, which provided comprehensive dental care facilities and services to its employees. Utilization rates ranged from 60 percent to 99 percent of the eligible employees. In 1981, the Council reported that the results of their survey of union-sponsored dental-care facilities revealed five unions that provided dental-care facilities. Estimates were that about 65 percent to 80 percent of the eligible members used these facilities. The Council observed in 1980 that most of the facilities were established for the convenience of the employees and as opportunities for increased goodwill on the part of the employers. In addition, most of these facilities appear to have been established before the rapid growth in prepaid dental plans. As a result, additional growth in the number of facilities appears limited since the Council (1980) noted that it seems "unlikely that many corporations are capable of expending, or willing to expend, the funds required to establish dental facilities" (p. 950).

And, as Cohen (1985) reported, "these are programs which appear to be idiosyncratic models for corporate America. Although interest abounds, there is no massive movement nor even a gradual trend for other major corporations to establish like facilities" (p. 20.s).

Diagnostic and Preventive Services at the Worksite

In addition to the comprehensive programs provided by a few corporations, several experimental programs also exist, which provide diagnostic and preventive services but do not compete with the private sector dentist (Cohen, 1985).

One such program was developed by Johnson and Johnson Dental

Products Company in 1980 (Meadow & Rosenthal, 1983). The model on which Johnson and Johnson's program is based essentially seeks to separate "well care" from "treatment". Diagnostic and preventive services are provided but patients with treatment requirements are referred to their own dentists or provided with a list of dentists in the area. This practice permits the group to focus on prevention and maintenance and assigns the patient the responsibility for obtaining treatment from the private dentist. The authors report that of those patients requiring dental treatment, 73 percent received the needed care within the first year of the program. Cohen (1985) reported that, although the program met initial resistance from the private sector, dentists now accept the program because of the demand generated at the worksite.

Capsule Summary

Initially, programs in the workplace seem to have been concerned with work-related injuries. Eventually they shifted to providing more comprehensive, nonwork-related treatment and, finally in the 1940s, to some preventive efforts. Insurance coverage also increasingly provided for dependent coverage. However, it must be remembered that these programs occurred in large companies (usually more than 500 employees) and thus were not available to a large segment of the work force. This is particularly significant when one recalls that of the nation's 57 million blue-collar workers, approximately 75 percent of them are employed in work settings of less than 100 people. Thus, although employers may be interested in the health of the worker, it is frequently not feasible or possible to provide dental services at the worksite.

Frequently in the worksite, it has been the professional and white-collar worker who used the services (Avnet & Nikias, 1967) more than the blue-collar and semi-skilled individuals, although this latter group would be expected to have greater dental treatment needs. The hourly worker is also at another disadvantage in that time lost from work because of the need to seek dental treatment also results in lost wages and earnings. If there exists an incentive bonus, then this is lost too.

The problem may be further exacerbated for the working mother with children who must take them to the dentist. Thus she must lose time not only for her own dental treatment but also for that of her children. Consequently, some writers have suggested that dental offices should provide expanded or nontraditional hours. For salaried workers, this does not appear to be a significant problem.

Dental Prepayment

For the sake of completeness, and because of their potential impact, let us consider briefly dental insurance or dental prepayment plans. Although frequently used, the term *dental insurance* is a slight misnomer since, in fact, there exist few dental insurance policies. The plans most commonly encountered are, in fact, prepayment plans.

Dental prepayment began in 1954 on the west coast when the International Longshoremen's and Warehousemen's Union-Pacific Maritime Association won a dental benefits program from the west coast shipping industry (Bishop, 1983). From that beginning, more than 50 million people had some type of dental prepayment program by 1980. The development of dental prepayment plans has been chronicled by Bishop (1983).

The effects of dental insurance on use of dental services and on oral health status have only recently been examined. At a symposium entitled "Dental Insurance and Oral Health," at the 1984 meeting of the International Association of Dental Research, data from the Rand Health Insurance Experiment was presented indicating that dental insurance increased use of dental services. In addition, the first evidence was presented that dental insurance had a major impact on improving oral health status (reduced caries and periodontal disease). The people who appeared to benefit most were people with more dental disease and who were less well educated.

It has been estimated that currently more than 23,000 different plans in existence provide varying degrees of coverage. Bishop (1983) and Striffler, Young, and Burt (1983) predict that the number of people having these plans will continue to increase.

Dental Health Promotion Programs in the Workplace

Although health promotion programs appear widespread in the workplace (as other chapters in this book will show), few worksites have dental health promotion programs. In the book entitled *Managing Health Promotion in the Workplace: Guidelines for Implementation and Evaluation* (Parkinson et al., 1982), only one of seventeen companies reporting health promotion activities had a dental awareness program (Sentry Life Insurance Company). Thus it appears that dentistry has not been considered in the development of most worksite health promotion programs.

MAINE WORKPLACE PROJECT

In 1978, the Bureau of Health Education and Audiovisual Services of the American Dental Association was approached by representatives of several large industries to develop a workplace educational program in dental health for their blue-collar workers.

Somewhat later, the state of Maine expressed an interest in promoting the dental health of industrial workers in that state, and again came to the Bureau for guidance in developing and tailoring a program suited to their specific needs. (Preliminary data available suggested higher prevalence rates for dental disease in Maine than for the United States in general. Contributing to the problem was the large number of communities that did not have the benefit of water fluoridation. Maine is one of five states nationwide that requires a public referendum to initiate fluoridation.)

As previously noted, little has been done in the area of education and dental health promotion among adults. Young (1971) reviewed the literature on dental health education of adults both in the United States and in other countries. Her review focussed on a variety of important variables. However, she was forced to conclude as follows:

> Too often studies are scattered and ad hoc, and are generally lacking in the requisite systemization, so that research questions asked are not meaningfully related one to another. Consequently, there is no cumulative progression of knowledge emerging from the research efforts. Nor are the research data derived from cost studies meaningful in relation to needed program expedience and ease of handling. (pp. 268–269)

Young's review appears to be the most recent one, and unfortunately, her 1971 statements about dental health education of adults still applies today.

Controversy continues regarding the effectiveness of educational interventions, and the issue has by no means been resolved. Albino (1980) argued that the failure of educational programs has occurred primarily because such programs have ignored the social values and salient concerns of the patients, an argument also made by Davis (1980). Legler, Mayhall, and Bradley (1972) have suggested that educational and health promotional efforts directed at disadvantaged groups should focus on the concept that dental disease can be controlled and prevented.

In addition, coverage for dental treatment is becoming an increasingly common phenomenon for an increasing number of people in the workplace. Presumably, the removal of financial barriers, particularly for those individuals of lower socioeconomic status, will result in increased use of health services and more attention to preventive services. While this appears to have been the case with medical services, such does not appear to be the case with the use of dental services (Albino, 1980; Avnet & Nikias, 1967). With dental services provided in prepayment plans, the professional and executive groups continue to overuse dental services as compared to semi- or unskilled groups (Avnet & Nikias, 1967). Avnet and Nikias felt that the individual's attitudes toward dentistry and dental care played the most important role in deciding to seek dental service and that personal motivation combined with the removal of financial barriers would increase use of services. This, of course, raises the question of how one increases motivation to seek dental services.

These observations and the need to respond to the lack of programs directed at blue-collar workers led to the design and development of a project to evaluate various health promotion interventions in the workplace. Thus a series of meetings were undertaken by the Bureau of Health Education and Audiovisual Services and the Bureau of Economic and Behavioral Research to formulate a strategy to meet these needs. Although the Bureau of Health Education and Audiovisual Services had developed an extensive collection of educational materials, none of them had been designed specifically for blue-collar workers and had not been evaluated for their appropriateness. The desire was to develop a package or set of guidelines that could be used by any industry or local dental society to implement a dental health education and promotion program that would have a direct impact on the dental health behaviors of the blue-collar worker and his family—a target group that appears to have been primarily ignored.

The following represents a project overview and a summary of some preliminary findings and impressions to date.

Project Development

Project Design.   With the aid of the Office of Dental Health of the Maine Department of Human Services, the Maine Dental Association, and volunteer dentists in Maine, a project was designed to study dental health promotion in workplace settings in Maine.

The project design necessitated the participation of at least six companies. A listing of the companies in Maine was obtained, and all companies with at least 200 employees were contacted to determine their interest in the proposed project. Of those companies indicating an interest, six were eventually selected and were located in different parts of Maine. They consisted of two shoe factories, one electronics circuitry company, one meat packing company, one lumber mill, and one printing and publishing company.

Three of the companies provided dental prepayment plans for their employees and families. Three did not. One company from each group (prepayment; no prepayment) was selected to receive traditional dental health education programs (including classroom education and a mass media approach). One company from each group also received only the mass media approach. This latter strategy was selected because of the data, which suggest that information on "how to" may be as effective as the more concerted efforts of traditional health education programs (Barnes, Hoeksema, & Ayer, 1973; Garrity, 1983; Kaplis, Drolette, Boffa & Kress, 1979; Ramirez et al., 1971).

Project Implementation

A resource manual was developed for the intervention groups with specific worksheets for each type of intervention and instructions on how to conduct the programs. Although the topic may have been oral hygiene, for example, the traditional groups received classroom education and mass media information on the importance of and reasons for good oral hygiene and practice in brushing and flossing with disclosing tablets.[1] The how-to groups received only the oral hygiene aids and mass media materials for review.

To carry out these interventions, it was necessary to develop "instructional teams" for the four intervention settings. Each team consisted of individuals selected by each company and usually coordinated by the personnel director. Local dental professionals also served as consultants and/or lecturers (depending on the type of intervention strategy).

The teams were trained by the investigators and prepared to handle the questions and issues raised by the participants. They participated in a workshop that provided information on the nature of dental disease and proper oral hygiene practices. Information was also provided regarding available resources in the community.

The initial interventions for both the traditional education groups and the information-only groups were scheduled to occur at weekly sessions during an eight-week period.[2]

The weekly topics, scheduled to cover eight areas identified as important, were as follows:

1. The use of fluorides.
2. Dental health versus dental disease.
3. Methods of plaque control.
4. The importance of fluorides.
5. The importance of regular checkups.
6. The care of children's teeth.
7. How to be a wise dental consumer.
8. Dental prepayment plans.

Mass media materials were the same for all companies, although traditional companies held classroom sessions (approximately one-half hour each) on topic areas.

Following these interventions, project staff met with the teams to determine the nature of the interventions for subsequent years. In other words, information obtained regularly from what had been successful and what had been unsuccessful in this phase of the program was used to determine the

---

[1]Disclosing tablets are tablets containing a harmless red dye that stains any plaque on the teeth. Proper brushing and flossing eliminate plaque, and the dye assists patients in finding areas they may have missed.

[2]The nature of the companies indicated that certain times of the year were more appropriate than other times for these sessions. For example, the months of February and March were better for conducting the program than were the months of June, July, and August, since the companies frequently closed for vacation during the summer months.

subsequent nature of the interventions. It may well be that more emphasis must be given to oral hygiene skills, or dental consumerism, and so forth. This approach represented an effort to determine the needs of the blue-collar workers and to respond to them, rather than inflict on them a program that might be inadequate.

Baseline Data Collection

Prior to any of the interventions, several dependent variables were selected for inclusion in the study. These variables included clinical examinations of the oral tissue and DMFS rate, a simplified periodontal assessment (Modified Russell's Periodontal Index), the Simplified Oral Hygiene Index (OHI-S), treatment needs, and prosthetic appliance needs. The clinical examinations were conducted in the workplace setting in an effort to maximize participation.

Permission was obtained from the participants (and their dentists) to obtain data from their dental records on dental visit behaviors and the types of treatments received. Permission was also obtained to collect the same type of information on their family's dental visit and treatment behaviors. Family member dental behavior and treatment type were considered important because it was felt that the effects of the dental health promotion interventions might be observed in the family.

Additional data collected included information on a Dental Health Knowledge, attitudes and behavior, sociodemographic data, absenteeism due to dental problems, perceptions about the educational and audio-visual material used, sociodemographic data, and perceptions about how the program could be changed to more accommodate the workers' needs.

This project was designed to be conducted during a four-year period, primarily because of the need to determine if any actual change in dental health occurred.

Preliminary Findings

Demographic Data.   Data were collected on 469 subjects. The average age of the subjects was 36.5 years. The sample was composed of 37.8 percent males and 62.2 percent females. Approximately 69.2 percent of the subjects were married, 19.5 percent were single, 9.6 percent were divorced, and 1.7 percent were widowed. The average number of children per family was 1.7. The average highest educational level completed was 11.6 years, which indicates that most workers in the present study had not completed high school.

DMF Rates.   Clinical examinations were conducted on the participants. The findings on DMFT (Decayed, Missing, Filled Teeth) are shown in Table 10.2 and compared with findings from the National Center for Health Statistics. The DMFT is considerably higher in Maine. If one examines the components of the score, then it can be observed that the increase is accounted for by the large number of missing teeth in the Maine Samples. In fact, some 13.6 percent of the workers examined were edentulous (with no teeth). These findings confirmed the impression that the oral health status of the sample was worse than that for the United States in general.

Table 10.2.   DMFT Baseline Data from Maine Compared with Data from National Center for Health Statistics (NCHS)

Component	Maine	NCHS
DMFT	19.6	14.9
D	1.6	1.7
M	10.3	4.9
F	7.6	8.3

Periodontal Status and Oral Hygiene.   Periodontal and oral hygiene assessments were made and the findings are shown in Table 10.3. Again, for the Periodontal Index (PI) it can be observed that the score for the Maine workers is almost twice that of the score reported by the National Center for Health Statistics. The debris and calculus indices were also higher in comparison to the data from NCHS, indicating that the participants' oral hygiene was worse.

Table 10.3.   Scores for Modified Russell's Periodontal Index (PI), Debris Index and Calculus Index for Maine Sample Compared with Findings from the National Center for Health Statistics (NCHS)

Index	Maine Sample	NCHS Samples
PI	1.26	.69
Debris	.79	.57
Calculus	.58	.36

Oral Hygiene Practices.   The participants were asked when they brushed their teeth, whether they used dental floss, satisfaction with their family's teeth, and the type of toothpaste they used. The findings show that most individuals brushed on arising or after breakfast and at bedtime. A large percentage of the sample (44.5 percent) reported that they did not use dental floss. Only 12 percent used floss daily. A larger percentage of people was dissatisfied with the condition of their family's teeth than with their family's health. Interestingly, almost all of the subjects used a fluoridated toothpaste.

Treatment Requirements.   More than 71 percent of the subjects were in need of periodontal treatment ranging from prophylaxis to full mouth extraction because of advanced periodontal disease.

Approximately 20.5 percent of the edentulous subjects needed new dentures or repairs for the upper jaw and 23.4 percent needed new dentures or repairs for the lower jaw. Approximately 4.3 percent of the sample was currently experiencing dental pain and required immediate attention. An additional 11.3 percent had dental conditions that were judged as requiring immediate attention. Thus an alarming 15.6 percent of the sample who presented with conditions at the screening examination required immediate attention and treatment.

Dental Prepayment.   When the data were analyzed according to whether the companies provided some type of dental prepayment, significant effects for prepayment were observed. Workers in companies without prepayment programs were seven times more likely to demonstrate conditions requiring immediate treatment for the relief of existing pain.

In addition, workers without prepayment plans demonstrated significantly more untreated disease as reflected in the DMFT scores and significantly higher Debris and Calculus scores (indicating poorer oral hygiene practices).

Interestingly, the effect of prepayment on periodontal treatment requirements was not significant, and approximately equal numbers in both categories were in need of specific types of treatment.

The availability of prepayment plans also exerted an influence on dental visit behavior. Employees with such programs were one and one-half times more likely to report having been to a dentist within the last year, and having been for the purpose of a checkup, cleaning (prophylaxis), and fillings.

Evaluation on Phase I

Participant Response to Phase I.   At the conclusion of the first intervention, the participants were asked to identify the areas they had found most helpful. The three areas deemed most helpful were methods of plaque control, the importance of regular checkups, and dental health versus dental disease. The areas on which they desired more information included the care of children's teeth, dental prepayment plans, and dental health versus dental disease.

When asked about the usefulness of the program, most (74 percent) reported the program as helpful or extremely helpful. The participants (73 percent) also reported that they took the pamphlets and educational materials home to family and friends who also found them useful. The observation was made by team members that few of the educational materials were discarded at the worksite.

When asked how the programs could be changed, the respondents suggested more meetings with dental professionals, the involvement of the entire family, and more educational materials. The dental screening examinations received positive responses from the employees and appear to have provided an important function by providing the employees with some knowledge of their oral health status. As a result, it may be that the employees then feel justified in seeking and making dental appointments for conditions that they know exist. Consequently, loss of wages may assume a less important role as a barrier. These findings suggest that it may be important for companies and local dental societies to consider screening at the worksite as a way to provide inducement to obtain care.

PRELIMINARY RECOMMENDATIONS

Company team members and the investigators reviewed employee survey comments, on-site employee responses, management needs, and baseline data to plan the second phase of the intervention strategies. Some of the major findings are summarized below.

The nature of the companies' work activities and schedules indicated that certain times of the year were more appropriate for conducting the interventions. The months best for the Maine participants were February and March, with the least acceptable months being June, July, and August.

Videoplayers and cassettes were placed in each company to permit employees to view topics during free time. Interestingly, the placement of these machines did not receive the enthusiasm the investigators had expected. The companies and the subjects viewed the equipment as expensive and capable of being stolen. Thus they were quite uncomfortable in having the equipment. In addition, the cassettes initially used were too long (7–15 minutes) and too technical. Suggestions were made to use shorter, simple, fun films that could be used in continuous playback units.

Posters were placed in common areas such as employee lounges or lunchrooms. Although most of the posters were viewed quite favorably, the employees did not like several posters and requested that they be removed. Again, suggestions were made to use simple, fun messages, deleting the use of more technical material.

Company team members felt the pamphlets and brochures selected for use were too numerous, making distribution to all employees difficult. The teams also questioned the mass distribution of all materials to all employees, since specific needs could vary (i.e., pamphlets on dental health and pregnancy were given to single people and grandparents, pamphlets on flossing were given to the edentulous, etc.). Recommendations were made either to provide pamphlets to people more discriminately or to place certain types of materials in racks to allow those interested in a topic to select the appropriate materials.

The topic area on the "Importance of Children's Teeth" (although not identified specifically by employees as of interest to them on the employee surveys) received many positive comments on site, and many requests for additional copies were made. Thus team members and the investigators felt this topic should be used as a major topic area in subsequent booster sessions. The assumption was that parents (grandparents, uncles, aunts, etc.) do focus on the needs of children and often learn about their own health care needs while trying to learn about those of children.

At this point in the study, the investigators have been forced to conclude that classroom sessions as part of the traditional dental health educational methods probably cannot be implemented in worksites like the settings used in this study. The settings are small, while workers' time is quite limited and regimented. Moreover, it appears difficult to conduct the programs either during work hours or outside of work hours. Given that workers in our study are also on incentive plans (over a base level of production, additional production means additional money), few were induced to take time away from their jobs. However, as we had hypothesized, providing information and resources constitutes another method for overcoming this problem. It would appear that combining mass screening examinations (using a dental health day) with the availability of company resource teams, as well as educational materials, may be the most effective plan.

FINAL PHASE OF THE PROJECT

The project is currently entering its final year. The ultimate measures of the program's impact must be assessed in terms of changes in dental disease, preventive and maintenance self-care activities, and dental visit behavior. During the final year, the participants will be scheduled for clinical examinations. Data on dental visit behaviors and types of treatments received by the employee and his or her family will be obtained.

A continuing problem seems to be that health promotion programs appear to be fragmented and almost impossible to integrate. Many of the health promotion efforts in other areas are equally applicable and germane to dental health promotion. Because of the small company sizes, it may be relatively impossible to implement integrated programs as we traditionally know them. However, because of the sheer size of the population working in these settings and the nature of the problems, it is important that we continue to investigate ways of doing so.

REFERENCES

Albino, J. E. (1980). Motivating underserved groups to use community dental services. In S.L. Silberman, & A.F. Tyron, (Eds.), *Community dentistry*. Littleton, MA: PSG Publishing Company.

American Dental Association, Council on Dental Care Programs. (1980). Survey of corporately owned and operated dental care facilities in the United States providing comprehensive dental care services. *Journal of the American Dental Association, 101,* 945—949.

American Dental Association. (1981). Special Report 2 of the Council on Dental Care Programs:

Union-sponsored dental care facilities. *American Dental Association Annual Reports and Resolutions*, pp. 74–75.

Avnet, H. H., & Nikias, M. (1967). *Insured dental care*. New York: Group Health Dental Insurance.

Ayer, W. A. (1982). The dentist-patient relationship. *International Dental Journal, 32*, 56–64.

Ayer, W. A., & Corah, N. L. (1984). Dental providers and oral health behaviors. In *Social Sciences and Dentistry* (Vol. 2). Berlin: Federation Dentaire Internationale, Quintessence Publishing Company.

Bailit, H. L., Beazoglou, T., Hoffman, W., Reisine, S., & Strumwasser, I. (1983, January). *Workloss and dental disease*. Report to the Robert Wood Johnson Foundation.

Balma, Dr., & Semeraro, D. (1964). Dental services in a large factory in northern Italy. *Dental Magazine, 81*, 202–206.

Bishop, E. (1983). *Dental insurance*. New York: McGraw-Hill.

Cohen, L. K. (1985). *Market and community responses to changing demands from the workplace*. Supplement to Community Health Studies, IX.1, 18.s–24.s.

Gift, H. (1984). Professional services and utilization. In *Social Sciences and Dentistry* (Vol. 2). Berlin: Federation Dentaire Internationale, Quintessence Publishing Company.

Davis, P. (1980). *The social context of dentistry*. London: Croom-Helm.

Glass, R., & Fleisch, S. (1981). Decreases in caries prevalence. In J. Hefferren, W. A. Ayer, & H. Kochler (Eds.), *Foods, Nutrition and Dental Health* (Vol. 3). Park Forest South, IL: Pathotox.

Glass, R., & Fleisch, S. (1983, Fall). *Health Care Financing Review, 5*, 7.

Jerge, C. R. (1980). Winston-Salem Dental Care Plan, Inc. *New Dentist, 11*, 18–22.

Kriesberg, L., & Treiman, B. R. (1962). Preventive utilization of dentists' services among teenagers. *Journal of the American College of Dentists, 29*, 28–45.

Legler, D. W., Mayhall, C. W., & Bradley, E. L. (1972). Behavioral characteristics of disadvantaged adult patients. *Journal of Public Health Dentistry, 32*, 15–21.

Liss, J., Evenson, P., Lowey, S., & Ayer, W. A. (1982). Changes in the prevalence of dental disease. *Journal of the American Dental Association, 105*, 75–79.

Meadow, D., & Rosenthal, M. (1983). A corporation-based computerized preventive dentistry program. *Journal of the American Dental Association, 106*, 467–470.

Newman, J. F., & Anderson, O. W. (1972). Patterns of dental service utilization in the United States: A nationwide social survey. *University of Chicago Research Series 30*. Chicago: Center for Health Administration Studies.

Nyyssonen, V., Rahala, M., & Paunio, I. (1983). State and development of dental care for employees in Finland. *Community Dentistry and Oral Epidemiology, 11*, 209–216.

Parkinson, R. S., et al. (1982). *Managing health promotion in the workplace: Guidelines for implementation and evaluation*. Palo Alto, CA: Mayfield.

Richards, N. D. (1971). Utilization of dental services. In N. D. Richards & L. K. Cohen (Eds.), *Social sciences and dentistry—A critical bibliography*. The Hague, Netherlands: Federation Dentaire Internationale, Sijthoff.

Stewart, P. H., & McLenaghan, J. E. (1957). Public health and dental programs in industry. *Journal of the California State Dental Association and Nevada State Dental Clinic, 33*, 12–21.

Striffler, D. F., Young, W. O., & Burt, B. A. (1983). Dentistry, dental practice and the community (3rd Ed.). Philadelphia: Saunders.

Young, M. A. (1971). Dental health education of adults. In W.D. Richards & L.K. Cohen (Eds.) *Social Sciences and Dentistry*. The Hague: Sijthoff.

11

Occupational Safety and Health Hazards and the Psychosocial Health and Well-Being of Workers

JEANNE M. STELLMAN and BARRY R. SNOW

Occupational diseases have been known from ancient times when pyramid builders were diagnosed as having contracted silicosis. The first identified human cancers were among Czechoslovakian uranium miners. It was industrial carnage, rather than the need for pure food and drugs, that spurred Upton Sinclair to write *The Jungle*, which ironically led to the establishment of the first pure foods legislation. But almost another half-century passed, before the Occupational Safety and Health Act, promulgated in 1970, ensured that Americans would be guaranteed a workplace "free from recognizable hazard."

Despite the legislative advance, more than one and one-quarter million people are injured and thousands are killed in the workplace annually, according to estimates made by the federal government (Bureau of National Affairs, 1981). In addition, the vast majority of occupational diseases remain undiagnosed and their impact on the nation's health is not well estimated. This lack of recognition is due in part to the chronic nature of many occupational diseases; moreover, professionals, including behavioral scientists, are not usually trained to recognize diseases or dysfunctions related to the workplace setting.

In this chapter we will provide an overview of the extent and range of occupational hazards that exist and their potential health effects. The approach used will be a "whole person approach", which will emphasize the importance of considering both the physical environment of work and the psychosocial environment as determinants of occupational health and well-being. Specific examples will be drawn from two major employment sectors: the office environment and health care industries. The neglected area of women's occupational health will also be considered. Possible preventive strategies for reducing or eliminating occupational health hazards will be discussed as well.

EXTENT OF OCCUPATIONAL ILLNESS AND INJURY

In 1980, the US Department of Labor recorded more than 1.25 million cases of reported occupational injuries (Bureau of National Affairs, 1981). According to the National Safety Council, more than 245 million workdays were lost due to workplace injury. The cost to the nation of these injuries was estimated to exceed $30 billion dollars (National Safety Council, 1981). These numbers represent only a fraction of the actual incidences, since many injuries are never reported to the state and federal agencies that keep the records.

Estimates of the extent of occupational illnesses are still more difficult to make. Some estimates are, however, available. In 1980, the Department of Labor reported that during 1977 there were 62,366 illnesses, which included cancers, respiratory diseases, skin ailments, and mental illnesses, among others (Root & Daley, 1980). Similar to the data on occupational injury, these data underestimate the actual incidences of occupational illness, since the majority of occupational illnesses are neither diagnosed nor reported. For example, during 1977, only 243 malignant neoplasms were reported. This is in stark contrast to the estimates made by Bridbord et al. (1978) that between 30 percent and 50 percent of the total cancer burden can either in whole or in part be attributed to occupational causes.

Similarly, the Department of Labor recorded 4,292 illnesses attributable to radiation exposure, while the National Academy of Sciences has estimated that a minimum of 614,000 people are occupationally exposed to significant doses of ionizing radiation annually (National Academy of Sciences, 1972). These figures suggest that workers in a wide range of settings may be exposed to occupational health hazards.

In Table 11.1 we have provided an outline of the major causes of occupational injury and illnesses and typical occupations where they may be found. It can be seen that hazards are not limited to any one group of occupations but can be found in virtually all types of work and worksites. Occupational hazards are thus relevant to virtually all the nearly one-hundred million men and women who are gainfully employed in the United States. The pervasiveness of occupational hazards contrasts with the commonly held notion that only workers in heavy industries are exposed to health hazards on the job.

Table 11.1. Overview of Occupational Health Hazards and Examples of Associated Work Settings

Hazards	Examples of Typical Work Sites
Chemical hazards	
Carcinogens (e.g., asbestos, benzene, radon, nickel)	Chemical industry, hospitals, metal mining and smelting, welding, agriculture
Pulmonary toxins (e.g., cotton dust, silica, cigarette smoke, indoor air pesticides)	Textile industry, chemical industry, coalminers, construction workers, office workers
Reproductive hazards (e.g., solvents, lead pesticides)	Health care workers, chemical workers, textile workers, laboratory technicians, agricultural workers, microelectronics industry
Skin irritants and sensitizers (e.g., formaldehyde, spices, dyes, metals, photographic chemicals)	Food handlers, health care workers, office workers, household workers
Physical hazards	
Ionizing radiation (e.g., X-rays, alpha, beta, and gamma rays)	Health care workers, nuclear power plant workers, airline flight crews
Non-ionizing radiations (e.g., microwaves)	Radar operators, diathermy machine operators, food processors
Noise and vibration	Airport workers, factory workers, computer operators, woodworkers, power tool operators
Heat and cold	Meat handlers, bakery workers, glassworkers, smelter and mill workers
Safety hazards	
Electrical hazards	Electricians, health care workers, office workers
Fire	All workers
Biomechanical stressors (e.g., lifting, uncomfortable working position)	Video display terminal operators, health care workers, truck drivers, hand tool operators
Slipping and falling hazards	Construction workers, cleaners, and maintenance workers
Psychological hazards (e.g., work overload, organizational structure, job insecurity, interpersonal relationships)	All workers

INTERACTION OF PHYSICAL AND PSYCHOLOGICAL HAZARDS

Any model of occupational health must consider both physical and psychological factors in the environment. This perspective is necessary because an individual worker is likely to be exposed to several of the hazards described in Table 11.1. A hospital worker may, for example, be exposed to hazards from infectious agents, to anesthetic gases and to the stress of dealing with sick patients within a rigid hierarchical structure. A coal miner may be exposed to the hazards of silicacious coal dusts and to working in uncomfortable or even physically dangerous surroundings, while at the same time facing the threat of potential unemployment and working at intrinsically dissatisfying tasks.

It should also be noted that occupational health hazards are not limited to blue-collar or service workers. White-collar workers, such as office workers, can also be subjected to psychological and physical hazards. Automation of the office has, for example, introduced the widespread use of video display terminals, which can cause eyestrain and eye irritation, musculo-skeletal complaints, as well as increased job dissatisfaction and feeling of irritation and anxiety (Stellman et al., 1984).

Not only is it likely that individuals will be exposed to multiple hazards while on the job, but it is also likely that these hazards will interact with each other and have a greater health effect on the employee than they would if they were to occur alone. This synergism has been observed by House, for example, in his studies of factory workers (House, et al., 1979). House found that as the number of physical exposures, such as dusts and fumes, increased, the relationship between psychological stresses and signs of physical disease became stronger. The impact of psychological stressors was, however, minimal in work settings where physical stressors were negligible.

Psychological stressors can also alter the effect of physical workplace hazards. Various studies have shown that the physical effects of noise can be exacerbated when there is a degree of unpredictability about when the noise occurs (Glass & Singer, 1972). Similarly, Moos (1981) has suggested that psychological stressors may increase the impact of low levels of workplace contaminants. These levels may have previously been regarded as too minute to have a significant impact on worker health. In addition, another way for physical and psychological factors to interact must be considered: in essence, this mechanism refers to the principle that physical occupational health hazards can inherently be viewed as potent psychological stressors on the job (Snow, 1982). Recent data collected after the Three Mile Island incident, for example, indicate that workers at nuclear power plants who did not experience crises were also more likely to report various physical and psychological complaints after the TMI event, because of the heightened awareness of workplace hazards (Kasl, 1981). These effects are not considered when psychological descriptions of the workplace use a narrow focus, which includes only those factors that describe the type of work being performed.

Finally, a comprehensive model of occupational health must, therefore, also consider the psychological and behavioral effects of physical workplace hazards. These effects can be measured in the form of performance deficits and other signs of neurotoxic hazards. This perspective broadens the range of outcome variables beyond the mental health and disease end points traditionally used. Indeed, the relatively new field of behavioral toxicology seeks to define the extent of this psychological effect (Weiss, 1983).

We have outlined five possible ways whereby physical and psychological factors may have an impact on worker health and well-being. Despite the important nature of this interaction, little empirical work has been done to evaluate these relationships. The next section of this chapter describes three work areas where such an integrative approach is especially needed.

HAZARDS IN HEALTH CARE: A SPECIFIC EXAMPLE

More than seven million people were employed in health care facilities in the United States in 1981. Employment in the health care sector represented approximately 8 percent of the total employment in the United States. (US Bureau of Census, 1973) Evidence gathered from investigations of workers and hospital working conditions suggests that many jobs in the hospital are potentially hazardous.

Among the potential health hazards in hospitals are several recognized human carcinogens, including ionizing radiation and asbestos. Highly mutagenic and reactive substances, which have been found to be carcinogens in laboratory animals, are also present. Ethylene oxide, a gas used for room temperature sterilization, is an example of such a substance. Substances and conditions toxic to normal human reproduction, such as anesthetic gases and some infectious agents, can be expected to be found in most hospitals. Table 11.2 describes some of the occupational health hazards in hospitals.

Epidemiological and clinical data support the hypothesis that some types of hospital work can be hazardous. Radiologists were among the first occupational groups to have been found to be at risk for occupational cancer. Excess leukemia in this group was observed and attributed to ionizing radiation exposure (March, 1944). Cohort studies on another at-risk group— nurse anesthetists and anesthesiologists—have found an elevated rate of spontaneous abortion, congenital abnormalities in offspring, and hepatic and renal disease. Some studies, moreover, have suggested that there is excess leukemia and lymphoma in this occupational group, while other studies have found no elevated risks for cancer (Cohen et al., 1974; Edling, 1980). Additional cancer risk may also derive from halogenated ether anesthetics, which have been found to be mutagenic in in-vitro assays (Low, 1979). Increased mutagenic activity in the urine of nurses administering anticancer chemotherapeutic agents has also been documented, indicating the possibility of an inadvertent occupational exposure to these potent drugs, albeit at low doses in comparison to the patient (Falck et al., 1979). Industrial hygiene work by our group has shown that the chemotherapeutic agents can be spread to work and body surfaces during administration (Stellman et al., 1984).

Table 11.2. Hazards of Health Care

Examples of Hazards	At-Risk Groups
Infection	
Viral hepatitis	Kidney dialysis personnel, dentists and dental technicians, clinical laboratory workers and those in contact with drug abusers
Tuberculosis	Prevalent in inner-city or impoverished areas or among some immigrant groups; therefore, associated workers at risk
Herpes simplex virus	Can come from contact with infected sputum, especially tracheotomy patients
Rubella virus (German measles)	Potential problem in pediatric wards
Skin disorders	
Nail infections	Dishwashers, nurses' aides, and others who must keep hands wet for prolonged periods
Allergic (contact dermatitis)	Food handlers; aides and nurses handling medications
Dermatitis from irritation	Those handling disinfectants and soaps
Chemical hazards	
Sterilizing agents (ethylene oxide, formaldehyde, furfuraldehyde)	Maintenance and supply workers, some professional staff
Anesthetic gases	Operating and recovery room personnel
Laboratory chemicals	Laboratory workers
Chemotherapeutic agents	Oncology nurses
Physical hazards	
Microwave radiation	Diathermy machine operators, operating room personnel
Ionizing radiation	X-ray technicians and radiologists, cancer nursing personnel
Heat stress	Laundry workers
Safety hazards	
Back injuries, puncture wounds, cuts, abrasions and injuries, electrical shocks	Patient care personnel, nursing personnel, laundry workers, maintenance workers, equipment operators

Source: Adapted from Stellman (1978)

* Although many hospitals and other facilities take precautions to isolate infected patients, there will still be contact with undiagnosed cases. In many instances, routine handling of potentially infectious objects and patients may be inadequate.

Health care workers are also exposed to various psychological stressors. These stressors may be present in the specific nature of the job and in the organizational structure in which they work. Examples of these stressors include low levels of staffing, which present workers with unacceptably high levels of work and can limit their ability to deliver optimal care to their patients. Factors inherent in the organizational hierarchy, which impose restraints on independent decision-making authority of professionally trained health care workers, are also a source of stress.

The nature of work, which requires the delivery of services to people in need on a regular basis, has also been associated with a syndrome of psychological and physical exhaustion known as "burnout" (Maslach, 1976). Recent studies have indicated that workers in these settings are likely to report high levels of such symptoms as fatigue, insomnia, depression, irritability, anxiety, cynicism, gastrointestinal disturbances, headaches, and other complaints (Freudenberger, 1974).

The health care setting can present a multitude of safety hazards to health care workers as well. These hazards include fire, explosion, electrical shock, and slips and falls. Puncture wounds from needles and sharp instruments occur commonly and increase the risk of infection among the injured workers. A common injury associated with a safety hazard in health care does not derive from an artifact of the environment, however, but rather from unaided patient lifting. Lewy reports data on accidents that he has observed at a major medical center: contusions and bruises, 25 percent; back injuries, 16 percent; lacerations, 12 percent; strains and sprains, 12 percent. Lewy also found that nurses account for 60 percent of the reported incidents, but only represent 33 percent of the hospital work force (Lewy, 1981). This may be related to the greater likelihood of nurses seeking medical assistance than the lower-level staff who may not have the opportunity to leave their jobs readily and visit the employee health services. It may also be related to the fact that patient care and patient services areas may be at higher risk for injury.

Unfortunately, there has been little or no research on the interaction between the various factors discussed above. Furthermore, most of the research has been directed toward higher level professionals so that minimal data are available on the tens of thousands of hourly employees in health care industries.

HAZARDS IN OFFICE WORK: A SECOND EXAMPLE

Forty-two percent of the U.S. labor force—approximately 45 million people—work in office buildings. Although the extent of exposure to toxic chemicals and to physical hazards is in general more limited than in health care or in heavy industry, the office worker can be subjected to various physical, chemical, safety, and psychological hazards. These have been extensively reviewed elsewhere (Stellman & Henifin, 1984).

Recent work by our group has helped to further define specific physical factors that are related to office worker health and well-being. These factors, which derive from a cross-sectional study of more than 2,000 office workers in four different workplaces, are summarized in Table 11.3. They include noise, air quality, extensive machine use, limited privacy, poorly designed workstations, machine-paced work, and work which is meaningless or provides little opportunity to use one's skills and discretion. Of particular interest is the fact that indoor air quality was found to be highly correlated both to symptoms of upper respiratory tract distress and to satisfaction with the office environment. Co-worker and supervisor support were similarly significant and strongly related to outcome measures of health, well-being, and satisfaction. Noise, moreover, was found to be the strongest univariate correlate to job satisfaction in each of the four workplaces studied. These findings highlight the need discussed earlier to consider a wide range of both psychological and physical disease outcomes for both psychological and physical stressors.

The results of the multivariate analyses of these physical and psychosocial conditions on the health and well-being of the workers who participated in the study add more evidence to the need for considering both physical and psychological factors in a workplace analysis. A multiple regression analysis showed that both the physical and the psychological factors were highly correlated to the health outcomes studied. For example, when levels of irritation, anxiety, and fatigue, as measured by reliable self-reported indices, were assessed with respect to both the physical and psychosocial factors, it was found that poor air quality was the single most important working condition associated with these reports. The strength of the correlation observed for these physical factors was as great as that observed for organizational factors, such as hostility from either co-workers or the public or with perceptions of a good job future.

Similar findings were obtained for other outcomes. The most important physical factors, which related to the study participants' overall job satisfaction, were found to be *ergonomic comfort*, defined as responses to a scale consisting of items on convenient furniture and worksurface arrangements, as well as chair height and adjustability (reliability = 0.78) and overall pleasant appearance of the workplace. Psychological factors, such as supervisor support and the opportunity to learn new things, were also strong correlates of this measure.

The variability of the outcomes measured could be equally and strongly attributed to both the physical and psychological stressors that were surveyed. These results argue strongly for the whole person approach presented earlier.

WOMEN'S OCCUPATIONAL HEALTH: A NEGLECTED AREA

More than half of the adult female population was employed in 1982, and indeed, women represented almost 40 percent of the total labor force in the

Table 11.3. Combined Effects of Psychological and Physical Working Conditions among 2074 Office Workers

Health Effect	Combination of Factors Found to be Most Related to Health Effects			
	"Psychological"	Correlation (r)	Physical	Correlation (r)
Satisfaction with office	Good job future	.37	Pleasant appearance	.53
	Supervisor support	.30	Poor air quality	-.53
	Hostility	-.28	Privacy	.52
	Work "makes sense"	.26	Noise	-.51
	Co-worker support	.18	Ergonomic comfort	.47
	Repetitious work	-.14	Windows	.25
Job satisfaction	Good job future	.47	Pleasant appearance	.39
	Supervisor support	.42	Ergonomic comfort	.38
	Decision latitude	.32	Lighting too bright	-.27
	Repetitious work	-.31	Noise	-.27
	Work "makes sense"	.31		
	Able to take break	.31		
	Learn new things	.27		
Anxiety	Hostility	.19	Poor air quality	.25
	Supervisor support	-.16	Ergonomic comfort	-.23
	Ease of finding new job	-.14	Noise	.21
	Decision latitude	-.11	Difficulty communicating with co-workers	.19
Fatigue	Hostility	.32	Poor air quality	.30
	Repetitious work	.18	Ergonomic comfort	-.29
	Able to take break	-.18	Noise	.27
	Ease of finding new job	-.11		

Source: Adapted from Stellman et al. (1984).

United States (U.S. Department of Labor, 1982). It is thus natural that the growing awareness of workplace health issues and the large presence of women in the paid work force should make workplace environment health issues of direct relevance to the majority of women in the United States. This is particularly true since much of the research, training, and occupational medical efforts tend to be focused on male-dominated, high-risk industries.

Despite the major growth in employment of women workers, women by and large still hold the jobs traditionally associated with women's work. In the blue-collar sector they are employed in the textile industries and in the assembly of small machines and electronics. They are heavily engaged in health and other service industries. They are waitresses, housemaids, child-service workers, and retail clerks. And, of course, clerical work on all but the managerial levels is the major employment sector for women workers.

Several comprehensive reviews of the health hazards of women workers are available (Hunt, 1979; Stellman, 1978). Table 11.1 includes many of the chemical and physical hazards associated with the major employment areas of women's work. These hazards include potential exposure to cancer-causing chemical and physical agents, such as ionizing radiation and formaldehyde in jobs such as oncology nursing and laboratory work. They include exposure to infectious agents, such as Hepatitis B virus, in patient contact and clinical laboratory health care work. They involve biomechanical stresses, such as wrist and arm injury from assembly and sorting jobs. And they involve accidental injury, such as back injury from lifting in health care, cleaning, or waitressing.

In addition to hazards associated with toxic chemicals and physical agents, a major occupational health problem, indeed perhaps the major occupational health problem, for women workers is the effects of stress. Women workers are particularly likely to be exposed to stressful conditions both because of the nature of the work that they do and because of the multiple social roles that the majority of women workers fulfill.

There are many sources of stress for most women workers. These include dead-end employment patterns; routine, boring work, often with underutilization of skills; burnout in jobs involving direct care of patients, clients, children, and so on; low pay; long hours, particularly in combination with family responsibilities; role conflicts associated with multiple roles; sexual harassment; and physical hazards (Stellman, 1978).

The relationship between stress factors on the job and the lifestyle risk factors for chronic disease, such as smoking and drinking, is still not established but there is mounting evidence that women workers in stressful occupations may be at particular risk for engaging in this practice as part of a complex reaction to their work and home roles (Stellman & Stellman, 1980). Such interactions will, of course, have profound adverse implications for the health and well-being of women in the paid work force.

It is extremely important to include full consideration of the multiple roles and responsibilities of women workers. Such considerations should also include investigation of the presence or absence of social support systems, such as child care, flexible working hours, and so on, and their role as stress mediators. The reader is referred to other sources for additional in-depth analysis of the problem (Piotrkowski, Stark, & Burbank, 1983).

FUTURE DIRECTIONS

The previous portions of this chapter focused on a detailed analysis of the environmental hazards that affect the psychosocial well-being of employees. These hazards have been associated with the range of chronic diseases that are the major causes of morbidity and mortality in the United States today. Thus the public health relevance of occupational health hazards in the 1980s can be considered analogous to the importance of infectious agents as a public health concern in earlier decades in the industrialized world. Nevertheless, professional attention does not place the same priority on environmental control of health and safety hazards as were accorded to infectious agents. Instead, person-centered strategies, which stress alterations in health habits and lifestyles thought to be associated with the diseases, are emphasized rather than control of hazardous agents at their source.

Smoking-cessation and stress management programs are two examples, which seem to be straightforward and beneficial health promotion strategies. Taken alone, however, they are in general not reflective of the "whole person" approach to health and well-being, which places emphasis on the need to consider how personal lifestyles themselves may also be affected by environmental factors. Smoking, drinking, and eating habits can be viewed in whole, or in part, as maladaptive responses to stressors that serve as the focus of intervention efforts. This section of the chapter reviews the clinical, research, and educational projects that must be pursued to further develop this approach.

Clinical

Various workplace modifications can be introduced to improve the psychosocial well-being of workers. These strategies emphasize the redesign of jobs and work practices so that worker exposure to toxic substances and to physical and psychological stressors is either reduced or eliminated. A specific example of a practical set of workplace interventions is the elimination of exposure to anticancer chemotherapeutic agents among workers mixing and administering these medications. These drugs are potent, toxic substances to which nonmedicinal exposure should be strictly controlled and all unnecessary body contact or inhalation eliminated. Evidence shows, however, that health care personnel are exposed to the drugs (Stellman, 1984).

Specific environmental control technologies, such as the use of protective gloves, glasses, and laboratory coats, can be used to reduce the exposure to the hazard. Administrative measures that focus on the redistribution of work so that a worker has adequate time, training, and space to work effectively and safely can also be taken (Women's Occupational Health Resource Center News, 1983).

There is a need for empirical work as well. Behavioral scientists can play a major role in studying the behavioral correlates of changes such as these. Comparative studies of the efficacy of intervention with psychological stressors, physical stressors, and both types of stressors combined can be mounted. These approaches would consider the impact of the total workplace on the worker and thus may be more effective than other techniques, which limit their focus to solely the physical or psychological hazards, or may not consider the workplace at all, but only the lifestyle behavior of the worker.

Research

Research into the multifactorial nature of chronic disease must continue. Of particular importance is the inclusion of work-related factors and occupational histories into major ongoing research. Extensive investigation into the relationships between job demands, occupational hazards, and other work-related factors and lifestyle factors, such as smoking, alcohol and drug abuse, and nutritional status, is of urgent importance. This urgency is demonstrated by the considerable amount of information currently available about the relationship between smoking and cancer and other diseases without the concurrent knowledge of the contribution of occupation. Such knowledge is essential for the development and application of prevention programs.

In addition, specific research efforts are needed in the following areas: workplace stress and mental and physical health; health hazards in health care and office work; and targeted research on the effects of chemicals in use in female-dominated industries, such as microelectronics, laundry and dry cleaning, meat handling, and sewing and stitching.

Basic and applied research into biomechanics and workplace design in the office and in industry, particularly in the areas of video display terminals and in small hand tools, are research priorities. Standards for personal protective equipment for women in nontraditional jobs and in traditional work require immediate attention.

The tools to conduct some of this basic and applied research also merit professional concern. Federal, state, and local governments have long been concerned with development of recommended public health forms (e.g., birth and death certificates). There is an urgent need for further form development, which will more accurately reflect occupational history, particularly of women who are often recorded as "homemakers" or "housewives". Registration of birth defects, possibly at entry into the school system, together with parental work history is needed since the majority of developmental defects are not obvious at birth. Such registration could provide the database for investigation of reproductive hazards in the workplace.

A research program into the linkage of records currently maintained by federal and state programs on places of employment, the chemical and physical agents in use, and the health of employees is urgently needed. Such data are contained within various agencies, such as the Social Security Administration (employee/employer links and vital status), State Workers' Compensation (accidents and injuries), Environmental Protection Agency (substances manufactured), NIOSH (hazards priorities), Veterans Administration (disabilities among veterans who were also employed), and so on. Feasibility and pilot studies could establish and test potential methods of linkage. The linkage would provide a sufficient number of people to answer many of the research questions that are currently difficult to carry out because of the small number of subjects available for study and the difficulties of cohort identification.

Education and Training

To provide for the needs of behavioral medicine in industry, targeted programs to increase the knowledge and competence of general practitioners in medicine, nursing, the behavioral sciences, social work, engineering, and chemistry are needed. Grant support for curriculum development, academic awards, and special stipends for skills improvement in the physical and psychological aspects of occupational health are potential mechanisms for this training.

Specialized programs to develop professionals, such as the NIOSH supported Educational Resource Centers, must be expanded. Additional programs on the high school and vocational school level should be encouraged through specific grant programs so that teachers and future workers become aware of potential hazards and how to minimize their effects. Many personal exposures can be avoided through appropriate personal and organizational work practices. In addition, such information may allow students to choose career paths more intelligently and to become informed consumers of health care.

Just as record linkage is needed for research database development, skills linkage in educational institutions must be fostered. Specific programs, which seek to combine expertise and interests of business schools, schools of public health, engineering schools, and schools of social work to focus on the health and well-being of workers, are needed. A nation that trains its business administrators in ergonomics and stress, its chemists and engineers in toxicology and industrial hygiene, and its social and health professionals in the recognition and treatment of occupational disease may well be able to avoid repetition of past mistakes and safeguard the health of future workers.

CONCLUSION

In conclusion, it can be seen that much remains to be done with regard to the broadening of the concepts and practices of behavioral medicine in industry. Careful examination of the data and recommendations reveals, however, that implementation of the programs discussed here would be of significant public health benefit. A "whole person" approach to behavioral medicine empirical studies and applied interventions would no doubt increase their effectiveness. Similarly, increased knowledge about the hazards, increased training, and proliferation of information would serve to improve health in the workplace as well as to make the job of the behavioral scientist more easily accomplished. Since work and the workplace are in fact the central focusing factor in industrial life, determining the places we live, the hours we keep, and the structure of our lives, attention to the multiple factors in the workplace and how they together interact on health and well-being is an essential requisite of any comprehensive behavioral science program.

REFERENCES

American College of Obstetricians and Gynecologists. (1977). *Guidelines on pregnancy and work.* NIOSH Contract 210–76–0159. Washington, DC: U.S. Government Printing Office.

Baetjer, A. (1946). *Women in industry: Their health and efficiency.* Philadelphia: Saunders.

Bridbord, K., Decoufle, P., Fraumeni, J. F., et al. (1978). *Estimates of the fraction of cancer in the United States related to occupational factors.* Bethesda, MD: National Cancer Institute, National Institute for Environmental Health Sciences, and the National Institute for Occupational Safety and Health.

Bureau of National Affairs. (1981). Occupational injury, illness rates for 1979–1980, tabulated by the Bureau of Labor Statistics. *Occupational Safety and Health Reporter, 11,* 500–504.

Cohen, E. N., et al. (1974). Occupational diseases among operating room personnel: A national study. *Anesthesiology, 41,* 321–340.

Edling, C., (1980). Anesthetic gases as an occupational hazard. A review. *Scandinavian Journal of Work, Environment and Health, 6,* 85–93.

Falck, K., et al. (1979). Mutagenicity of nurses handling cytostatic drugs. *Lancet,* 1250–1251.

Freudenberger, H. J., (1974). Staff burnout syndrome. *Journal of Social Issues, 30,* 159–165.

Glass, D. C., & Singer, J. E. (1972). *Urban stress: Experiments on noise and social stressors.* New York: Academic Press.

House, S. J., Wells, J. A., Landerman, L. R., McMichael, A. J., & Kaplan, B. H. (1979). Occupational stress and health among factory workers. *Health and Social Behavior, 20,* 139–160.

Hunt, V. R. (1979). *Work and the health of women.* Boca Raton, FL: CRC Press.

Kasl, S.V., Chisholm, R. F., & Eskenazi, B. (1981). The impact of the accident at the Three Mile Island on the behavior and well-being of the nuclear workers. *American Journal of Public Health, 71,* 484–495.

Lewy, R. (1981). Prevention strategies in hospital occupational medicine. *Journal of Occupational Medicine, 23,* 109–111.

Low, S. (1979). Mortality experience among anesthesiologists. *Anesthesiology, 51,* 195–199.

March, H. C. (1944). Leukemia in radiologists. *Radiology, 43,* 275–8.

Maslach, C. (1976). Burned out. *Human Behavior, 5,* 16–22.

Moos, R. H. (1981). *Creating healthy human contexts: Environmental and individual strategies.* Los Angeles, CA: American Psychological Association.

National Academy of Sciences. (1972). The effects on populations of exposure to low levels of ionizing radiation. Washington, DC: National Research Council.

National Safety Council. (1981). *Accident Facts.* Chicago: National Safety Council.

Piotrkowski, C., Stark, E., & Burbank, M. (1983). Young women at work: Implications for individual and family functioning. *Occupational Health Nursing, 31*(11), 24–29.

Root, N. & Daley, J. R. (1980). Are women safer workers? A new look at the data. *Monthly Labor Review,* 3–10.

Snow, B. R. (1982). Safety hazards as occupational stressors: A neglected issue. *Occupational Health Nursing, 30,* 38–41.

Stellman, J. M. (1978). *Women's work, women's health: Myths and realities.* New York: Pantheon.

Stellman, J. M., & Andrews, L. R. (1982). The assessment of toxic exposures in the workplace. *Toxic Substances Journal, 2.*

Stellman, J. M., Aufiero, B. M., & Taub, R. N. (1984). Assessment of potential exposures to antineoplastic agents in the health care setting. *Preventive Medicine.*

Stellman, J. M., & Henifin, M. S. (1984). *Office work can be dangerous to your health.* New York: Pantheon.

Stellman, J. M., Klitzman, S. K., Gordon, G. C., & Snow, B. R. (1984). *The health and well-being of video display terminal operators.*

Stellman, S. D., & Stellman, J. M. Women's occupations, smoking and cancer and other diseases. *CA, A Journal for Clinicians, 31,* 29–43.

U.S. Bureau of the Census. (1973). *Census of population: 1970 detailed characteristics, Final Report PC(1)–D(1).* Washington DC: U.S. Government Printing Office.

U.S. Department of Labor. (1982). *Employment and earnings, 29,* 5. Washington, DC: U.S. Government Printing Office.

Weiss, B. (1983). Behavioral toxicology and environmental health science: Opportunity and challenge for psychology. *American Psychologists, 38,* 1174–1187.

Women's Occupational Health Resource Center. (1983). *Handling Chemotherapeutic Drugs, 5,* 2. New York: Columbia University School of Public Health.

12

Cancer Control at the Community Level: The Modification of Workers' Behaviors Associated with Carcinogens

SANDRA M. LEVY, BILL HOPKINS, MARGARET CHESNEY,
KNUT RINGEN, PETER NATHAN,
and VERNON MacDOUGAL

In any discussion of worker behavior relevant to occupation and cancer, two fundamental questions must be addressed: Does workplace exposure to carcinogens affect cancer incidence? If so, does behavior and behavioral change make any difference related to this cancer risk? The affirmative answer to the first question in human populations has been provided through epidemiological examination of the occupational cancer evidence; the answer to the second question is significant only if the answer to the prior question is affirmative.

PROPORTION OF CANCER INCIDENCE ATTRIBUTABLE TO OCCUPATIONAL EXPOSURE

Doll and Peto (1981) discussed a number of "avoidable" causes of cancer, among which they list occupational exposure to carcinogens. Using a conservative approach, these authors estimated an overall 4 percent of cancers in the United States attributable to occupational exposure, with attributions of varying proportions for separate cancer sites. They concluded their report noting that, although 4 percent overall is a fairly small proportion of cancer deaths directly attributable to occupational exposure, this amounts to approximately 8,000 deaths in the United States every year.

In addition, occupational cancer tends to be concentrated among particular groups of individuals (such as minority workers) and "such risks can usually be reduced, or even eliminated, once they have been identified" (Doll & Peto, 1981, p. 1245). Their conclusion sounds optimistic, since environmental carcinogens are, or should be controllable. Nevertheless, several complex issues, which can affect such a reduction, must be addressed through careful, systematic research. To date, the issues have not been examined.

SETTING THE INVESTIGATIVE PARAMETERS

In a 1977 NIOSH document, "The Right to Know," and in a 1977 OSHA publication, "Informing Workers and Employers about Occupational Cancer," only vague reference is made to "notification." But how to inform effectively (so that the information is perceived and believed, and the individual is motivated to act in the most appropriate, health-enhancing fashion) is not even addressed as a potential issue. Nor are the behavioral effects of being informed of risk status discussed in any substantial way. The last section of this chapter will address behavioral consequences of past exposure in former workers being screened for cancer incidence.

For workers who are currently in potential or actual hazardous environments, other issues must also be considered. For example, OSHA and the relevant unions have been concerned with environmental controls of toxic substances, but the behavioral environmental implementation of protection has never been systematically assessed: that is, through Federal regulation, allowable exposure levels have been determined—and to varying degrees enforced—but the effective reduction in exposure linked to specific worker practices has not been addressed.

Presently, 18 agents or chemical processes are known to be associated with occupational cancer. Examples of known carcinogens are asbestos and the arsenic compounds. In addition, literally hundreds of other agents are suspected carcinogens or promoters of cancer initiation. One rational way to select hazardous substances and associated occupations, which should be the target for such behavioral-environmental studies, would be to select cancer sites with the highest proportion of occupationally related mortality, and then study the behavior of workers exposed to agents known to be associated with higher risks for cancer in those sites. Referring to Doll and Peto's (1981) data, one might then select cancers of the mesentery and peritoneum, plura, nasal sinuses and other respiratory sites, lung, bladder, blood (leukemia), and liver. Major agents associated with these sites include substances such as asbestos, the arsenic compounds, aromatic amines, and the widely used industrial solvent, trichloroethylene.

Whatever the agent chosen to which exposure is to be modified, the basic position is conservative. Sufficient evidence exists that at least some of these industrial substances, such as vinyl chloride and asbestos, are carcinogenic in humans, and others, such as formaldehyde, are probably so. The assumption here is that it is prudent to reduce exposure to the fullest extent in this range of agents.

The answer to the first question, related to increased cancer risk in occupationally exposed individuals, is clearly yes. The next question that should be considered is whether the individual worker's behavior contributes to carcinogenic exposure. Of course, if it does not, then there would be no place for behavioral science in this research arena, nor would there be any role for public health efforts aimed at individuals or for educational shaping of individual responses.

However, a range of controls seems to exist in the workplace, extending from *engineering devices* to *specific work practices,* where as yet no engineering controls are available. Clearly, behaviors related to this range of controls would also differ. Where engineering controls are available, behaviors concerning proper maintenance, responses to emergencies and equipment failures, and so forth, would be the most relevant to consider. Specific, protective work practices would be particularly important whenever engineering controls allow workers to have frequent, direct contact with hazardous substances. For example, a 1978 special occupational hazard review of trichloroethylene (TCE), published by NIOSH (1978), concludes by stressing the importance of improved work practices. Among preventive measures recommended were use of protective clothing, institution of practices for clean up of spills, establishment of good general housekeeping and sanitation procedures (e.g., food consumption should be avoided in areas where TCE is handled), institution of practices for safe entry into confined spaces such as tanks, and long-term establishment of medical surveillance.

Related to various work practices, these behavioral procedures are typically arrived at by common sense and consensus. The practices themselves are rarely systematically measured (e.g., quantifying extent of behavior, with independent observers rating degree of perfection). Nor, in many cases, are the specific practices validly correlated with actual exposure.

Despite this lack of systematic behavioral measurement, however, we can still ask, Is there a problem in behavioral cooperation from the worker? Anecdotally and clinically, some disagreement exists. Some people report that there is no problem, management simply has the protective equipment installed, and then successfully educates the workers to use it.

On the other hand, others anecdotally report that an industry and worker compliance problem does exist. Few hard data on actual plant and worker violations in exposure levels and recommended practices are available, but observations have been made of improper equipment usage and total failure to use protective equipment in particular plants until "two or three days before we sent a hazard review team to their place. The workers told us that they never had used the equipment up until then."[1]

Our assumption is that noncompliance rates probably differ by industry,

---

[1]Personal communication, NIOSH official.

plant size, socioeconomic status of workers, and so forth, and our educated assumption is that there is a problem. We know from the general health promotion literature that people do not comply. For example, only a 40 percent compliance rate with long-term, prophylactic, health-maintenance regimens has been recently reported in the review of the compliance literature (Haynes, Taylor, & Sackett, 1979).

In essence, we do not have the systematic data on the extent of the problem, that is, the prevalence of noncooperation with optimal work practice. In fact, data do not exist to inform us about the extent of compliance with the work practices proposed, nor the correlation of practices with actual exposure levels. Does the worker behavior occur or not occur and to what extent (in what workers, in what occupations, etc.) does it occur? If the behavior occurs, how can we increase its occurrence, what difference does it make in the rate of exposure reduction? Reasons for this lack of empirical data include the methodological difficulty of conducting carefully controlled behavioral studies in the workplace; the historical antipathy between labor and management and the consequent impediments to collaborative research in the industrial setting; and the technical difficulty of measuring personal exposure levels of toxic agent absorption to measure accurately the effects of worker behavior modification. But these basic research questions could be addressed by systematic investigations in this area.

BEHAVIORAL LEVELS OF EXAMINATION

There is an entire range of issues that could be investigated in the area of worker behavior and cancer risk, but perhaps we could view them on two levels: (1) Public health and effective communication levels; and (2) The behavioral level, that is, work practice enhancement and alteration of other, detrimental discretionary behaviors by various interventions, linking that change to risk reduction.

At the first level of communication research, we build worker educational programs. But public educators must build on knowledge of their audience, then use techniques that will effectively influence those listeners. They have to understand the social setting, what is feasible in terms of cost/benefit for the listener, the motivation and values inherent in population subgroups, the range of possible responses to an educational message, and so forth.

Focusing on the second, behavioral level, several other factors, which probably affect the quality of worker response to educational and supervisory input, must also be understood to develop effective worker training strategies. For example, both the historical cohort to which a worker belongs, and the time on the job at a particular occupation could affect willingness to change habitual patterns of behavior or could affect receptiveness to educational input. Other variables, such as ethnicity, general health status, race, and migratory status, may all potentially contribute to work practice variance in the industrial environment and must be understood in order to build more effective intervention strategies in these subgroups.

To date, only one study has systematically intervened to modify workers' behaviors in a high-risk environment. This research discussed below can be viewed as a prototype of behavioral intervention, superimposing behavioral shaping onto ordinary work-practice routine in an industrial setting.

AN EXAMPLE OF BEHAVIORAL PROCEDURES FOR REDUCING WORKER EXPOSURE TO CARCINOGENS: A RESEARCH CASE HISTORY

In August 1977, the National Institute for Occupational Safety and Health (NIOSH) issued a request for proposals entitled, "Behavioral Procedures for Reducing Worker Exposures to Carcinogens." The proposal had originally recommended work with both hexavalent chrome and styrene. During negotiations for the contract, it was decided that styrene alone would be the target substance, because carcinogens were usually regulated to exposures at the lower end of the range in which they could be measured. Therefore, any changes in exposures to carcinogens might not be measurable. In addition, styrene was known to be in use in manufacturing processes in many plants in the geographical study area. Styrene was regulated only to a level that allowed a wide range of measurable exposures, and researchers would likely encounter levels of exposure such that any systematic changes could be detected.

Manufacturing Processes

Manufacturing of reinforced, laminated plastic products, such as those manufactured from styrene-containing resins, typically consists of a series of operations. A mold that has the converse shape of the desired product is cleaned and waxed and then moved to the gelcoat sprayer, who sprays a mixture of pigmented polyester resin and styrene monomer onto the mold with the compressed air sprayer. The compressed air sprayer is similar to a paint sprayer and is constructed to mix a catalyst, such as methyl ethyl ketone peroxide (MEK-p), with the resin-styrene mixture as it leaves the gun.

When styrene is used as a diluent-reactant, it polymerizes with the resin after application to the mold. During this curing process, the mold is set aside to allow the gelcoat layer to harden. After hardening, the cured gelcoat is given a reinforcing lamination of fibrous-glass. The lamination is applied with a second spray gun that shoots a mixture of chopped fibrous-glass, resin-styrene mixture and catalyst. The operator of this machine is called the chop sprayer.

Immediately after application of the reinforcing lamination, additional reinforcement may be built into the part by the integration of wooden or metal members or woven fibrous-glass mats soaked in a resin-styrene-catalyst mixture. The reinforcement is typically bonded to the part with a light spray of the chopped fibrous-glass mixture.

In the next operation, workers using rollers, similar to those employed for painting, roll the newly applied lamination to remove gas bubbles from the mixture and to ensure that the resin and fibrous-glass are thoroughly compressed and mixed. These workers are called rollout people. The molds and parts are set aside to cure. After curing, the parts are removed from the mold, and the mold is inspected and repaired if necessary. The person who performs this latter operation is called the mold repair person.

Although other finishing operations may be performed, and plants differ with respect to floor plans, engineering controls, storage, and equipment, the above described steps are typical and characterize the industry.

Plant Mapping. While the project staff was becoming familiar with the manufacturing processes, the project industrial hygienist took numerous grab tube samples with a Bendix, Model 400 Gastec pump and Fink styrene detector tubes to identify the processes, jobs, and plant areas that involved relatively high exposures to styrene. If plants had functioning exhaust ventilation, the following results were relatively common in most plants.

High exposure areas, with momentary concentrations ranging from 110–280 ppm, were two kinds of spray booths in which the gelcoat mixture and the resin-chopped fiberglass mixture were sprayed onto the molds. These two processes apparently introduced a majority of the styrene into the plant because styrene was vaporized by the spraying process and came to evaporate at a high rate because the surface area for evaporation became large and the addition of the catalyst to the mixture produced heat.

The rollout and curing areas of the plant also yielded relatively high momentary levels of styrene, ranging from 70–170 ppm, resulting from the action of the catalyst. The two spraying jobs and the job in which the resin-fiberglass mixture was rolled out involved not only the greatest momentary exposures for workers, but also the greatest total time of relatively high exposure.

Several jobs, such as repairing molds and touching up blemishes in gelcoat surfaces, occasionally introduced relatively small quantities of styrene into the air. Many jobs, such as those involved in grinding flashings from parts, polishing parts, and moving parts from one area to another, introduced either no or negligible amounts of styrene into the plant. Workers in these jobs were exposed to styrene as a result of the ambient concentrations produced by the styrene-introducing processes described above. These momentary ambient concentrations typically ranged from 2–20 ppm.

Plant Surveys. During the familiarization visits in the plants, once high exposure jobs and processes had been identified, project staff observed how work was being done and speculated about ways in which job tasks could be accomplished so that workers would receive less exposure to styrene. At times, workers were asked briefly to try out a work procedure thought to be useful and to provide the staff feedback on the ease with which the procedure could be used, and whether any change in exposure could be subjectively noticed. At other times, grab tube samples would be collected to determine momentary styrene exposures as the workers carried out a procedure in their accustomed way and in the manner suspected to be more useful to reduce exposures.

Interviews with Workers and Management. During the plant familiarization visits, informal interviews were conducted with 38 workers and lead people and with 14 members of management at the various plants. During these interviews, project staff explained their interests and the reasons behind them, and the plant employees were also asked to make suggestions. These suggestions were sometimes tested for practicality, and momentary exposure differences were measured as described above.

One obvious suggestion, for example, was that workers should always engage the exhaust ventilation when carrying out the spraying or rollout jobs in a booth. Momentary grab tube samples indicated that this work practice could reduce exposures. Similarly, if workers were spraying or doing rollout work in a ventilated booth, momentary samples would differ considerably, depending on whether the worker stood on the upwind or downwind side of the part being produced.

The above searches produced a compendium of possible work practices that were then built into the workers' repertoire of behaviors during the period of this study. Examples of work practices and procedures are the following:

1. Exhaust ventilation should be turned on any time a worker is present in the work area. (Work areas were defined for each of the plants in relation to the ventilated booths in and near which nearly all spray, layup, and rollout work was done. This practice provided for dilution of styrene concentrations, removal of evaporated and vaporized styrene from the work area, and reduction of styrene introduced into the ambient plant air.)

2. Workers should position themselves with respect to parts and materials so that their breathing zones remained 18 inches or more from sources of vaporizing or evaporating styrene. These sources particularly included the nozzles of spray guns, containers of raw styrene, and parts on which there were curing resins. (Concentrations of airborne styrene in general decrease with distance from the source. Therefore, the greater the distance from the source to a worker's breathing zone, the less styrene that person might inhale.)

3. Workers should not allow styrene or styrene-containing resins to touch any body surfaces. (Styrene is readily absorbed through the skin. Rollout and layup workers, particularly, often worked with styrene containing resins with their bare hands. Reducing skin contact might reduce the amount of styrene absorbed into the body and also

reduce a proximate source of evaporating styrene, which could then be inhaled.)

4. Workers should always work on the upwind side of any source of airborne styrene. (Presumably, airborne exposures would in general be less when workers kept their breathing zones in the relatively clean air being drawn through the work area by the exhaust ventilation rather than in the styrene-contaminated air that would exist between the source of styrene and the ventilation exhaust.)

5. All booth filters should be in place. (If filters were omitted or improperly seated in their frames, the exhaust ventilation system could soon become sufficiently blocked by gelcoat, resin, and chopped fiberglass, and its efficiency would be reduced.)

Observational Definitions.   Eleven work procedures and 20 housekeeping practices were converted into observational definitions, and the definitions were revised until project staff could use them, while observing conditions and workers in a plant, and obtain a high degree of interobserver agreement on the occurrence and nonoccurrence of the specified events.

Research Methods

Subjects.   Some data were collected on 41 subjects at Plant A, 9 subjects at Plant B, and 10 subjects at Plant C. The workers ranged in age from 19 to 52 years and in duration of employment at a plant from 1 day to 15 years. All subjects participated voluntarily after being informed about the nature of the research and signing consent forms.

Data Collection Procedures

Behavioral Data.   Paid observers were trained to use observational definitions. This training consisted of familiarizing observers with the definitions and codes, and using videotape sequences of workers carrying out their jobs to familiarize observers with the behaviors as they occurred in plants. Observers practiced using the definitions and observational codes, practiced collecting data from videotapes, and practiced on-the-job rating until a 90-percent agreement (defined as total agreement on occurrences and nonoccurrences of specific behaviors) occurred between all observers and project management staff.

Training

Training consisted of the following elements:

1. Short meetings during which the subjects viewed the videotapes in which the recommended work procedures and housekeeping practices were presented and in which brief discussions of the procedures and practices were held;

2. On-the-job rehearsal of the procedures and practices;

3. On-the-job behavioral tests measuring the extent to which the workers engaged in the recommended procedures and practices;

4. Immediate feedback on the results of the tests;

5. Social approval and praise for good performances;

6. Opportunities to rehearse again and retake tests when a subject did not meet the criteria for passing.

Nine meetings for viewing tapes and discussing procedures and practices were scheduled at the convenience of the department manager and the workers in each of the plants. These meetings were usually held at the beginning of a shift or during coffee breaks. Meetings averaged about 20 minutes in length and the longest meeting required 30 minutes. All meetings were held in the lunch areas of the departments.

During a meeting, the trainer showed a videotape to the workers, stopping the tape to answer any questions. At the conclusion of the tape, the trainer asked the workers if they had any questions about the procedures and practices shown. Furthermore, the trainer encouraged the workers to discuss ways in which the procedures and practices might cause them problems and ways in which they could share responsibilities for housekeeping requirements. At the end of each session, the trainer gave the workers sheets of paper describing the criteria they would have to meet to pass the on-the-job tests. These criteria were to engage in the specified work procedures 100 percent of the time and to meet each of the specified housekeeping conditions at least 90 percent of the time. If a subject missed a training session because of an absence from work, a makeup meeting was held for this person during the next day of attendance at work.

The nine training sessions were spread over 4 weeks in Plant A, 4 weeks in Plant B, and 5 weeks in Plant C.

On-the-Job Tests.   On-the-job tests were given both as the last step of training, to ensure that workers had learned what was intended by the videotapes, and as a maintenance procedure, to ensure that they continued to use the procedures and practices once they had learned them.

Social Approval and Correction

From the beginning of training until the termination of data-collection in a plant, the trainer provided social approval whenever subjects were observed to be following the work procedures, or work area conditions were observed to meet criteria for housekeeping practices. The trainer was instructed to keep the approving comments informal and to use the vernacular of casual conversation in the plant. Examples of typical comments would be: "That's the way to stay upwind of your work!"; "I see you have on your gloves. That'll help keep the resin off your hands!"; and "You guys are doing a good job keeping the overspray off the walls. This place really looks neat!"

Whenever workers failed to engage in the recommended work procedures, the trainer corrected them. Corrective comments were again informal and not negative and they always included a statement of explanation. Typical comments would be: "George, move the mold over here and the exhaust will draw the fumes away from you better"; "You need to spray in this direction only so the fumes won't blow back up in your face"; and "The overspray is getting too deep on the floor. That will make it harder to turn the molds." The trainer was instructed to follow up on each corrective comment by visiting a worker again within a few minutes and provide approving comments if the suggested corrective action was being taken.

The trainer was instructed to adjust the frequency of approving and corrective interactions with workers throughout the course of posttraining data collection. The frequency should be low at the beginning of training because workers were responsible for few procedures and practices. The frequency should then gradually increase during training and remain high until the work procedures and housekeeping practices were occurring as desired. The frequency of approval and correction should then gradually decrease but should remain high enough to maintain acceptable use of the procedures and practices.

Several monetary incentives were also provided to each worker for co-operating with the research and for meeting certain criteria. These incentives were announced to the workers in advance and were paid to the workers in cash as soon as the specified event occurred.

Results

The training and motivation programs were significantly effective in changing the subjects' work procedures and housekeeping conditions in all three plants. This conclusion held, not only for the mean data, averaged for subjects, work procedures, plant areas, and housekeeping practices, but also for all individual workers and plant areas and for all the recommended practices and procedures. Thus it is quite safe to assume that the major objective of the research was met in every respect. Work procedures and housekeeping practices, selected for their relevance to reducing exposures to workplace toxic substances, can be readily and dramatically changed in prescribed directions.

In addition, these changes in work procedures and housekeeping practices were not transient. They were maintained in both Plant B and Plant C for more than 120 workdays with no systematic tendencies to revert to baseline levels. Although the research was not intended to address the question of necessary and sufficient causes for the long-term maintenance of changed worker behavior, it is likely that some of the relatively unique motivation methods were responsible for these prolonged effects. Instead of treating needed changes in behavior as a problem for one-shot training or education, this intervention provided not only training but also included ongoing feedback, social approval, and motivation. This approach is similar to many engineering approaches that not only install intended controls, but also build them and adjust and maintain them as necessary to produce the desired result. In this context, it is interesting that improvements in the use of the recommended work procedures and housekeeping practices continued to occur in all three plants long after training had been completed.

It is perhaps also significant that the approach to behavior control did not simply consist of the presentation of information about the suspected carcinogen and about general forms of work procedures and housekeeping practices. Rather, the objectives of the training and motivation methods were specific behaviors and conditions. Much of the training was carried out on the job. Many of the workers and their supervisors had been involved in the development of the recommendations and the workers were involved in developing the practical details of how they would carry out the recommendations. The recommendations were not selected by management and imposed on the workers or vice versa. Rather, they were developed jointly by the workers, management, and the research staff.

It is important that this approach to controlling exposures be viewed as a supplement to other control methods. Specifically, the research staff speculated that the procedures and practices particularly useful to reduce exposures were those that advantageously used existing engineering controls. Engaging exhaust ventilation, placing work to take advantage of exhaust airflow, and positioning oneself to take advantage of airflow and dilution were particularly important behaviors. All of these were supplemental to the engineering controls typically provided in this industry. However, it should also be noted that if these behaviors are as important as hypothesized, the engineering controls would be relatively less effective unless the proper behaviors also occurred.

It is perhaps axiomatic that the usefulness and importance of particular behaviors will vary with the physical properties of hazardous substances and their routes of possible entry into the body. It is also likely that behaviors having to do with use and maintenance of engineering controls and proper responses to emergencies and/or equipment failures will be relatively more important in those industries that have relatively complete engineering controls. Behaviors that are directly involved in production and handling of substances will be more important in industries in which engineering controls still allow frequent, direct contact with hazardous substances.

Methods for selecting those human behaviors that must be controlled to achieve good exposure control should still be viewed as tentative. The present research methods, including the identification of the areas of a plant and the processes that generated exposure, observations of ways in which people behave with respect to hazardous agents, the development of alternate behaviors that might reduce exposures, and the partial use of momentary samples to estimate the value of the alternate behaviors to reduce exposures, seem rational. But this process can certainly be improved through further investigation.

Other Lifestyle Contributors to Risk in Toxic Work Settings

The interaction of other lifestyle factors, such as smoking, alcohol, and dietary patterns, that could affect increased risk for some groups and could certainly be considered as outcome variables to be modified should be considered. Certainly, for cancer risk, smoking is the single most important risk factor, both alone and in synergistic or additive association with environmental carcinogens such as asbestos. (However, to avoid redundancy with other chapters in this volume, tobacco use in the workplace will not be further considered here.)

BEHAVIORAL IMPACT OF INTERVENTION PROGRAMS FOR HIGH-RISK EXPOSED COHORTS

The last section of this chapter will address psychosocial and behavioral consequences of past exposure, notification of risk status, and subsequent intervention to screen for early signs of cancer.

Approaches to notification of high-risk occupational groups is fraught with methodological, ethical, organizational, and economic difficulties that have as yet to be addressed adequately. Considerable concern has been raised about the behavioral impact of worker notification and intervention programs (Schulte & Ringen, 1984). These concerns derive from the recognition that people place an extremely high value on their health, and that cancer, in particular, strikes fear in people. But the concerns relate to more abstract problems as well, including the difficulty of conveying the concept of risk to the general public (risk is often equated with disease), and the possibility that individuals labeled to be at risk might become stigmatized in their communities.

The problem of worker notification is linked closely to many other areas of concern regarding technological change and environmental causation of disease, such as hazardous waste dumps and nuclear plant accidents. All of these areas have one common element: the acknowledgment that the problem through notification is *acute*, whereas the disease for which the people involved are found to be at risk is *chronic*.

In 1979, the National Institute for Occupational Safety and Health (NIOSH) and the Workers' Institute for Safety and Health initiated cooperation in the planning of possible high-risk notification and intervention programs. In the course of conducting occupational research, NIOSH has identified many cohorts whose participants may be at high risk, and was interested in the feasibility of conducting notification of results on a routine basis. The Workers' Institute, under a grant from the National Cancer Institute, was interested in developing model programs of intervention in high-risk cohorts subsequent to their notification that could be implemented by the labor movement in the future.

From the outset, it was recognized that certain organizational criteria would have to be met if a successful model approach were to be developed. *Foremost,* a bootstrap approach would be required to be acceptable within the limited resources available to labor unions. *Second,* the approach would have to be replicable and sufficiently simple so that the predominantly nonmedical staff members and officers of labor unions might be capable of adopting it. *Third,* the approach would have to be capable of addressing four fundamental flaws observed in previous notification/intervention attempts: failure to use existing networks of communication, particularly those of the appropriate labor organizations; reinforcement of attitudes and activities that perpetuate dependence on others to make vital, personal decisions in health maintenance; ignorance of the long-established fact that if behavioral control is justified, it will occur most effectively through peer groups removed from the hostility of the workplace and with family involvement and community support; and failure to enhance the ability of the worker and his or her family to manage the legal, financial, and psychological problems of lifelong surveillance, intervention, and treatment, and premature death (Samuels, 1980).

Given these considerations, in 1980 NIOSH and the Workers' Institute began three demonstration projects. Only one of these, concerned with an asbestos-exposed cohort in Port Allegheny, Pennsylvania, will be discussed here.

Port Allegheny, Pennsylvania Asbestos Health Program

Approximately 1200 members (all white male) of the Flint Glass Workers' Union in Port Allegheny, Pennsylvania are at high risk of developing cancers associated with workplace exposure to asbestos at the Pittsburgh Corning Corporation's glass and insulation production plant. NIOSH identified this asbestos hazard in 1971 and informed the company and the union. At about that time, the use of asbestos in production at the plant ended. Subsequently, after lengthy discussions, a nonprofit community program was established. The governing board contains representation from the union, corporation, community groups, and medical providers. With advice from the Workers' Institute and Mt. Sinai School of Medicine, New York, this program provides education and information, maintains an office for scheduling of activities and recording of results in an ongoing cohort registry, contracts for periodic medical surveillance, and generates financial support for ongoing activities. This program has been extended to family members of exposed workers as a result of potential family risk due to secondary exposure. Observers have characterized this program as a model of community collaboration to address a serious work-related health program.

Intervention Methods

Because controling various risk factors can exacerbate the action of asbestos, the Port Allegheny program (PAAHP) was designed as a health education and medical surveillance program.

Health Education. Smoking-cessation classes, including materials and clinical counseling, were offered because of the highly synergistic interaction of asbestos and smoking in lung cancer. Information on occupational hazards that could advance pulmonary obstruction (such as coal mining, exposure to silica, etc.) was provided. Importance of regular shots to prevent influenza and pneumonia was emphasized and other measures were provided to protect individuals with pulmonary obstruction from the debilitating (and often deadly) effects of a superimposed respiratory infection (PAAHP offers free shots).

Medical Surveillance. Lung function tests to determine the degree of pulmonary obstruction were administered. Induced sputum production for cytology to identify cell transformation indicative of neoplastic development, and chest x-ray examinations (P.A. and Lateral) to examine obstructions visually and to identify the nature of pulmonary disease were conducted. Referral of positive cases for follow-up was also an integral part of the program.

Approaches to Program Evaluation

The project reported here was a demonstration project, and the evaluation design was directed at the primary objective related to notification of risk and the development of the intervention program.

The extent to which the project reached and recruited the cohort members into the intervention programs was the primary objective to be evaluated. To promote participation and enhance coping ability, it was felt that family involvement would be central. Three measurements were the focus of this evaluation. The primary evaluation measurement of project effectiveness was participation rates in the primary intervention programs. Participation rate was defined as number of participants as a percentage of identified eligible program participants. The goal for participation was set at 70 percent, which is high compared to most community—medical screening programs. Second, to evaluate the extent to which the educational messages penetrated the target populations, rates of compliance with key program components were evaluated. In Port Allegheny, smoking-cessation rate was a major compliance indicator.

Results

This project is only now completing the implementation phase and beginning to enter the evaluation phase. Presented here are some highly preliminary findings from the types of data that have been monitored regularly during the implementation phase to monitor progress, and are limited to impact on the workers.

In 1978, 1979, and 1981, the medical program was offered by field examination medical teams from Mt. Sinai School of Medicine, New York City. By mid-1982, medical services became offered by specially trained community physicians in the Port Allegheny area, initially on a modest scale of about 10 eligible participants per week. To date, 54 percent of the eligible workers have been screened. However, this relatively low rate is misleading because every screening session has been fully subscribed and there is a long waiting list. What characterizes the Port Allegheny program at this time is a well-structured and organized approach that will probably lead to acceptable participation. The program has as yet to address on a routine basis the household residents of the workers who are eligible for program participation, and they are not expected to be included until most of the workers have been recruited into a routine system of medical surveillance.

Some data have been collected on smoking cessation rates, a key compliance indicator. Of 355 workers screened in 1979, 39.4 percent had already quit smoking. To what extent publicity about the asbestos hazard since the plant was identified as a serious potential health problem in the late 1960s contributed to this rate is unknown, and it is now being investigated. A small sample of workers ($n = 83$) were examined for a second time in 1982 (i.e., these were workers who had gone through either the 1979 or the 1981 screenings). It is important to note that 7 individuals had quit smoking during this interval, which means that a compliance incidence rate of 26.9 percent was achieved for the 26 participants who were smokers at the inception of this period. Of course, the important point is that only 19 individuals, or 22.9 percent of the total reexamined population, remain smokers. This smoking prevalence is well below half the rate of smoking in the general blue-collar U.S. population.

Smoking cessation in the Port Allegheny population is especially interesting, and it will be investigated intensively. If the smoking cessation rate of 39.4 percent prior to the first major medical screening in 1979 is found to be generalizable to the entire population of workers at risk, and if it is found to be attributable to the fact that the workers determined they were at high risk, it could have a major bearing on future approaches to worker notification and intervention. Simply stated, such a finding might indicate that under certain circumstances the act of notification by itself might be adequate on the grounds that the workers at risk would seek out additional information and fend for themselves. Certainly, such a finding would obviate the current opposition to worker notification that is grounded in the belief that such action would be a great psychosocial disservice to the workers at risk. It also would be important to study the smoking-cessation effect during the interval between the first medical examination and reexamination. The formal smoking-cessation program had not been developed to any extent during this period (only a total of a dozen or so workers agreed to participate in three smoking-cessation programs organized by a worker who had received training from the American Cancer Society as a smoking-cessation facilitator). However, the examining physicians gave strong counsel to each clinic participant about the importance of smoking cessation. To what extent such counseling resulted in cessation remains to be determined, but it might be an important function of the physician in programs directed at the blue-collar workers, where cessation successes have been generally poor.

The program reported here demonstrates that notification and intervention programs for high-risk occupational cohorts are feasible within the existing structures of labor–management relations, medical-care delivery and community services, and social networks. It also demonstrates that much remains to be learned about our responses to risks carrying relatively low probabilities and chronic disease effects with long latency periods.

CONCLUSION

Clearly, more questions than answers exist related to the enhancement of health protective behavior in a carcinogenic environment, interaction with other lifestyle factors that further contribute to cancer risk, as well as management of the behavioral aftermath of past exposure notification. Health risk reduction in the workplace—particularly the lethal risk that has been under discussion here—is a priority research area and a fertile field of inquiry for behavioral medicine investigators to pursue.

REFERENCES

Doll, R., & Peto, R. (1981). The causes of cancer: Quantitative estimates of avoidable risks of cancer in the United States today. *Journal of The National Cancer Institute, 66*, 1191–1308.

Haynes, B., Taylor, D., & Sackett, D. (1979). *Compliance in health care*. Baltimore: Johns Hopkins University Press.

NIOSH, CDC, PHS document. (July, 1977). *The right to know*. Report prepared at the request of the Subcommittee on Labor, Committee on Human Research, and the U.S. Senate.

NIOSH, (January 1978). *Special occupational hazard review of Trichlotoethylene*. DHEW Publication No. 78–130. Washington, DC: U.S. Government Printing Office.

OSHA and the Committee on Public Information on the Prevention of Occupational Cancer, National Research Council. (1977). *Informing workers and employees about occupational cancer*. Washington, DC: National Academy of Sciences.

Samuels, S. (1980). Workers at high risk. In J. Last (Ed.), *Public health and preventive medicine*. New York: Appelton-Century-Crofts.

Schulte, P., & Ringen, K. (1984). Notification of workers at high risk: An emerging public health problem. *American Journal of Public Health, 74*.

13

Promoting Job Safety
and Accident Prevention

JUDITH L. KOMAKI

A quick glance at some alarming facts about workplace accidents indicates why occupational safety continues to be a major national concern:

Almost two million disabling work injuries occur each year.

Work-related deaths occur at a rate of more than one an hour, for a grisly total of 11,200 deaths a year.

Annual losses due to work-related accidents are estimated to be more than $31 billion (National Safety Council, 1983).

To the chagrin of many observers of the safety scene, these figures do not represent any substantial decline from levels reported before the passage of the landmark Williams-Steiger Occupational Safety and Health Act more than a decade ago.

This chapter will illustrate a constructive, feasible solution for reducing unsafe acts, the primary cause of occupational accidents. The guidelines are based on the principles of the experimental analysis of behavior (Keller, 1969; Skinner, 1974) and on the experience gained while successfully implementing behaviorally based programs in businesses and industries.

Many thanks to Milton Blood, Thomas Coates, Anthony DeCurtis, and Beth Sulzer-Azaroff for their constructive help and supportive encouragement of my work.

WORK SAFETY IS NO ACCIDENT

Injuries had jumped sharply at the large wholesale bakery. For three prior years, the injury frequency rate had hovered around 35 disabling injuries per million man hours worked. Then it jumped to 53.8—more than double that of the bakery industry generally and substantially higher than in such hazardous occupations as mining and meat packing. Management was naturally alarmed, particularly as its Workman's Compensation premiums were rising as well. What to do?

An initial investigation of the bakery convinced Ken Barwick and Larry Scott, both students in Management at Georgia Tech, and the author, that safe practices were not being maintained because workers received little, if any, positive reinforcement for performing safely. Nor were they trained to avoid unsafe practices.

Thus a system of specification, measurement, and reinforcement was put into effect that was much different from the usual approach of posting signs and admonishing workers to be careful. The result: A remarkable improvement in performance and a dramatic reduction in accidents.

First, a safety performance instrument was developed wherein specific worker practices could be identified as safe, unsafe, or not observed. Next, workers received instruction in the desired safety definitions and in avoiding unsafe practices, and they were encouraged to improve their safety performance in order to lift the plant out of the cellar in the parent company's safety ranking.

Most important, feedback was provided. Workers were told their safety performance would be observed and recorded on a graph for all to see. A departmental goal of 90 percent safe performance was agreed to by the employees. Supervisors also were asked to recognize workers when they performed selected activities safely and to make specific comments.

After this program had been in effect 11 weeks in the wrapping department and 3 weeks in the makeup department, the observers stopped observing and providing feedback. To assess the effect of this reversal phase, observations were reinstituted five weeks later for a period of four weeks.

The results: See Figure 13.1. From performing safely 70 percent and 78 percent of the time, employees in the two departments substantially improved their safety performance to 96 percent and 99 percent, respectively. Consequently, it was concluded that the program, particularly the feedback, was effective in improving safety performance.

During the reversal phase, performance dropped back to the original baseline (71 percent and 72 percent) and management was finally convinced arrangements for plant personnel to observe, record, and post the safety level were necessary.

Within a year there was a fivefold decrease in the number of lost-time accidents, from over 50 accidents per million hours worked to less than 10, a

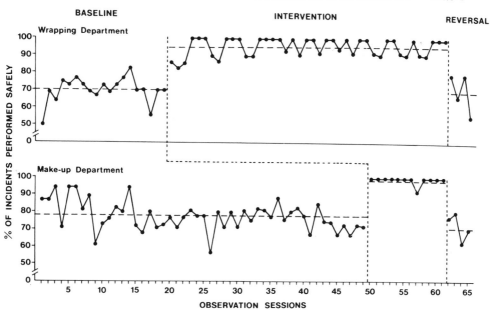

FIGURE 13.1. Results of a behavioral safety program introduced in two food manufacturing departments. (From Komaki, Barwick, & Scott, 1978. Copyright 1978 by the American Psychological Association. Reprinted by permission.)

relatively low figure. The plant moved from last to first place in the company standings and received a safety plaque from the company "in recognition of successfully working 280,000 hours (over ten months) without a disabling injury."

(Reported in *Journal of Applied Psychology*, August 1978, pp. 434–435, and in *Wharton Magazine*, Winter 1979, pp. 8–10.

This example (Komaki, Barwick, & Scott, 1978) is only one of a series of recent studies documenting the effectiveness of a behaviorally based approach to occupational safety. In 1978, Smith and Anger at the National Institute for Occupational Safety and Health, along with an independent contractor, reported one of the first successful attempts to reduce eye accidents in a ship-building organization (Smith, Anger, & Uslan, 1978). In 1978 and 1980, Sulzer-Azaroff and her associates at the University of Massachusetts published two articles, showing that an informational feedback package resulted in sharp reductions in hazardous conditions in a university research laboratory facility (Sulzer-Azaroff, 1978) and in a private manufacturing organization (Sulzer-Azaroff & de Santamaria, 1980). In 1980,

Zohar and his colleagues at the Israel Institute of Technology documented the increased use of personal protective equipment, again using informational feedback (Zohar, 1980; Zohar, Cohen, & Azar, 1980). In 1982 and 1984, teams of investigators at the University of Kansas (Haynes, Pine, & Fitch, 1982) and Louisiana State University (Reber, Wallin, & Chhokar, 1984) showed once more the efficacy of a behavioral approach to promoting safety performance.

POTENTIAL OF THE BEHAVIOR ANALYSIS APPROACH FOR PROMOTING JOB SAFETY

This section illustrates why the behavioral approach is particularly suitable for facilitating safety and preventing accidents. Its three salient characteristics are described, and these characteristics are then contrasted with traditional approaches to safety.

Salient Characteristics of the Behavior Analysis Approach

Action Orientation.   The aim of virtually all behavior analysis research is to enhance behaviors of importance in a meaningful way. Thus investigators actively intervene in work settings. They develop programs with the aim of producing desired changes on the job, and they introduce and evaluate these programs to see whether they are successful.

Emphasis on Behavioral Measurement.   Another noteworthy characteristic of the behavior analysis approach is its focus on behavior. Investigators are typically interested in workers' actions. When addressing the problem of reducing accidents, for example, the focus is on what workers can feasibly do to avoid having accidents in the future.

In line with the emphasis on workers' actions, behavioral psychologists have developed a sound way of assessing behaviors in applied settings. These techniques, referred to here as applied operant measurement because of their widespread use by *operant* psychologists in *applied* settings, have been successfully used to assess and improve the behaviors of clients, students (Ciminero, Calhoun, & Adams, 1977; Hersen & Bellack, 1976; Mash & Terdal, 1976), and workers (Komaki, Collins, & Thoene, 1980).

In assessing safety performance, for example, trained observers actually watch workers on the job. Self-reports, reports from secondary sources, or accident records are not sufficient; the behaviors themselves must be sampled. To enhance the "fairness" of the information obtained, only those behaviors and outcomes that workers can directly control are tapped; the observations are made frequently during an extended period of time to reflect both the peaks and the valleys of performance; and the observations are planned in advance to ensure they are representative and unbiased. To enhance the accuracy of the information obtained, the definitions of safe performance are unambiguously defined until two independent observers consistently agree; the observers are trained until they reach a set criterion of interobserver agreement; they record their findings immediately; and they continue to assess interobserver agreement throughout the formal data collection.

Moreover, recent evidence suggests that these applied operant measures of safety performance are related to occupational injury rates (Reber & Wallin, 1983). Applied operant measures of safety were developed for 12 departments of a farm machinery manufacturing company, and data were collected. The behavioral measures of safety were then correlated with departmental injury incidence rates. A significant inverse relationship was found between behavioral performance and the overall injury rate, and between behavioral performance and the lost-time injury rate: that is, as the safety performance decreased, the accident rate climbed, and as safety performance increased, the accident rate declined. This way of measuring performance holds considerable promise for appraising, and ultimately improving, safety performance.

Focus on Performance Consequences. The third characteristic of the behavioral approach concerns its view of human behavior. Worker behavior, for example, is thought to be best understood by focusing on the consequences of that behavior, that is, those events that occur *after* the behavior. Thus, if you wished to understand why workers perform unsafely, you would look at the consequences of safe *and* unsafe behaviors. You would examine what happens after safe performance: what type of comments, if any, were made; what types of activities occurred; and whether accidents were always avoided. You would also examine what happens after unsafe performance: whether reprimands were issued; what types of activities occurred; and whether accidents regularly occurred. In general, when positive consequences consistently occur after defined performance, it has been found that workers will be motivated to perform safely. Conversely, if there are few consequences and if workers are inadvertently rewarded for performing unsafely, workers will probably not be motivated to perform safely for long.

An analysis of work environments reveals that the consequences for performing safely are usually either nonexistent or negative. Co-workers rarely make favorable comments to one another about lifting properly or using safety glasses, for example. Furthermore, supervisors rarely recognize workers for following safety practices. Because a safe action seldom results in a tangible product, both superiors and employees have trouble accurately assessing performance levels. Safety committees, whether they be composed primarily of workers or supervisory personnel, union or management, rarely note improvements made since the last inspection; no news is usually the best news. Some safe actions (mopping spills, putting on safety goggles) require workers to take extra time or to add steps to their usual sequence of performance. The net effect is that when one is safe, there is often less time for other work (or nonwork) activities. Moreover, accidents sometimes occur even when workers perform safely: a truck lid flies open in the face of an inspecting mechanic; pans fall down onto the head of a passerby. Just because workers act safely all the time does not mean that workers will necessarily avoid having an accident. Safe performance generally has the following consequences:

1. Few co-worker comments.
2. Little management recognition.
3. Infrequent safety committee mention.
4. Less time for other activities.
5. Occasional accident.

For unsafe actions, however, the story is the same, and few consequences typically occur, as follows:

1. Few reprimands.
2. More time for other activities.
3. Relatively few injuries.

Negative sanctions rarely occur when workers perform unsafely. Unsafe acts, like safe acts, leave no trace, unless an accident occurs. Even then the accident must be fairly serious for it to be apparent. Supervisors and co-workers sometimes indirectly encourage workers to neglect safety practices and take shortcuts so workers can spend more time on other activities. Accidents, so-called natural consequences of unsafe behavior, seldom occur when workers perform unsafely. Employees can perform unsafely hundreds of times and yet remain injury-free. When accidents do not occur after unsafe acts, then workers essentially learn that unsafe acts are not necessarily unsafe.

The above analysis of the safety environment helps explain why workers perform unsafely, despite compelling arguments to the contrary. The work environment provides little, if any, positive reinforcement for safe performance and the natural consequence of performing unsafely, the occurrence of accidents, is frequently, albeit fortunately, missing.

This focus on performance consequences not only helps to explain why workers perform unsafely but it also provides a way to motivate workers to perform safely. The key again is performance consequences. Study after study has shown that performance will be enhanced and will maintain itself for extended periods of time when performance consequences are judiciously arranged and are contingent on performance (Frederikson, 1982; O'Brien, Dickinson, & Rosow, 1982).

Thus, to motivate people to do their jobs safely, positive consequences should be arranged to follow desired performance. In work settings, four types of consequences are possible:

1. Informational consequences, such as feedback.
2. Social consequences, such as recognition.
3. Activity consequences, such as time off.
4. Organizational consequences, such as promotions, pay increases, training opportunities, and job assignments.

Informational consequences such as feedback are particularly recommended. They have been successfully and extensively used to improve safety practices (e.g., Komaki et al., 1978, 1980, 1982; Sulzer-Azaroff, 1978; Sulzer-Azaroff & de Santamaria, 1980; Zohar, 1980; Zohar et al., 1980). Feedback is readily acceptable to both workers and management, costs little, effects steady improvements, and is probably more likely than other methods to be continued for extended periods of time. Furthermore, employee reactions to feedback programs have been found to be quite positive (e.g., Komaki et al., 1978). When feedback is introduced, workers often check the feedback graphs to see how their units are doing. When performance falls below 100 percent, employees typically inquire as to which item(s) had been scored unsafe. Workers are often seen reminding one another about being safe in the future. Healthy informal competitions frequently occur among units; employees and supervisors check to see how adjacent units do. In fact, recent evidence suggests that employees prefer a system that allows them to see how they are doing rather than one that provides only rules and reminders; three-quarters of the employees in one plant, when polled, noted they preferred the feedback to the reminders (Komaki et al., 1982). Interest in these programs also typically remains high. In the above plant, performance remained high and steady for at least six months.

In short, the three characteristics of the behavioral approach—its applied orientation, its emphasis on behavioral measurement, and its focus on performance consequences—lend themselves to the successful promotion of job safety and accident prevention. These characteristics are in distinct contrast to traditional ways of approaching safety described next.

Traditional Approaches to Occupational Safety

Predominance of Descriptive Studies. The active orientation of the behavioral approach differs dramatically from traditional safety efforts, most of which are descriptive. A venerable and long-standing way for psychologists to study safety, for example, is to examine the relationship between accidents and such factors as age, experience level, and personality characteristics of the work force (e.g., Verhaegen, Vanhalst, Derijcke, & Van Hoecke, 1976). Although these descriptive studies may add to our knowledge of individual differences, the results do little to alter practices on the shop floor. Even if workers were willing, workers cannot readily change their personality, age, or experience level.

Studies conducted by safety researchers are also primarily descriptive. Accident categorizations are common, with accidents catalogued by the type of equipment and the work condition (e.g., Institute of Occupational Health, 1981). Several safety researchers have attempted to identify the factors important to the functioning of an effective safety program; they have compared the safety efforts of high and low accident firms matched for type of industry, size, and geographical location (e.g., Cohen, Smith, & Cohen, 1975; Shafai-Sahrai, 1971; Simonds, 1973). Rarely do the investigators involved take the next step, however, and design and implement an accident prevention program, based on their findings. It is not sufficient to know that the employment of safety officials in top positions and the discussion of safety in plant publications is related to low accident rates. These ideas should be introduced on an experimental basis to see whether they do, in fact, result in safe performance and prevent accidents.

Focus on Accident Rate. In contrast to the behavior analysis approach, which focuses on performance and performance measurement, injury frequency rates or accident rates are typically used as an *indirect* index of safety performance: (1) disabling injuries (commonly referred to as lost-time accidents), which include deaths, permanent total disabilities, permanent partial disabilities, and temporary total disabilities; and (2) medical treatment injuries, which include first-aid cases and those injuries that do not qualify as disabling but require a doctor's services.

The problem with these indirect indices is that it is questionable whether they fairly reflect workers' behaviors on the job. Workers can perform safely and still experience an accident due to events entirely beyond their control (e.g., a roof caves in). In such cases, it would not be judicious to rely on injury rates alone as an index of worker performance. Even when no injuries occur, say for a month, one cannot necessarily assume that workers performed safely all that time; workers frequently act unsafely hundreds of times before incurring an injury.

Further methodological problems hamper both types of injury frequency indices. It is difficult to reflect sensitively the level of safety performance on a week-to-week or even month-to-month basis, using lost-time accidents, because of their relative infrequency and severe fluctuations, called the *rare events* phenomenon. In the wholesale bakery (described above), there was an average of less than one lost-time accident per month in a plant of 200 employees with an extremely high accident rate. During one five-month period, no lost-time accidents occurred, whereas in another month four accidents occurred.

Medical treatment injuries, although they occur more frequently, are also not a reliable indicator of worker performance because they are particularly subject to reporting and recording inaccuracies. If a decline in medical treatment injuries is reported, it is often not clear whether this type of injury actually decreased, or whether injuries were reported or recorded less often. When workers incur these relatively minor injuries, peers sometimes persuade them to forget reporting accidents, particularly if a large safety prize is at stake. In addition, the recording of relatively minor accidents is apt to be lax.

Because of the methodological problems associated with lost-time accidents and medical treatment injuries, accurate and fair measures of safety performance are a particularly welcome addition to the field of safety.

Training as the Panacea. In-house attempts to improve safety rely almost exclusively on the provision of safety training and the posting of rules and reminders (e.g., Anderson, 1975; Laner & Sell, 1960; Leslie & Adams, 1973; Milutinovich & Phatak, 1978; "Operator Training," 1975). The assumption is that workers who know what to do will automatically conduct themselves in a safe manner for extended periods of time regardless of the consequences on the job. This emphasis on antecedent strategies, which occur prior to the behavior, is in distinct contrast to the behavior analysis focus on performance consequences.

Although safety training is no doubt essential, the results of a recent study suggest that it alone is inadequate, that more attention should be devoted to the provision of consequences for desired performance, and that feedback is an effective and readily accepted motivational strategy (Komaki, Heinzmann, & Lawson, 1980). In this research, performance did not improve when vehicle maintenance employees received training in the form of a slide presentation, verbal explanations, and written rules. In fact, it was not until feedback was provided that performance significantly improved. Thus it was concluded that training alone is not sufficient to improve performance substantially, even though desired practices were objectively defined and examples were tailored to specific job situations.

The next section describes how to implement a behavioral safety program, using feedback as the primary consequence.

SUGGESTED METHOD FOR PROMOTING JOB SAFETY

Two major tasks are involved in implementing a behavioral safety program:

1. Identifying and assessing desired safety practices.
2. Providing feedback.

Each task is described in the section below and then an example is presented.

Identify and Assess Desired Safety Practices

Specify Desired Safety Practices. The first step in developing a behavioral safety program is defining desired work practices or safety definitions. The definitions of safe performance establish performance requirements and describe what people should be doing.

In generating the definitions, we found that two sources of information are most valuable and should be consulted: (1) the accident/injury experience; and (2) interviews with supervisors and employees. The following steps are recommended:

1. Identify the accidents that have occurred during the previous three to five years. Group them by type (e.g., those involving lifting, house-keeping).

2. Determine whether each type of accident is preventable, that is, whether workers could avoid a similar incident if they altered their behavior. Eliminate the unpreventable accidents, that is, those that are not related to workers' behaviors per se but are the result of extraneous factors such as roofs caving in or truck lids flying open.

3. Discuss preventable accidents with on-site personnel to determine what actions could have been done to prevent such accidents from occurring again. Ask supervisors and employees to describe, or better yet to model, what an employee should do to prevent the accident. When vague descriptions are given (e.g., Workers should be "more careful" or "less rushed"), ask personnel questions such as the following: What should people do or say? What does the safe person do that the unsafe one doesn't? Solicit descriptions in terms of verbs, such as "turn off", "release", and "look forward", rather than adjectives, such as "careful". At the same time, note the conditions under which the action is to occur (e.g., when, with whom).

4. Make a preliminary list of the desired safety practices and the conditions under which each practice should be conducted for each department or unit (15 to 20 practices maximum suggested per unit).

5. Ask on-site personnel to add to the list and to identify dangerous acts and conditions that may have not yet caused an accident. At the same time, ask them to eliminate from the list or to modify overly fussy actions.

At the end of this phase, you should have a list of approximately 10–20 safety practices per unit. Each definition should contain an action (e.g., look toward knife being sharpened) or an outcome (e.g., no liquid spills on floor); include the conditions under which the action is to occur (e.g., when sharpening); and be phrased positively, indicating what workers are supposed to do (e.g., walk), rather than what they are not supposed to do (e.g., do not run).

Prepare Recording Materials. The next step involves recording information about the level of safety performance. Normally, an observer goes to the worksite, observes, and records on a data sheet whether the practice was conducted safely, unsafely, or was not observed. "Safe" actions are those that are completed as defined. "Unsafe" is recorded any time a practice is performed unsafely, regardless of the number of times it was done safely. "Not observed" is recorded any time the conditions under which the practice is done have not occurred (e.g., no mechanics were working underneath vehicles, so the observer would not be able to record whether proper eye protection was worn).

A preliminary data sheet should be drafted. It should include, at a minimum, a list of the desired safety practices and spaces to indicate whether each practice was done safely, unsafely, or was not observed during the observational period. Refer to Table 13.1 for an illustration of a portion of a data sheet that was used in the wholesale bakery operation. Room can be left on each data sheet to calculate the safety score as follows:

$$\text{\% of practices performed safely} = \frac{\text{no. safe}}{\text{no. safe} + \text{no. unsafe}} \times 100 = \text{safety score}$$

Table 13.1. Definiing Desired Safety Practices: Sample Safety Definitions and Data Sheet Used in a Wholesale Bakery Operation

Definitions of Sale Performance	Safe	Unsafe	Not Observed
1. When pulling dough trough away from dough mixer, hands are placed on the front rail of the trough and not on the side rails.	_____	_____	_____
2. When cutting wire bands from stacks of boxes or spacers, employee cuts with one hand and holds the metal strap above the cut with the other hand.	_____	_____	_____
3. There are no cardboard spacers (defined as cardboard 30mm square or larger) on the floor	_____	_____	n.a.

Source: Komaki, Barwick, and Scott (1978). (Copyright 1978 by the American Psychological Association. Reprinted by permission.)

Decide on Observational Procedures. Making decisions about the observational procedures are next. One can observe on a worker-by-worker basis or by physical location or area. One can observe workers in all positions or workers in only certain positions, or one can observe all areas of a department or only select areas. One can observe each worker or a sample of workers, or one can observe the entire department, each section within a department, or a sample of sections. The choices made depend on the situation and the accident pattern.

The decision to observe on a worker-by-worker basis is in general made when each worker is more or less stationary and does the same job repeatedly, such as poultry-processing employees working in the same spot on the same job on a line. The decision to observe by physical location is usually made when workers move freely within a given area and do a variety of different tasks during a particular workday, such as vehicle maintenance employees who diagnose problems and repair vehicles located at different places within a corporation yard.

The decision to observe workers in only certain positions or in certain locations is usually made when workers in certain positions (e.g., gizzard cutters) have èxperienced accidents and workers in other positions (e.g., gizzard pullers) have not, or when accidents almost always occur in one location or locations (e.g., maintenance area) and rarely occur in other locations (e.g., plant entrance).

The decision to observe all or a sample depends on the total number of workers and locations involved, and on the amount of time the observer can devote to collecting data. If all pertinent workers and/or locations can be observed each time, that is ideal. If only a sample can be observed, then the sampling should be arranged so that it is random and so that all cases are periodically observed.

Another set of decisions involves whether to observe for a specified period of time (e.g., five minutes in sections A, B, and C of the wrapping department; five minutes per worker) or to observe based on the frequency of occurrence of certain events (e.g., five hens passing on the line; five widgets completed). In general, when frequency cues are present in the work environment and these are fairly regular (e.g., the line typically moves at 60 hens per minute), then the frequency cues are preferred. The main consideration is that each observation session be comparable; that is, the worker should have the same number of opportunities to demonstrate that a given practice was performed safely on observation 3 during the first week as that worker would have on observation 29 during the tenth week.

A final decision involves the duration of a single observation period. In general, it is suggested that the duration be as short as possible. Each observation period should probably be no longer than 30 minutes per department, with 5 to 15 minutes preferred. The observation duration is kept short, because it is more likely that the safety information will continue to be collected frequently over time. The collection of safety information on a frequent, extended basis is integral to the success of a behavioral safety program using a relatively benign consequence such as feedback.

Fieldtest and Refine Measurement Instrument.   Once you have generated the desired safety practices, prepared a preliminary data sheet, and made initial decisions on the observational procedures, then you are ready to fieldtest and refine the instrument. The test of interobserver agreement plays a significant role during this refinement stage. It provides the instrument developer feedback about the objectivity of the definitions and the feasibility of the observational procedures.

To conduct a test of interobserver agreement:

1. Two people go to the worksite and *independently* assess workers at the same time, using the same list of safety practices and the same observational procedures.

2. Afterwards, each safety practice score is examined and noted as either an agreement or a disagreement. If both agree on scoring, this constitutes an agreement; if they do not, a disagreement is scored. Refer to Table 13.2 for a sample interobserver agreement form.

3. The percentage agreement score is then calculated as follows:

$$\frac{\text{no. of agreements}}{\text{no. of agreements} + \text{disagreements}} \times 100 = \text{percentage agreement}$$

The definitions and observational procedures of the instrument are not considered acceptable until agreement is reached on the scoring of virtually

Table 13.2. Refining Terms of Behavioral Safety Instrument: Sample Interobserver Agreement Form

Code Item	Observer A			Observer B			Agree
	Safe	Unsafe	Not Observed	Safe	Unsafe	Not Observed	
1.	x	___	___	x	___	___	+
2.	___	___	x	___	___	x	+
3.	___	x	n.a.	___	x	n.a.	+
4.	___	x	___	x	___	___	−
5.	___	___	x	x	___	___	−
6.	x	___	___	___	x	___	−
7.	x	___	___	x	___	___	+
8.	x	___	___	___	___	x	−
9.	x	___	___	x	___	___	+
10.	x	___	___	x	___	___	+

$$\text{percentage agreement} = \frac{\text{No. of agreements}}{\text{No. of agreements} + \text{No. of disagreements}} = \frac{6}{10} \times 100 = 60\%$$

Note: n.a. = not applicable

all safety practices all the time: that is, two independent observers must agree on the "safeness" of all practices before formal data collection begins. Iteration of these reliability checks and revisions of definitions and observational/recording procedures continue until 100 percent agreement is reached.

For example, numerous revisions in a recent study were entailed to meet this criterion of 100 percent agreement (Komaki et al., 1980). Originally, one safety item read, "There should be no oil spills." After observer disagreements, the item was amended to include both oil and grease spills and the appropriate remedy, rice hull or grease compound. Another change involved noting the location of the spill; walkways were differentiated from spaces beneath equipment. Observer differences mandated still another revision to include a definition of inside and outside walking areas. Eventually, the item read: Any oil/grease spill larger than 8×8 cm (3×3 in.) in an interior walking area (defined as any area at least 30 cm, 1 ft, from a wall or a solid-standing object) or an exterior walking area (designated by outer white lines parallel to the wall and at least 30 cm, 1 ft, from the wall) should be soaked up with rice hull or grease compound.

In short, it does not matter how objective the practices appear or how clear the procedures seem. The proof is in the agreement between the independent observers. When criterion is achieved, then and only then are the terms considered objectively defined. Following this unambiguous definition of terms, observers are then trained to criterion and the formal data collection begins.

Recruit and Train Observers. Various people can conduct the observations: on-site personnel, such as the safety director, supervisors, the employees themselves, or outside personnel. In selecting observers, the primary consideration is that people be available to be trained to criterion, and that they be able to collect information frequently for an extended period of time. At least two people should be recruited to observe so that they can conduct periodic interobserver agreement checks.

To ensure that people are properly trained, they should conduct interobserver agreement checks and obtain scores of 90 percent or better on at least three consecutive, representative occasions before formally collecting data. Only when observers meet this criterion are they considered trained and can they formally collect data.

Make Frequent and Unbiased Observations. To determine how well employees are performing initially, observations should be made at least three times each week over a period of three to six weeks. Once the program is introduced, data should continue to be collected at the same rate, in essence, at least three times each week for the duration of the program.

Observation times should be carefully planned and scheduled in advance. Observers can be scheduled on randomly or proportionally selected days of the week and times of the day. Or observers can be asked to collect data three times a week at their convenience, making sure that they change the days from week to week and the times from day to day so that the observations are varied and unpredictable.

Assess Reliability. During the formal observation period, checks on interobserver agreement should be conducted regularly. Once every 10 observations is suggested.

Provide Feedback

Once the desired safety practices have been specified and the initial level of safety performance assessed, the next task is to provide feedback to the workers about their desired performance. No matter how carefully information is collected, it is irrelevant, unless communicated as feedback. Feedback should be provided in the following manner:

1. **Presented on Graphs.** Post percentage scores for each department or group on a graph, similar to the one in Figure 13.1. People can see at a glance how their group has done and how they compare to their previous performance.

2. **Posted Publicly.** Display graphs prominently in the work area so that all relevant parties can see how the group has done. The public posting of the information also makes it possible for people to compare how they are doing with other groups within the organization, thus fostering a healthy competition among the groups.

3. **Provided Frequently and Immediately.** Make sure feedback is provided at least three times a week immediately or shortly after the information is collected. Workers can then have sufficient opportunities to demonstrate what they can do and to take corrective actions before feedback is provided again.

4. **Discussed with People Doing the Work and Their Supervisors.** Encourage supervisory and upper-level management personnel to acknowledge improvements and the maintenance of safe performance.

The program is typically announced during a brief meeting (30 to 50 minutes) of all relevant employees. During the meeting, the following procedure is recommended:

1. Show pairs of slides in which employees are posed performing an action both safely and unsafely. After the unsafe scenes, ask what the problem is. After the safe scenes, confirm what behaviors are desired.

2. Describe how the baseline information has been collected and how the group safety score has been computed. Then ask employees to estimate the group's safety performance level *before* showing them the baseline data.

3. Display a large graph with the baseline data plotted on it. Compare their estimates with their actual baseline level. Point out the overall safety level, designated as the mean, and note where the actual level

was higher or lower than the estimated level. Note those items consistently performed safely.

4. Encourage performance improvements: for the employees' own benefit, to aid the company, to gain recognition for the group.

5. Tell employees that their safety scores will be posted on the unplotted portion of the graph and that the graph will be posted in a conspicuous place for everyone to see.

Immediately after the meeting, post the graph and a list of the safety definitions in a prominent place in the work area, and arrange for data to be collected and the scores posted on the graph. Thereafter, encourage supervisory and management personnel to recognize improvement and to refer to the graph, a source of objective information about the workers' safety level.

STEP-BY-STEP EXAMPLE OF A BEHAVIORAL SAFETY PROGRAM

To illustrate the steps involved in designing and establishing a behavioral safety program, one of the author's recent studies is presented as an example (Komaki, Heinzmann, & Lawson, 1980).

This study was conducted with 55 employees in the vehicle maintenance division of a large western U.S. city's department of public works. The department had one of the highest accident rates in the city, recording 84.2 lost-time accidents per million man hours. The only units with higher accident rates were in the police department and water pollution control plant. (It should be noted, however, that the accident rate was comparable to the average rates of other vehicle and equipment maintenance units in the United States.)

The vehicle maintenance division was responsible for equipment repair and maintenance of the majority of the city's rolling stock. It was organized into four sections: (a) Sweeper repair constructed and repaired equipment, provided welding services, and maintained the city's line of street sweepers; (b) preventive maintenance fueled, washed, and lubricated vehicles; (c) light equipment repair provided a ready line of police cars and other light vehicles (pick-up trucks, motorcycles); and (d) heavy equipment repaired maintenance vehicles weighing 1500 pounds and over.

Identification and Assessment of Desired Safety Practices

Specification of Desired Safety Practices.   Accident logs from the previous 5 years were examined. When it was found that an accident occurred after a worker fell off a jack stand, for instance, an item was included regarding the proper use of jacks and jack stands. Supervisors were also asked to suggest items they believed to be important. Throughout the developmental stage, both supervisors and workers were invited to prune excessively scrupulous items. Examples of desired safety practices were:

1. **Proper Use of Equipment and Tools.**   When reaching upward for an item more than 30 cm (1 ft) away from extended arms, use steps, stepladder, or solid part of vehicle. Do not stand on jacks or jack stands.

2. **Use of Safety Equipment.**   When using brake machine, wear full face shield or goggles. When arcing brake shoes, respirator should also be worn.

3. **General Safety Procedures.**   When any type of jack other than an air jack is in use (i.e., vehicle is supported by jack or off the ground), at least one jack stand should also be used.

Eventually, between 15 and 22 objectively defined practices were written for each section. These definitions formed the basis for the observational procedures and the feedback.

Observational Procedures.   Trained outside raters, identified as interested in the area of safety, served as nonparticipant observers. Observations took place from three to five times each week. The exact times varied; no observation was carried out at the same time of day on any two consecutive visits.

The observers, in full view of the workers using the list of safety practices, rated each practice as being safe, unsafe, or not observed. If an employee in light equipment repair was seen working under a vehicle, for instance, it was observed whether he was wearing eye protection. If the person was, that item was scored "safe"; if he was not, the item was scored "unsafe." If no work was being done under a vehicle during the observation period, the item was marked "unobserved." The percentage of incidents performed safely was calculated as the number of safe items observed divided by the number of total incidents observed (the number of items checked as being either safe or unsafe) and multiplied by 100.

The total observation duration for each section ranged from 15 to 20 minutes. Observers spent 5 minutes in each of three shops in the sweeper repair section (15), 5 minutes in each of four locations in the preventive maintenance section (20), 10 minutes in the heavy equipment repair section (10), and 10 minutes in one location and 5 minutes in a second location of the light equipment repair section (15).

Reliability Assessment.   A total of 165 observations was conducted over a

45-week period. Interobserver agreement was assessed 21 times, for an average of one reliability check every 7.5 observations. Primary and secondary observers consistently obtained high percentages of agreement throughout the study, averaging 94.8 percent overall.

To minimize the possibility of observer bias, approximately half (52 percent) of the observations were made by three people (of seven) who were not told about the order or the timing of the phases. A total of 15 (of 21) interobserver agreement checks were conducted with at least one of the naive observers. Interobserver agreement here was also consistently high, with an average of 94.4 percent, indicating that the changes were not likely to be a function of observer bias.

Provision of Feedback

Supervisor-provided Feedback.   Each supervisor collected information on his section's safety level and posted the data on a graph in his section. One experimenter taught the supervisors how to observe and provide feedback to their sections. Using the same observational procedures as the observers, the supervisors practiced collecting data with one of the experimenters until they agreed consistently over 90 percent of the items.

Introductory Meeting.   Workers attended an introductory session that lasted from 30 to 45 minutes during their regular workday. During the meeting, the men were shown a series of 35-mm transparencies. Each slide portrayed a safe and an unsafe condition or practice as described in the observational instrument. The slides had been taken by the city safety officer in the employees' own work areas, showing equipment they used, with a fellow worker as a model. One slide, for instance, showed an employee working under a vehicle without eye protection. (*Note:* The pictures of employees performing unsafely were carefully posed, and this announcement was made clear to the viewers.)

After seeing an unsafe action portrayed, the employees were encouraged to discuss the hazards portrayed. Slides were then shown depicting the dangerous condition corrected or the act performed safely. The employee working under the vehicle now wore eye protection, for instance. As each safe scene was shown, the relevant safety rule was stated. The men were then given a copy of the safety rules, and were encouraged to pay special attention to rules that were overlooked most often.

The section supervisor then explained that he would be making a randomly timed daily safety observation and that he would post the results on a graph. The supervisor used a specially prepared graph to point out the mean safety level during the baseline and drew attention to the line demarcating the section's goal. The supervisor emphasized that the goal was realistic, that it had been set in conjunction with the superintendent, and that it was based on the employees' previous safety performance. He made it clear that performance at the 100 percent level was neither expected nor required. After the meeting, the graph was posted where all the employees would be able to see how they were doing in comparison with their previous performance and the goal.

Thereafter, supervisors collected information and posted data on graphs in their sections. Periodically, they conducted interobserver agreement checks, averaging over 96 percent.

Results

In this study, the safety level improved significantly when the supervisors provided feedback regularly.

SUMMARY

The motivation to perform safely seems obvious: to avoid immediate injury and long-term suffering. A closer, and more behavioral, analysis of the situation, however, reveals why these seemingly compelling arguments do not necessarily result in sustained safety on the shop floor. The behavior analysis approach helps not only explain why workers perform unsafely, despite cogent arguments to the contrary, but also provides a constructive way to enhance and upgrade workplace safety. The essential steps of a behavioral safety program—the identification and measurement of desired safety practices, and the provision of feedback—are detailed in this chapter. Examples and guidance are given so that the reader can implement a behavioral program in his or her own organization and begin to accrue the benefits of a positive, effective safety effort.

REFERENCES

Anderson, C. R. (1975). *OSHA and accident control through training.* New York: Industrial Press.

Ciminero, A. R., Calhoun, K. S., & Adams, H. E. (Eds.). (1979). *Handbook of behavioral assessment.* New York: Wiley.

Cohen, A., Smith, M., & Cohen, H. H. (1975, June). *Safety program practices in high versus low accident rate companies* (DHEW Publication No. NIOSH 75-185). Washington, DC: U.S. Government Printing Office.

Frederiksen, L. W. (Ed.). (1982). *Handbook of organizational behavior management.* New York: Wiley.

Haynes, R. S., Pine, C., & Fitch, H. G. (1982). Reducing accident rates with organizational behavior modification. *Academy of Management Journal, 25,* 407–416.

Hersen, M., & Bellack, A. (1976). *Behavioral assessment: A practical handbook.* New York: Pergamon.

Institute of Occupational Health. (1981). *Current research projects: 1981.* Helsinki: Institute of Occupational Health.

320 REFERENCES

Keller, F. S. (1969). *Learning: Reinforcement theory.* New York: Random House.

Komaki, J. L., Barwick, K. D., & Scott, L. R. (1978). A behavioral approach to occupational safety: Pinpointing and reinforcing safe performance in a food manufacturing plant. *Journal of Applied Psychology, 53,* 434–445.

Komaki, J. L., Collins, R. L., & Penn, P. (1982). Comparison of antecedent and consequent control approaches to work motivation. *Journal of Applied Psychology, 67,* 334–340.

Komaki, J., Collins, R. L., & Thoene, T. J. F. (1980). Behavioral measurement in business, industry, and government. *Behavioral Assessment, 2,* 103–123.

Komaki, J., Heinzmann, A. T., & Lawson, L. (1980). Effect of training and feedback: Component analysis of a behavioral safety program. *Journal of Applied Psychology, 65,* 261–270.

Laner, S., & Sell, R. G. (1960). An experiment on the effect of specially designed safety posters. *Occupational Psychology, 34,* 153–169.

Leslie, J. Jr., & Adams, S. K. (1973). Programmed safety through programmed learning. *Human Factors, 15,* 223–236.

Mash, E. J., & Terdal, L. G. (1976). *Behavior-therapy assessment: diagnosis, design, and evaluation.* New York: Springer-Verlag.

Milutinovich, J. S., & Phatak, A. V. (1978). Carrying the safety training load—tips for all managers. *Industrial Engineering, 10,* 24–32.

National Safety Council. (1983). *Accident facts, 1983 edition.* Chicago: National Safety Council.

O'Brien, R. M., Dickinson, A. M., & Rosow, M. (Eds.). (1982). *Industrial behavior modification: A learning-based approach to business management.* New York: Pergamon.

Operator Training. (1975, October). OSHA's hidden failure: How you can correct it. *Modern Materials Handling,* pp. 45–60.

Reber, R. A., & Wallin, J. A. (1983). Validation of a behavioral measure of occupational safety. *Journal of Organizational Behavior Management, 5*(2), 69–78.

Reber, R. A., Wallin, J. A., & Chhokar, J. S. (1984). Reducing industrial accidents: A behavioral experiment. *Industrial Relations, 23,* 119–125.

Shafai-Sahrai, Y. (1971). An inquiry into factors that might explain differences in occupational accident experience of similar size firms in the same industry (Doctoral dissertation, Michigan State University, 1971). *Dissertation Abstracts International, 32,* 2247A. (University Microfilms No. 71-23, 242.)

Simonds, R. H. (1973). OSHA compliance: "Safety is a good business." *Personnel, 50,* 30–38.

Skinner, B. F. (1974). *About behaviorism.* New York: Vintage.

Smith, M. J., Anger, W. K., & Ulsan, S. S. (1978). Behavioral modification applied to occupational safety. *Journal of Safety Research, 10,* 87–88.

Sulzer-Azaroff, B. (1978). Behavioral ecology and accident prevention. *Journal of Organizational Behavior Management, 2,* 11–44.

Sulzer-Azaroff, B., & deSantamaria, M. (1980). Industrial safety hazard reduction through performance feedback. *Journal of Applied Behavior Analysis, 13,* 287–296.

Verhagen, P., Vanhalst, B., Derijcke, H., & Van Hoecke, M. (1976). The value of some psychological theories on industrial accidents. *Journal of Occupational Accidents, 1,* 39–45.

Zohar, D. (1980). Promoting the use of personal protective equipment by behavior modification techniques. *Journal of Safety Research, 12,* 78–85.

Zohar, D., Cohen, A., & Azar, N. (1980). Promoting increased use of noise through information feedback. *Human Factors, 22,* 69–79.

THREE

A Compendium of Health Promotion Programs in the Workplace

14

STAYWELL:
Evolution of a Behavioral
Medicine Program
in Industry

MURRAY P. NADITCH

Control Data's STAYWELL program, initiated in 1979, was one of several programs that aimed to control rapidly spiraling health care costs and to affect productivity. STAYWELL was conceived in the Control Data tradition of making a significant investment in the solution of a social problem affecting the company, and then realizing a positive return on that investment by marketing the program to other companies.

Design and development were based on an empirical evolutionary perspective. The strategy was to develop a state-of-the-art program in 1980, test empirically the efficacy of the program for a number of years, and continue to modify STAYWELL until it evolved into first an effective vehicle of health behavior change and then into a cost-effective program.

This evolutionary perspective has yielded some interesting results both in terms of what has, and what has not, worked in a workplace behavioral medicine program. This chapter intends to share some of those successes and failures of the STAYWELL program with other companies and scientists in the field who are developing programs.

PEOPLE WITH POOR HEALTH HABITS COST THE COMPANY MORE MONEY

The primary focus of the program was on behavioral risks related to chronic degenerative diseases, such as cardiovascular disease and lung cancer. The basic premise was that it would be less expensive to prevent these diseases than to pay for them after they had already developed. Consequently, it was expected that any return on investment from the program would derive from a slowdown in the rate of chronic degenerative diseases several years after the program had been implemented.

In the course of evaluation, however, some interesting data came to light that suggested that there may also be significant potential for short-term as well as long-term cost savings. Like most major companies, Control Data is self-ensured. Unlike most companies, however, Control Data also processes its own health-care insurance claims. This data base gave the STAYWELL program a unique opportunity to examine the association between health behaviors and actual health care costs, as well as the potential to examine the effect of behavior change related to the program on health care costs. Employees who smoked, were sedentary, were overweight, or who were hypertensive were compared to employees lower in behavioral risk factors in terms of health care cost claims data and the number of average days spent by each employee per year in the hospital. These data are shown in Table 14.1.

Table 14.1. The Relation between Habits, Health Care Cost Claims, and Hospital Stays

Health Habit	Average Total Dollars Paid	Number of Average Days in Hospital
Smoking		
Current smokers and those who quit less than five years ago N = 2,376	$390.87	0.60
Never smoked or quit more than five years ago N = 3,193	$313.27	0.28
Significance Level	P<.03	P<.05
Exercise		
Sedentary N = 1,219	$436.92	0.57
Some or vigorous exercise habits N = 4,350	$321.01	0.37
Significance Level	P<.01	P<.19
Overweight		
Greater than 20% overweight N = 1,637	$362.42	0.61
20% or less overweight N = 3,932	$339.71	.33
Significance Level	P<.55	P<.22
Hypertension		
Greater than or equal to 160/95 N = 300	$692.95	0.53
Less than 160/95 N = 5,269	$326.65	0.41
Significance Level	P<.02	P<.51

Note: Population is non-HMO employees who took Health Risk Profile in 1980.

Employees who smoke cost the company considerably more in health care claims than those who do not. Smokers also spend more than twice the number of days in the hospital per year as employees who do not smoke. Employees who report being engaged in a physical fitness program have much lower health care costs than those employees who are sedentary. Employees who exercise also had a lower average number of days in the hospital each year, although this latter relationship was not statistically significant. As would be expected, employees who were hypertensive had much higher health care claims than those who were not, and there was also some trend for hypertensives to have longer hospital stays. There was also a suggestion that employees who were more than 20 percent overweight had higher health care costs and longer hospital stays than those employees who were not, although these latter relationships were not significant.

In a separate analysis, relations between health habits and some subjective measures of absenteeism and productivity were assessed. The results of that analysis indicate that there may also be some relations among health habits, absenteeism, and productivity. Figure 14.1 shows the relation between health

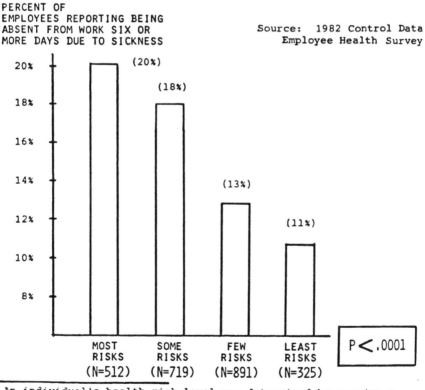

PERCENT OF
EMPLOYEES REPORTING BEING
ABSENT FROM WORK SIX OR
MORE DAYS DUE TO SICKNESS

Source: 1982 Control Data
          Employee Health Survey

P<.0001

An individual's health risk level was determined by summing a person's risk status in five STAYWELL related areas: Weight, stress, fitness, nutrition and smoking.

FIGURE 14.1.   The relation between health habits and absenteeism.

habits and percent of employees reporting being absent from work six or more days per year due to sickness. Health habits are assessed in terms of behavioral risk for chronic degenerative diseases. An individual's health-risk level was determined by summing a person's risk status in the five areas in which STAYWELL is focused: weight control, stress management, physical fitness, nutrition, and smoking cessation. These data show a clear linear relation between behavioral risk factors (health habits) and reported work absence. This subjective reporting of work absenteeism is a better measure than actually recorded sick days. Many employees take sick days as vacation days, and some classifications of employees are not required to report the number of sick days taken.

The relation between health habits and work limitations due to illness are shown in Figure 14.2. There is a clear linear relation between health habits

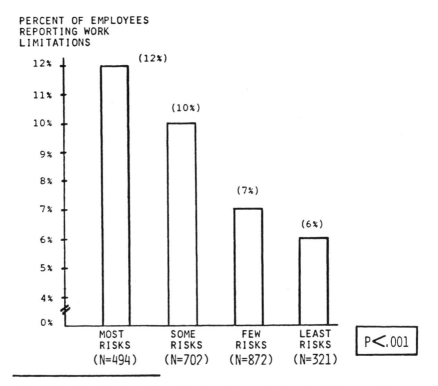

An individual's health risk level was determined by summing a person's risk status in five STAYWELL related areas -- weight, stress, fitness, nutrition, and smoking.

Source: 1982 Control Data Employee Health
          Survey.

FIGURE 14.2.   The relation between health habits and work limitations due to illness.

and reported work limitation because of health. Although this is a subjective measure of productivity, it does suggest a relation between health habits and productivity. This is important because productivity is particularly difficult to measure in any form. The relation shown in these data is reasonably robust, linear, and statistically significant.

Taken together, these data indicate that there may well be some short-term cost savings involved in assisting people in modifying their health-related behaviors and that these behaviors may also be related to absenteeism and work limitations.

PROGRAM DESCRIPTION

The STAYWELL program is currently being delivered to approximately 22,000 Control Data employees and spouses in 14 American cities, as well as being implemented in a number of other companies. At Control Data, the program is offered as a free corporate benefit to all employees and spouses.

The program is based on the premises that lifestyle and behavior have an important impact on health, and that people can be helped to understand and modify health behaviors. Emphasis is on long-term changes in health behavior facilitated by providing people with awareness, skills, and a work-place environment conducive to the initiation and maintenance of positive lifestyle behaviors. The program focuses on smoking cessation, weight control, fitness behavior, stress management, and nutritional practices related to the reduction of cholesterol, salt, and sugar.

STAYWELL consists of orientations for employees and management, a behavioral health screening and health hazard appraisal, behaviorally oriented courses in each program area, and an extensive program using employee volunteers to modify positively and affect aspects of the work environment and workplace norms related to health behaviors. The program is offered on employee time, except for the orientation and the screening, which are given on company time.

Program implementation in new sites begins with a two- to three-month preparation phase that includes an extensive communication program with local management. The objective is not only to inform managers, but also to enlist active management support, to use managers as role models, and to have the program minimize work disruption and to accommodate the unique needs of each facility.

Employee enrollment has ranged from 65 percent to 95 percent. The relationship between program enrollment and employee demographics was analyzed for education, age, sex, and income. Surprisingly enough, none of these relationships was statistically significant. There were trends for employees with less than a high school education to enroll less, for people 60 years and over to enroll less, and for men to enroll somewhat less than women. These data run counter to the findings of many studies in which there has been a significant linear relation between enrollment in prevention programs and education. Enrollment was also lower at worksites where people did more traveling, and consequently spent less time at the workplace.

Employees who enroll attend a health screening in which data on health lifestyle, height, weight, and blood pressure are collected. Data from the health screening are used to generate a confidential health-risk profile for each employee.

The health-risk profile is a computer-generated analysis of each participant's most significant health risks. Each participant receives a report comparing their chronological age with their risk age (how old they are in terms of risk) and their achievable age (how old they could be if they decide to make the appropriate changes in their health-related behaviors).

Control Data has a strict confidentiality policy. No one in the company can have access to any individual data. Only aggregate data are used for evaluation purposes. The only exception is in the case of an abnormal blood or blood pressure profile. Abnormal profiles are reviewed by a physician. That review is not shared with the company, and participants decide when they register for the program whether they would like possible abnormal findings sent to their personal physician. Follow-up letters encouraging participants to review abnormal profiles with their physicians are sent.

Health-risk profiles are interpreted for employees at group interpretation workshops. Workshops help people identify their health risks and provide immediate opportunities for initiation of specific health actions, such as registration in courses or in the employee voluntary participation group program.

Lifestyle change courses are offered in smoking cessation, stress management, weight control, nutrition, and fitness. Courses are taught by health educators, psychologists, exercise physiologists, and other health professionals trained by Control Data to deliver the program. Where the program has been purchased by other companies, Control Data has either implemented the program and provided instructors or has trained the company's existing personnel.

Employee volunteer participation groups are the primary vehicle through which the program attempts to initiate and maintain health behavior using sociocultural processes. The purpose of participation groups is to modify the worksite environment in a manner that will affect norms and expectations of positive health-related activities.

During the program preparation phase, the names of informal leaders are obtained during management interviews. Informal leaders and other employee volunteers are invited to special orientation sessions in which they are encouraged to take an active leadership role in the program and are given direct opportunities to initiate and lead action teams. Action teams focus on changing some aspect of the work environment, or on the formation of support groups in which employees assist one another in changing health-related behaviors.

Action teams focus on activities, such as obtaining more nutritious food for vending machines, initiating low-calorie cooking classes, clubs or activities related to running, walking, hiking, bicycling, aerobic dance, or any other activity related to health behavior promoting activities.

Support groups form during the final phase of lifestyle change courses. Support groups assist members in implementing, continuing, and maintaining the difficult job of changing lifelong habits related to health behaviors. Support group activities include sharing of experiences, innoculation against failure techniques, and other activities that provide a supportive, sympathetic environment in which to continue or initiate health habit changes.

Participation groups give people opportunities to interact socially, extend opportunities for program entry beyond the formal entry phases, and facilitate program momentum. Participation group activities have been positively received by employees, and there tends to be more participation in these activities than in formal courses.

PROGRAM EVOLUTION BASED ON EMPIRICAL DATA

Significant changes have been made in the content and process of the program based on the evaluation data. The following vignettes illustrate the process:

MIDDLE-MANAGEMENT SUPPORT

When the program was first introduced in a pilot form in 1979, there was strong top-management support and considerable employee support for the program. At the pilot sites in New York and San Diego, however, initial implementation of the program was difficult. Evaluation data indicated that these problems were due to a lack of lower- and middle-management support for the program. Although we had been successful in selling top management, lower-level managers had serious reservations about how this program would affect their operations. A special orientation session for lower- and middle-level managers introduced in 1980 effectively solved this problem.

BLUE-COLLAR PARTICIPATION

Initial program participation by employees with less than a high school education was disappointing. Further analysis of the data indicated that employees who were working in production facilities had difficulty in taking the program because of a simple issue related to time structure. The program was offered on employee time, and the courses all had sessions that were an hour long. Blue-collar workers had only half-an-hour for lunch in many sites and consequently would have had to take the program before or after work to enroll in the courses. This problem was solved by developing a new set of courses whose sessions were either 20 minutes or a half-hour in length. This enabled blue-collar workers to take the courses during their lunch hours, and it resulted in a considerable increase in blue-collar enrollment.

A MORE "HANDS-ON" APPROACH

Both blue-collar and other workers were unhappy with the initial courses introduced in 1980. Although these courses were state-of-the-art, employees felt that they were too academic. People who enrolled in the program wanted a more "how to do it" approach, didn't want to feel like they were back in school, and wanted a less academic content. Major revisions were made in the weight, nutrition, fitness, and smoking-cessation courses as a result of this feedback. The revised courses were more popular, had less dropout, and resulted in more behavior change.

PROBLEMS IN COURSE COMPLETION

There was a high course dropout rate in the initial courses introduced in 1980 and 1981. Some of these problems were solved through course revision, but others were solved by directly addressing the dropout problem. An incentive system was developed so that people would have a higher incentive to attend class and complete the courses, as well as to develop group norms about course completion. In the weight-control program, for example, all the participants in a course could receive a gourmet, low-calorie cookbook. The cookbook is given to *everyone* in the course, but only if the *average* level of attendance is above a certain level. Consequently, course participants encouraged each other to attend class and not to drop out.

PROGRAMS FOR PEOPLE WHO CAN'T OR WON'T ATTEND COURSES

The revised program was still not reaching a significant subset of employees, even after the program revisions. Evaluation data indicated that these employees either had tight or unpredictable work schedules, were involved in extensive traveling, or did not wish to share their problems with other employees. This problem was addressed by developing a set of correspondence-like courses. These courses allow employees to receive assistance in changing their health-related behaviors, while pursuing the program on their own schedules.

CHANGES IN HEALTH BEHAVIORS

Self-reported health behavior change was compared among employees at STAYWELL and a series of matched non-STAYWELL control sites. These data were collected in a 10-percent representational sample survey ($N =$ approximately 5,000) of domestic Control Data employees. These data enable comparison of responses by employees in STAYWELL sites with employees in matched non-STAYWELL control sites and comparison of responses of employees at varying levels of program participation at the STAYWELL sites. Five levels of STAYWELL participation are differentiated: nonactive employees at STAYWELL sites, participation limited to the health hazard appraisal (HRP), any program activity beyond the HRP, and enrollment in lifestyle change courses.

Health behavior change was defined as whether employees reported significant changes during the past year in relevant health behaviors. The percentage of employees who reported any or substantial improvement in each area of health-related behavior is shown in Table 14.2. Participants in each of the lifestyle change courses report significantly more improvements in the five relevant health behavior areas than either STAYWELL employee groups at lesser levels of participation or employees at non-STAYWELL control sites. The strongest effects are in the area of smoking cessation. Fifty-eight percent of respondents indicate some change in smoking habits, and 35 percent indicate a substantial improvement in smoking behavior. Only 15 percent report any smoking improvement at control sites. These data are consistent with other data collected on the smoking-cessation program, showing that smokers who enrolled in the course smoked on an average of 1.6 packs per day. Twelve months after the course, 30.3 percent were not smoking, and another 43.5 percent were smoking less than 1 pack a day.

The data shown in Table 14.2 include all people who participated in each course, not just course completers. The inclusion of noncompleters and dropouts in these statistics represents a more accurate assessment of the effect of the courses on all those people who enrolled and gives a more conservative estimate of effects.

Additional data collected in the 1981 and 1982 Health Attitude Surveys suggest that these results are not due to self-selection bias. The 1981 Health Attitude Survey results indicate that employees who enroll in lifestyle change courses were not people who would have made these changes without course participation. Even fewer felt less capable of changing health lifestyle on their own than did those who failed to enroll in lifestyle change courses. The 1982 Health Attitude Survey results reveal that course participants are more likely to agree with items indicating that they are more likely to be motivated by co-workers, encouragement from doctors, and opportunities for group activity in changing their health lifestyle than are other employees. Citing reasons for why people do not change their behaviors, course participants are also more likely than noncourse participants to identify insuf-

Table 14.2.   Percentage of Employees Reporting Improvement in Health-Related Behaviors by Subject Area and Level of Program Participation

Subject	Control Sites	Not Active	HRP Only	Any Post HRP Activity	Course Partici- pation	Signifi- cance level P<
			Staywell Sites			
Weight Control						
% Improved	30	29	33	39	62	.0001
% Substantially Improved	8	11	8	11	18	.05
n	1688	223	226	169	74	
Smoking Cessation						
% Improved	15	15	13	11	58	.0001
% Substantially Improved	8	9	6	6	35	.0001
n	1661	216	218	204	31	
Exercise						
% Improved	40	27	34	53	62	.0001
% Substantially Improved	11	9	11	16	29	.01
n	1678	220	229	200	42	
Stress Management						
% Improved	34	28	32	31	62	.0001
% Substantially Improved	7	8	5	5	9	n.s.
n	1676	220	228	149	92	
Nutrition						
% Improved	31	29	33	39	52	.01
% Substantially Improved	8	7	7	9	15	
n	1684	220	228	197	48	n.s

ficient individual willpower, lack of support from family and friends, and the need for help to make behavior changes. These data suggest that participants were more, not less, likely to depend on social facilitation to initiate health behavior changes, and would not have shown as much positive health change without course participation.

Normative Change

An important premise underlying the STAYWELL program is that long-term health behavior change will be facilitated in a normatively conducive environment. The primary focus of the participation group program is on creating positive changes in health-related norms. Changes in health-related norms were measured by the 10 percent sample of all full-time employees. Normative change is measured by asking respondents the extent to which they perceive that *other* employees at the worksite are making positive changes in specific health-related behaviors. Three-year trend data reflecting

Table 14.3. Percentage of Employees Reporting Positive Health-Related Changes at the Worksite by Subject Area and Level of Program

| Subject Area | Level of Program Participation | | | Significance Level P< |
| | Control Sites | Staywell Sites | | |
		Non-Participants	Participants	
Number of over-weight Employees				
1980–1981	13	20	19	.0005
1981–1982	13	27	32	.0001
Smoking during work				
1980–1981	25	26	32	.015
1981–1982	20	30	32	.0001
Nutritious food availability				
1980–1981	19	26	31	.0001
1981–1982	19	26	32	.0001
Physical fitness				
1980–1981	12	30	44	.0001
1981–1982	13	40	54	.0001
Effective stress management				
1980–1981	11	13	20	.0001
1981–1982	12	20	24	.0001
Energy level at work				
1980–1981	8	12	15	.0001
1981–1982	9	13	18	.0001
Alcohol-related problems				
1980–1981	11	10	10	n.s.
1981–1982	13	16	12	n.s.
Coffee consumption				
1980–1981	7	8	8	n.s.
1981–1982	9	11	8	n.s.
Sample size				
1981	1436	378	412	
1982	1464	338	563	

responses to the positive changes on the part of other employees in the areas of physical fitness, overweight, smoking, stress, nutrition, energy level, coffee consumption, and alcohol use are shown in Table 14.3.

These data indicate that STAYWELL participants observe more change than do nonparticipants at STAYWELL sites or controls. Nonparticipants of STAYWELL sites also consistently observe more normative change than do people at control sites. Furthermore, these data are interpreted as suggesting that the STAYWELL program has been successful in initiating positive normative change in targeted areas, that these changes are perceived by nonparticipants as well as by participants, and becomes greater during the program's second year.

Two items were included to test for positive response bias and Hawthorne effects. These items were perceived changes in coffee consumption and alcohol problems. These socially desirable areas of change are not part of the STAYWELL program. There were no significant differences in these areas in either STAYWELL or control sites during the two years of this analysis. Normative change was limited to those specific areas targeted in the STAYWELL program.

RETURN-ON-INVESTMENT MODEL

A major premise underlying workplace health promotion programs is that these programs will be cost-effective for the companies who sponsor them. Control Data is in a unique position to assess cost-effectiveness because the company is not only self-insured, but also processes its own health care insurance claims.

The health care claims database has been included as part of the STAYWELL evaluation process, and the relation between behavioral health risks and health care costs and the relation between changes in health behaviors and health care costs are being examined. This evaluation is being done in the context of a mathematical simulation model. The model uses data in the STAYWELL evaluation as well as epidemiological and health economics data to estimate present and future potential cost-savings.

The model begins by estimating the percentage of people who smoke, have high cholesterol, high blood pressure, are overweight, and are not physically fit using demographic data on age, sex, and education, and normative data from Control Data's health hazard appraisal results. The extent to which the estimated population of employees at risk can modify their behaviors is calculated by using program participation rates based on the STAYWELL experience, STAYWELL evaluation data related to behavior change, and other health behavior change results from studies in the behavioral medicine literature. These behavior changes are used to determine estimated changes in chronic degenerative diseases over the long run using the Framingham and other epidemiological data, as well as to estimate short-term changes in health care and employee performance costs. Potential health savings estimated are discounted to current rates and compared with the cost of the program to estimate potential return-on-investment.

The model is useful because it makes assumptions explicit, provides a forum for scientists to critique the validity of studies underlying the model and focuses research on the development of more scientifically vigorous studies, allows users to vary assumptions and examine sensitivity of outcomes as a function of the model's assumptions, and facilitates use of operations research maximization models that can be used for optimal program impact planning by estimating such things as which potential target populations will yield maximum return on investment. The model is an organizing force for planning the program and for using evaluation results as they become available.

PLATO STAYWELL: A COMPUTER PROGRAM OF HEALTH BEHAVIOR CHANGE

Control Data has the largest library of computer-based instruction programs in the world. In early 1982, Control Data decided to examine the extent to which this computer technology could address the problem of health behavior change. The result is the STAYWELL PLATO program. That program is a computer-based and administered program of health behavior change. The natural evolution of an evolutionary empirical approach, PLATO STAYWELL "gets smarter" as more people go through it. It is not an artificial intelligence program. What it does is gather data to facilitate a continued improvement in the program's efficiency and to place people into an individualized program that has the highest probability of working for that person.

The program is extremely individualized. For example, the weight program contains an array of different approaches to control weight. Based on individual difference characteristics, people are matched to the programs that have the highest likelihood of being effective for them. The efficacy of that matching in accounting for variance in outcomes is assessed by a regression-based computer model. The program is also modified while a person is in it as a function of his or her behavior change and compliance. The branches that comprise these individual program modifications are also assessed by the computer-based model and updated as more people enter the program.

The STAYWELL program has a population large enough to establish the basic parameters required for an individualized program. Program users complete a behavioral profile prior to entry into the program. This behavior profile contains a listing of the major variables that have been hypothesized to relate to individual differences in response to programs in the clinical literature. For example, in the weight-control area such variables as knowledge about nutrition, the degree of social support at home, the degree of overweight, the number of programs a person has been in previously, sex, and other demographic characteristics are included in the profile.

In the initial iteration of the program, subjects are randomized across a number of intervention approaches. When a sufficient sample of people has run through the program, the individual difference variables are examined to determine their efficacy in predicting outcomes at program end and 12 months after program completion. Individual difference variables that are useful predictors remain in the model and those that do not account for significant variance are deleted. Variables whose main or interactive effects account for significant variance are used to match individuals to program paths in the next intervention. This procedure is repeated with each iteration (approximately every time 250 people complete the program), so that the system is able to make increasingly accurate predictions about the effects of matching people to program paths.

Each program path includes branches so that individuals who are not doing well may move to an alternative intervention, have the intervention enriched with adjunctive material, or repeat certain aspects of the intervention. Each branch point is treated and tested as an alternative experimental intervention. The efficiency of branch points are evaluated and reconsidered with each new cohort of people comprising one of the iterations in the evaluation process.

Through this approach, the computer-managed model can be considered as a paradigm for the evolution of empirically based theory that could yield cumulative scientific knowledge in this area. This method is consistent with the structural equation modeling paradigm using theory construction proposed by Blalock (1969).

The computer-managed program is individualized further by having a friendly, supportive tone, using the subject's name, remembering statements made by the subject earlier in the program, providing quantifiable and graphic feedback on the user's progress, facilitating comparison of the individual progress with other similar people in the program, and allowing users a wide latitude of choice.

The emphasis is on behavior change rather than limited to education. Each user receives only the information relevant to his or her specific problem. Users have opportunities to use new information in computer-managed simulation situations, make specific goals and objectives for behavior during the week following the session, and receive individualized computer-generated feedback on their success or lack of success in applying the techniques between lessons.

THE STAYWELL PROGRAM IN OTHER COMPANIES

The STAYWELL program has also been implemented in a number of other companies. These implementations have taken various forms, and consequently afford the opportunity to analyze the effects of some other program combinations and permutations.

In implementing the program in other companies, Control Data adopts that program to other company's policies and existing programs, if any exist. In some cases, Control Data has implemented a turnkey type of program, and in other cases Control Data has trained a company's employees to implement all or part of the program.

During the last year, the program was implemented at another company in which a small token fee was charged to each employee who enrolled in the program. The program at Control Data is given as a free corporate benefit to all employees and spouses. These different approaches afforded the opportunity to compare the results of payment on program outcomes. The company in which a token payment was charged had an initially lower enrollment level, but had significantly higher rates of both course completion and behavior change. The data from the other company were matched with a sample of Control Data employees for city size, type of work organization, and other variables that may have accounted for these differences. The differences remained significant, and suggest that a token payment may be useful in increasing behavioral change and compliance.

REFERENCES

Blalock, H. M. (1969). *Theory construction: From verbal to mathematical formulation.* Englewood Cliffs, NJ: Prentice-Hall.

15

The Johnson & Johnson LIVE FOR LIFE™ Program: Its Organization and Evaluation Plan

CURTIS S. WILBUR, TYLER D. HARTWELL, and PHILIP V. PISERCHIA

In recent years, many companies have initiated employee health promotion programs at the worksite (Alderman & Davis, 1976; Bjurstrom & Alexiou, 1978; Fielding, 1980; *New York Times*, 1981; Rundle, 1981). Expectations are high as new programs are launched amid promises of increased productivity, improved morale, decreased absenteeism, and attenuated illness care costs. Unfortunately, no formal studies have been designed and conducted with sufficient scientific rigor to support such claims. Moreover, it is not even clear whether such worksite programs have a significant and meaningful impact on the employee health and lifestyle characteristics. In short, great expectations are being created for worksite health promotion programs, but little evidence is being collected to substantiate whether such programs actually work as intended, much less achieve tangible "bottom line" benefits.

Employee health promotion programs require rigorous evaluation to establish themselves permanently in the business world. Evaluation establishes program accountability, an especially important feature at budget planning time. More important, an evaluation system can build the framework for systematically relating program inputs (e.g., physical fitness participation and improvement) to outputs (e.g., work performance, absenteeism). Only with such a system can employee health promotion programs function as the human resource management tool they are often promised to be.

338

Johnson & Johnson is conducting an epidemiological evaluation of its health promotion program, LIVE FOR LIFE™, using a formal research design. This is a unique effort to assess the overall health impact of a wellness program on an entire employee population. Simultaneously, an attempt will be made to relate employee illness cost data to those employee health and lifestyle characteristics collected in the epidemiological study. The overall evaluation objectives are to (1) assess whether the LIVE FOR LIFE™ Program has a meaningful impact on employee health and lifestyle characteristics and (2) determine the likely economic consequences of such changes. This chapter first describes the LIVE FOR LIFE™ Program in detail, then describes the epidemiological evaluation effort and presents selected baseline and first-year epidemiological results.

THE LIVE FOR LIFE™ PROGRAM

The Johnson & Johnson LIVE FOR LIFE™ Program (Figure 15.1) is a comprehensive health promotion program for all employees. By the end of 1984, LIVE FOR LIFE™ was available to more than 25,000 Johnson & Johnson

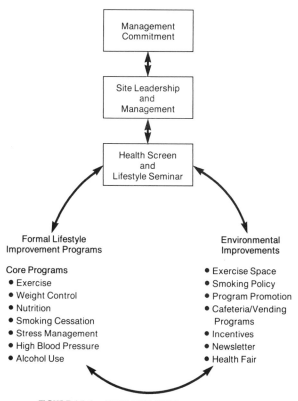

FIGURE 15.1.   LIVE FOR LIVE™ process at a plant.

employees at 43 locations in the United States, Puerto Rico, Canada, and Europe. It is specifically designed to improve employee health by encouraging healthful lifestyles. The long-term objective of the program is to help contain illness care costs attributable to unhealthy behavior and lifestyles that are amenable to modification in the work setting. Shorter term objectives include improvements in the quality of life, performance, and attitudes of Johnson & Johnson employees. Specific program objectives include improvement in physical fitness, nutrition, weight control, stress management, smoking cessation, health knowledge, blood pressure control, alcohol use, and proper use of medical programs.

LIVE FOR LIFE™ Corporate Staff

Primary responsibility for coordinating and delivering LIVE FOR LIFE™ Programs to the family of Johnson & Johnson Companies rests with a small corporate staff. They provide the technical assistance and support to a company as it builds its own LIVE FOR LIFE™ Program. Its primary expertise resides in supplying participating companies with the consulting expertise, training, core program components, professional services, and promotional materials necessary for program success. LIVE FOR LIFE™ staff are also responsible for program development and evaluation.

A program coordinator from the LIVE FOR LIFE™ staff is assigned to each participating company. The program coordinator functions as an account executive and is the essential link between the company and the full range of available LIVE FOR LIFE™ programs and services. The program coordinator is critical to program success. Although the program coordinator has no authority or responsibility for the day-to-day operations of a company's LIVE FOR LIFE™ Program, skillful consultation and selling ability with company leadership and management work to ensure that the LIVE FOR LIFE™ Program process is properly planned, organized, and implemented in a way that satisfies the health enhancement needs of the entire base.

Presentation to Company Mangement

The LIVE FOR LIFE™ Program process begins with a presentation to a company's management board. The management board is asked to make three commitments: to give financial resources, management time, and responsiveness to employee requests for environmental improvements. Finding sufficient financial resources is often the easiest commitment to fulfill. Management time is a more scarce resource. Nevertheless, management board members are made aware that the success of their LIVE FOR LIFE™ Program is linked to their willingness to recruit talented middle- to upper-management people to serve as senior volunteer leaders during the initial start-up phase of the program. The leaders work closely with the person hired as the LIVE FOR LIFE™ administrator at that site. Both the volunteer leadership group and the LIVE FOR LIFE™ administrator will recommend to the management board a range of environmental improvements necessary to encourage and support widespread employee participation and lifestyle change. The management board must be responsive to these requests and be willing to take those steps that, in their business judgment, are appropriate and sound. Development of exercise facilities, a company smoking policy, and flexitime are examples of proposed environmental improvements to which management boards are often asked to respond.

Selection of a LIVE FOR LIFE™ Administrator and Volunteer Leaders

A management board member is typically assigned operational responsibility for the LIVE FOR LIFE™ Program. This person is often the vice-president of personnel, but in a number of companies the vice-president of finance is selected. His or her first task is to identify and recruit a LIVE FOR LIFE™ administrator. This person is primarily responsible for organizing and marketing the LIVE FOR LIFE™ Program to employees, working closely with the senior group of volunteer leaders. The management board representative then selects the two or three senior managers (LIVE FOR LIFE™ Program cochairpersons) as leaders. The cochairpersons, in turn, select five to seven managers to lead task forces organized around such key areas as exercise, smoking cessation, stress management, weight control/nutrition, publicity, and general health knowledge. The task force leaders, in turn, recruit a cross section of employees as task force members.

The site LIVE FOR LIFE™ administrator and the volunteer leadership members work closely to execute the following management and administrative functions.

1. Develop and submit to management yearly marketing plans, participation objectives, and budget requirements.
2. Recommend environmental improvements to the management board.
3. Develop, execute and manage a strategy for achieving program plans and participation targets.
4. Request health enhancement services and materials from the LIVE FOR LIFE™ Program corporate staff and other vendors as required.
5. Prepare periodic reports to management on program results, including participation achievements and employee health and lifestyle improvements.

Health Screen

For most employees, the LIVE FOR LIFE™ process begins with participation in the LIVE FOR LIFE™ Health Screen. The Health Screen is promoted as a unique opportunity for all employees to find out how healthy they are and to learn about health in the process. The Health Screen is offered on company time and takes about one hour to complete. All employees who wish to participate in LIVE FOR LIFE™ exercise programs must complete the Health Screen. Health and lifestyle variables collected and analyzed during the Health Screen include biometric (e.g., blood lipids, blood pressure, body fat, height and weight, and estimated maximal oxygen uptake); behavioral (e.g., smoking, alcohol use, physical activity, dental hygiene practices, nutrition practices, coronary-prone behavior pattern, stress management skill use, sleep quality, human relations, and human potential); and attitudinal measures (e.g., general well-being, satisfaction with working conditions, relations with others, growth opportunities, ability to handle job strain, job involvement, job self-esteem, and organizational commitment).

Lifestyle Seminar

The LIVE FOR LIFE™ Program process continues with a *Lifestyle Seminar* that is usually scheduled about four weeks after an employee completes the Health Screen. The Lifestyle Seminar is the primary vehicle for introducing employees to their company's LIVE FOR LIFE™ Program. Personal responsibility for health is emphasized, especially within the context of a LIVE FOR LIFE™ Program in which the company has committed itself to creating a work environment rich in opportunities to improve and maintain personal health. Employee Health Screen results are returned at this point via an attractive document called a *Lifestyle Profile*. The Lifestyle Profile is positioned as a way to determine "how healthy you are," and provides the basis for taking action to improve health through LIVE FOR LIFE™ Program participation. Upcoming LIVE FOR LIFE™ Program health-enhancement opportunities are reviewed and promoted. The Lifestyle Seminar is offered on company time to groups of about 50 employees each, and takes about three hours.

Lifestyle Improvement Programs

An essential component of the LIVE FOR LIFE™ Program process is the ongoing opportunity for employees to improve or maintain their health through participation in exercise, weight control, nutrition, smoking cessation, stress management, high blood pressure, and alcohol use programs. The employees' participation is voluntary and is done on their own time. The set of core LIVE FOR LIFE™ Action Programs is described in Table 15.1.

In addition to Action Programs, a range of shorter educational and promotional programs are also available, built on such themes as breast self-examination, biofeedback, nutrition, blood pressure, and carbon monoxide analysis for smokers.

The strategy is to offer a combination of Action Programs that achieve high employee participation and, as a consequence, fundamentally restructure the social environment at work. Representative yearly participation targets are as follows:

Action Program Areas	Percentage of Eligible Employees
Weight control	25 percent of all overweight employees
Smoking cessation	20 percent of all employees who smoke
Stress management	20 percent of all employees
Exercise	40 percent of all employees
Total participation	60 percent + of all employees

Table 15.1. Descriptions of the LIVE FOR LIFE™ Program Care Action Programs

Action Program	Description
Smoking cessation	A 9-session program that teaches employees how to kick the habit
Weight control	A 10-session program that emphasizes eating fewer calories and burning up more through regular exercise
Exercise	A series of 12 week aerobic exercise groups
Applied stress management	An 8-session program designed to teach basic mental and physical relaxation skills
Yoga	A 12-session program that introduces employees to basic yoga exercises and skills
Personal Power	A 10-session program to teach employees how to manage stress through improved personal assertiveness
Nutrition	An 8-session program designed to cover the essential aspects of a prudent eating pattern
High blood pressure	A 4-session educational program and quarterly followup
Alcohol use	A 5-session program, which allows employees to explore the role alcohol and other substances play in their lives

Creating a Healthy Environment

The primary responsibility of the LIVE FOR LIFE™ administrator and volunteer leaders is to create a work environment that supports and encourages positive health practices among the greatest number of employees possible. Offering regular, convenient, and attractive Action Programs will substantially aid in creating a positive social environment based on high employee participation. Other aspects of the work environment are targets for improvements as well. The following examples describe several possible environmental improvements in key health and lifestyle areas.

Fitness:
> Shower and locker facilities on-site;
>
> Exercise facilities either on-site or rented from local organizations.

Weight control/nutrition:
> Scales in restrooms;
>
> Availability of convenient nutrition information where food is sold;
>
> Availability of nutritious foods in the company cafeteria and vending machines.

Stress management:
> Employee assistance program to provide professional treatment and referral services to troubled employees;
>
> Availability of management training programs designed to improve boss–subordinate relations;
>
> Flextime;
>
> Carpooling;
>
> Self-administered blood pressure equipment.

Smoking cessation:
> Smoking policy;
>
> Availability of LIVE FOR LIFE™ Program Thank-You-For-Not-Smoking signs.

Publicity and promotions:
> Incentive prizes and awards for participation in LIVE FOR LIFE™ activities;
>
> LIVE FOR LIFE™ Program newsletter;
>
> LIVE FOR LIFE™ Program bulletin board and information display area;
>
> Comprehensive recruitment brochure;
>
> Health fairs;
>
> Poster displays for upcoming programs.

The overall strategy is to surround the employee with information and incentives to spur healthier lifestyles. Long-term, this strategy seeks to create a corporate culture thoroughly supportive of good health practices.

EPIDEMIOLOGICAL EVALUATION DESIGN

To evaluate the impact of the LIVE FOR LIFE™ Program on a wide range of employee health and lifestyle characteristics, a two-year epidemiological study is underway. The epidemiological evaluation is being conducted jointly by the LIVE FOR LIFE™ staff and Research Triangle Institute. It is based on a quasi-experimental design in seven Johnson & Johnson plants in Pennsylvania and New Jersey. The design, designated as a nonequivalent control group design by Campbell and Stanley (1966), has four treatment plants and three control plants. This design was necessary in the present setting since employees could not be randomized to treatment or control within the same plant; and therefore, a randomized controlled trial, although preferable, was not feasible (Stolley, 1980). The design involves collecting Lifestyle Questionnaire and Health Screen data from employee volunteers in *both* types of plants (treatment and control) at baseline (before treatment), at one year, and at two years. After baseline data collection, the LIVE FOR LIFE™ Program is begun in the treatment plants but only a yearly LIVE FOR LIFE™ Health Screen is available in the control plants.

In addition to Health Screen volunteers, a random nonvolunteer sample was also selected at baseline in each of the treatment and control plants. This sample was a 53 percent sample[1] of all nonvolunteers in each plant as determined by employee rosters. The purpose of the nonvolunteer sample was to determine if (1) the volunteer and nonvolunteer populations were similar at baseline on the evaluation measures (e.g., Are smoking rates high among nonvolunteers?) and (2) to allow plant-wide estimation of parameters of interest. Thus the nonvolunteer sample allows one to examine such questions as whether only individuals with positive lifestyles become involved with the LIVE FOR LIFE™ Program. The employees in the nonvolunteer sample were contacted by trained interviewers and asked to fill out a shorter version of the Lifestyle Questionnaire (i.e., the "nonresponse questionnaire"), which required only 15 to 30 minutes of their time (this questionnaire was administered by the interviewer). Measures collected on the "nonresponse questionnaire" include smoking patterns, health knowledge, sick days, nutrition practices, self-care, general well-being, and physical activity.

Comparisons between respondents and nonrespondents indicated that the nonrespondents reported lifestyles as healthful as those of the respondents', with the exception that significantly more nonrespondents reported having smoked cigarettes and significantly more female nonrespondents currently smoke. Detailed results are described in Settergren, Wilbur, Hartwell, & Rassweiler (1980). The overall results indicate that the systematic biases found among nonrespondents in general population health surveys did not exist among this more homogeneous, employed population. In summary, the Health Screen volunteers provided a good estimate of the health status and lifestyle of the entire work force.

At year two of the epidemiologic evaluation, another nonvolunteer sample will also be drawn. This second nonvolunteer sample will include *all*

---

[1]This sampling rate was determined on the basis of minimizing cost to satisfy a minimum precision requirement.

individuals who gave data at baseline (volunteers or nonvolunteers) and did not volunteer data at year two. Thus an attempt will be made to obtain some data on the entire cohort of individuals who gave data at baseline. This second nonvolunteer sample will allow examination of the following questions: (1) Will only individuals with positive lifestyle changes return voluntarily at year two? (2) Does the rate of return at year two for individuals with positive or negative lifestyle changes vary between treatment and control plants? The nonresponse questionnaire at year two will collect data on smoking patterns, sick days, general well-being, physical activity, healthy heart behavior, alcohol use, stress management, and medical history. In addition, at year two a trained interviewer will aso collect blood pressure, weight, and height data on the nonvolunteers.

Results at Baseline for Selected Healthy Lifestyle and Attitude Measures

Baseline results from the seven Johnson & Johnson evaluation plants have been analyzed and are available for examination. Table 15.2 presents selected baseline results for volunteers by sex, as well as treatment or control.

Table 15.2 indicates that the volunteer Health Screen rate at baseline is approximately 77 percent with the rate somewhat higher at treatment plants (78 percent versus 75 percent). The Johnson & Johnson employee population at the seven evaluation plants is relatively young (average age for men = 34.9, and average age for women = 34.1), and the majority of men have at least a college education (60 percent). In general, the population at baseline

Table 15.2. Selected Baseline Characteristics of Health Screen Volunteers by Treatment and Sex

	Female		Male	
	Control	Treatment	Control	Treatment
Number of volunteers	751	1272	518	971
Volunteer rate (%)	73.5	79.3	77.3	77.7
Mean age (in years)	34.8	33.7	36.1	34.3*
Age distribution (%)				
18–24	24.3	22.4	8.1	11.1
25–34	35.1	41.0	43.6	48.2
35–44	14.7	16.8*	28.8	25.9
45–54	19.0	12.6	14.1	10.5
55	6.9	7.2	5.4	4.3
Total	100.0	100.0	100.0	100.0
Ethnicity (%) minority	9.8	16.4*	10.0	15.1*
College graduates (%)	11.6	20.7*	58.4	61.8

Table 15.2. (Continued)

	Female		Male	
	Control	Treatment	Control	Treatment
Smoking distribution (%)				
Never smoker	40.5	49.7	40.7	50.1
Former smoker	17.4	19.6	32.3	26.8
Current smoker	42.1	30.7*	27.0	23.1*
Total	100.0	100.0	100.0	100.0
Mean cigarettes per day (current smokers)	18.3	18.5	23.3	21.9
Mean total ethanol per week (oz)	2.4	23.8	4.6	4.0
Mean systolic blood pressure (mm Hg)	116.3	114.4	126.7	124.8*
Mean diastolic blood pressure (mm Hg)	75.1	73.3*	81.5	80.8
Hypertensive (% w/SBP 140 of DBP 90)	8.4	6.3	22.1	16.5*
Mean % above ideal weight [a]	15.1	11.7*	16.9	15.9
Mean total activity (kcal/kg/day) [b]	34.4	34.8	36.9	36.9
Mean general well being [c]	76.8	77.2	82.0	81.5
Mean Framingham type A behaviour scale [d]	6.1	6.0	5.3	5.3
Mean number of self reported sick days	6.1	5.6	2.7	3.8*
Mean job satisfaction with growth opportunities	59.5	61.0	63.8	62.9
Mean satisfaction with supervision	66.9	65.7	64.3	63.1
Mean satisfaction with working conditions	71.3	71.6	70.9	72.8

[a] Ideal weight is 5 pounds greater than the values in the Metropolitan Life Insurance tables

[b] Estimated from a previously validated, seven day physical activity recall interview (Blair et al, 1980)

[c] A scale based on a series of 18 questions developed from the National Center for Health Statistics (Frazio, 1977).

[d] A scale based on a series of 10 questions developed for the Framingham Study (Haynes, Levine, & Scotch, 1978).

*Difference between treatment and control with sex group is statistically significant at $p = .01$.

appears to be fairly healthy with overall systolic blood pressures of 125 for men and 115 for women. The percentage of smokers is 24.5 for men and 34.9 for women, and the mean number of self-reported sick days is 3.4 for men and 5.8 for women.

In general, at baseline, although some statistically significant differences exist, the three treatment and four control plants appear to have similar employee populations. The notable exception to the similarity of the populations is percent current smokers. For this variable, the treatment plants average only 26.9 percent, while the control plants average 35.0 percent. An examination of this phenomenon by plant indicates that two of the three control plants have a relatively high percentage of smokers.

Preliminary Results at One Year for Selected Health, Lifestyle, and Attitude Measures

Preliminary results on the cohort of employees attending both the baseline and Year 1 Health Screen are exciting. The evidence shown in Table 15.3

Table 15.3.   Preliminary One-Year Findings of the Epidemiological Study of the LIVE FOR LIFE™ Program

Health and Lifestyle Area	Health Screen Measure	% Change Baseline – 1 year		
		Treatment (n=1,223)[a]	Control (n=764)[b]	p
Fitness	Aerobic calories/kg/week	38	8	.001
Weight control	% above ideal weight	2	9	.001
Smoking cessation	% current smokers	−16	−4	.05
Stress management	General well being	5	2	.001
	% with elevated blood pressure (−140/90)	−26	−11	—
Employee attitudes	Self reported sick days	−9	13	.05
	Satisfaction with working conditions	3	−7	.001
	Satisfaction with personal relations at work	1	−3	.001
	Ability to handle job strain	0	−2	.001
	Job involvement	2	0	—
	Commitment to the organization	0	−2	—
	Job self esteem	0	−2	.05
	Satisfaction with growth	−1	−3	—

[a] 66% of employees screened at baseline and on-site
[b] 74% of employees screened at baseline and on-site

clearly indicates that the treatment cohort, in comparison with controls, consistently shows greater improvements in the major health, lifestyle, and attitude areas addressed by the LIVE FOR LIFE™ Program.

In summary, the preliminary baseline one-year comparison available at this point confirms that the LIVE FOR LIFE™ Program is capable of achieving significant and meaningful improvements in the health and lifestyles of Johnson & Johnson employees. The findings are consistent across the range of core program areas.

SUMMARY

LIVE FOR LIFE™ is in the business of selling health. It was created by Johnson & Johnson in response to the perception that a tremendous opportunity existed to improve the life quality of Johnson & Johnson employees significantly and, in the process, to realize an economic benefit. The LIVE FOR LIFE™ Program evolved as a process aimed at achieving high employee participation. Standardized operating procedures have been developed to ensure a reliable and consistent delivery of health promotion services wherever the LIVE FOR LIFE™ Program is offered. LIVE FOR LIFE™ Program services satisfy the needs of a diverse customer base including employees and management. Because the LIVE FOR LIFE™ Program is in the business of selling health to essentially healthy people, its marketing strategy emphasizes the positive benefits employees will experience as a consequence of participation in LIVE FOR LIFE™ Program activities. The LIVE FOR LIFE™ Program process begins with its acceptance and support by top management. A LIVE FOR LIFE™ administrator is recruited to manage a company's program, working with a group of volunteer leaders. Employees begin the process by participating in a Health Screen, and results are returned in the context of a Lifestyle Seminar. A series of core Lifestyle Improvement Programs are offered in areas such as smoking cessation, weight control, nutrition, fitness, stress management, high blood pressure, and alcohol use. Intermingled with these programs are efforts aimed at sustaining participation. Finally, environmental changes that support positive health practices are engineered by the LIVE FOR LIFE™ administrator.

The results of the LIVE FOR LIFE™ Program have been examined within the context of a rigorous epidemiological evaluation. Preliminary evidence clearly suggests that employees in the treatment cohort, in comparison with controls, consistently show greater improvements in the major health, lifestyle, and attitude areas addressed by the LIVE FOR LIFE™ Program.

The ultimate judges of the Program's effectiveness are the management and employees of Johnson & Johnson. The LIVE FOR LIFE™ Program must constantly satisfy the health enhancement needs of this diverse customer base. Although the preliminary evidence from the LIVE FOR LIFE™ Program epidemiological study indicates that customer response has been extremely positive, careful attention to customer needs must be maintained if the LIVE FOR LIFE™ Program is to prosper in time.

REFERENCES

Alderman, M. H., & Davis, T. K. (1976). Hypertension control at the work site. *Journal of Occupational Medicine, 18*(12), 793–796.

Bjurstrom, L. A., & Alexiou, N. G. (1978). A program of heart disease intervention for public employees. *Journal of Occupational Medicine, 20*(8), 521–531.

Blair, S. N., et al. (1980). Measuring physical activity by a seven-day recall method. *Crisis in the public sector* (program and abstracts, p. 232). Washington, DC: American Public Health Association.

Blair, S. N. (1984). How to assess exercise habits and physical fitness. In J.D. Matarazzo, S.M. Weiss, J.A. Herd, N.E. Miller, & S.M. Weiss (Eds.), *Behavioral health: A handbook of health enhancement and disease prevention.* New York: Wiley, pp. 424–447.

Campbell, D. R., & Stanley, J. C. (1966). *Experimental and quasi-experimental designs for research.* Chicago: Rand McNally.

Fielding, J. E. (1980). *Health and industrial relations.* Institute of Industrial Relations, University of California at Los Angeles.

Fielding, J. E., & Breslow, L. (1983). Health promotion programs sponsored by California employers. *American Journal of Public Health, 73*(5), 538–541.

Frazio, A. F. (1977). *A concurrent validation study of the NCHS General Well-Being Schedule.* DHEW Publication No. HRA 78-1347. Data Evaluation and Methods Research, Series 2, No. 73 (pp. 1–52). Washington, DC: U.S. Government Printing Office.

Haynes, S. G., Levine, S., & Scotch, V. (1978). The relationship of psychosocial factors to coronary heart disease in the Framingham Study: I. Methods and risk factors. *American Journal of Epidemiology, 107*, 362–383.

New York Times. (1981, August). New health plans focus on "wellness."

Rundle, R. L. (1981, July 27). Coors urging employees to belly up for fitness. *Business Insurance, 15*(30), 3, 29.

Settergren, S. K., Wilbur, C. S., Hartwell, T. D., & Rassweiler, J. (1980). Comparison of respondents and nonrespondents to a worksite health screen. *Journal of Occupational Medicine, 25*(6), 475–480.

Stolley, P. D., (1980). Epidemiologic studies of coronary heart disease. Two approaches. *American Journal of Epidemiology, 112*(2), 217–224.

16

Healthy People in Healthy Places: Health Promotion Programs in the Workplace

KENNETH R. PELLETIER

Historically, the involvement of large organizations in the health care of their employees has been minimal and consisted primarily of their investment in "health" insurance, which was estimated to be $42 billion in 1978 by the U.S. Commerce Department. Although a few instances of more direct, corporate involvement in medical care exist, these have been considered anomalies. After World War II, industrialist Henry J. Kaiser began to develop the first Health Maintenance Organization, a prepaid group practice, which emerged as the largest private, nonprofit, direct-service program in the world by providing care to more than four million people. Similarly, the Gillette Company began a health care program for its employees more than 30 years ago. With these few exceptions, most organizations limited their involvement to treating work-related accidents and created these programs in response to the labor union movement. This situation is changing rapidly. Clearly, the change in consciousness regarding health promotion in the workplace is not the dominant mode at present but is the trend of the future. During a 1981 speech to the National Industry Council for HMO Development, Robert Burnett, the chief executive oficer of the $100-million-a-year Meredith Publishing Company, stated, "What's the average CEO's information quotient on the subject of health care costs? Somewhere in the area of 0 to 5 on a scale of 100." This sentiment is echoed in the somewhat more optimistic tone of David A. Winston, chair of a Health and Human Services task force, when he stated, "On a scale of 1 to 100, I would rank corporate interest in health policy issues 25, but moving up rapidly" (Iglehart, 1982, p.120). This trajectory is most encouraging and two driving forces are behind this trend.

Most evident is the economic imperative as stated by Willis Goldbeck, quoted in a recent *New England Journal of Medicine:* "It is very important to recognize that you can purchase medical care like any other product ... tough policy and value decisions need to be made, and where economic leverage must be brought to bear ... remember that you are using your stockholders' money, and your profits to pay for health care that is not really needed. That's what it really comes down to, and in this era of limited resources none of us can afford that kind of waste" (Iglehart, 1982). Later in the same article, Robert Burnett reflected on his involvement in launching an HMO in Des Moines for his company. That effort failed primarily due to resistance by the medical community, but clearly Burnett and other top executives recognize the necessity of such reforms and intend to remain involved in this transition. According to Burnett, "the tragedy is that I can make more money for the corporation and its stockholders in the next three or four or five years ... by doing something effective in the way of cost control than I can by selling ... every dollar of health care costs that's saved goes straight to the bottom line." If this holds true, and there is every indication that it will, then it is increasingly likely that there will be an imperative for more organizations to take a serious role in health promotion for their own economic interests. For some communities, the organizations and corporations that employ a large number of men and women in those areas could emerge as the most potent providers of health resources in the future.

Before proceeding, let us examine the various ways in which this toll of overall disease has been related to specific impacts on employees and employers. Even beyond the appalling statistics themselves, the issue of even greater significance is that each instance is an example of the inextricable interactions among lifestyle, workplace, environment, and economic determinants of an individual's health and longevity. Furthermore, evidence already indicates that every instance can be significantly reduced by preventive and health promotion programs, which will be the focus of this chapter. It is essential to remember that these abstract statistics represent the accumulation of individual suffering, illness, and economic loss, as well as impaired productivity and economic loss for the organizations involved. To date, records indicate the following:

1. For alcohol abuse alone, the total costs were $44.2 billion, with $11.9 billion in direct costs and $32.3 billion in indirect costs. Alcoholic employees experience twice the average rate of absenteeism as other employees, which results in a significant loss for all concerned (DuPont, 1979; Erfurt & Foote, 1977).

2. There is a striking and often misquoted statistic regarding sickness

insurance in the automobile industry as one example. Among major corporations, General Motors discovered that it spent more money, $825 million per year, on health insurance and disabilities than on steel from U.S. Steel, one of its principal suppliers. Furthermore, these costs added more than $175 per automobile for its 1979 models. Disease treatment costs become a part of every service or product that is used or consumed (Parkinson et al., 1982; Sapolsky, Altman, Greene, & Moore, 1981; Stacey, 1980).

3. Similarly, Ford Motor Company estimated the health care per employee at $3,350 per year and $290 per automobile during the year 1980. On a positive note, Ford saved $2 million annually from the voluntary participation of 10,000 employees in a Health Maintenance Organization (HMO). That constitutes only 4 percent of its U.S. work force (Parkinson et al., 1982; *U.S. News and World Report*, 1980).

4. A survey by Pacific Mutual Life places the cost of poor nutrition at $30 billion annually and has estimated the cost of replacing a high-level executive between $250,000 and $500,000. This finding is echoed by Xerox Corporation, which spends $58 million annually on insurance and medical claims with the loss of an executive at age 41 costing from $600,000 to more than $1 million (*Behavior Today*, 1978).

5. Equitable Life Assurance has estimated the cost of one person with a chronic headache in 1979 at $3,394.50 per year. Consistent with their business, the costs were presented in a manner similar to an insurance premium (Manuso, 1978):

a. Visits to employee health center	$ 473.14
b. Time away from work	56.61
c. Work interference due to symptoms	2,206.95
Work interference affecting superiors	72.80
Work interference affecting co-workers	542.88
Work interference affecting subordinates	42.12

6. Finally, there is one often cited instance from a recent study by the National Institute on Occupational Safety and Health (1980), which estimated the cost of "executive stress" at $10 to $20 billion in the United States alone. That figure covers only the clearly measurable items as workdays lost, hospitalization, outpatient care, and mortality to executives, and was noted as the following:

Cost	Conservative Estimate	Ultraconservative Estimate
Cost of executive workloss days (salary)	$ 2,861,775,800	$ 1,430,887,850
Cost of executive hospitalization	248,316,864	124,158,432

Cost of executive outpatient care	131,058,235	65,529,117
Cost of executive mortality	16,470,977,439	8,235,488,720
	$19,712,128,338	$ 9,856,064,119

These are only a few examples of such innumerable statistics, which are tragically reminiscent of Everett Dirksen's famous quip, "A billion here, a billion there—pretty soon you are talking about real money!" Actually, there are innumerable ways in which these costs have been analyzed and presented. However, in most instances, if an individual or organization knows the costs, they have already taken action, and if these costs are not evident for a specific individual or organization, then elaborating them seldom leads to action; thus there is no need to belabor such statistics. Furthermore, preventive health care and holistic medicine concerns move beyond the quarter-of-a-million people directly involved in health care to address the 200 million who are not.

Spheres of Influence

Given the enormous and complex challenge of health promotion in the workplace, one person can initiate some actions, and more can be accomplished when people and organizations work together for mutual benefit. To clarify this issue, it is helpful to picture an individual at the center of a series of concentric spheres, like a three-dimensional bullseye surrounded by various determinants of health. This diagram represents "sphere of influence" to emphasize that it is possible for individuals and organizations to undertake an active role in influencing and changing each of these determinants for better or worse. We are not passive victims but have created organizational and environmental conditions that threaten the lives of their creators in a modern version of Mary Shelley's *Frankenstein*. One major cultural force at work today, as manifested in energy management, holistic health care, voluntary simplicity, and nuclear weapons control, is the empowerment of individuals to take a more active role in the determination of the course of their individual and collective destiny. In that sense, it is possible to consider certain vital fulcrum points or "Spheres of Influence," where actions can be effectively undertaken.

1. **Sphere I.**   At the center are personal health choices, such as stress management, "workaholic" behavior, "stress addiction," nutrition, physical activity, cigarette smoking, alcohol and drug abuse, and the entire range of personal health determinants. Among the personal lifestyle factors specific to the workplace are education in lifting techniques for prevention of back injuries, stretching exercises for secretaries and virtually all white-collar workers who spend long

hours seated at a desk, stress management programs, and eating healthier lunches in workplace cafeterias.

2. **Sphere II.** In the second sphere are factors that are primarily controlled by management, but which individuals and aggregates of workers must continue to influence, such as toxic chemical exposure, noise pollution, air quality, lighting conditions, monotonous jobs, reduction of unnecessary organizational stressors, improved employee with employer communications, introduction of modifications of Japanese "quality circles," educational programs in "medical self-care," and overall health promotion programs that extend personal health practices into the workplace.

3. **Sphere III.** Health policies and incentives determined primarily by management with worker input, such as regulation and elimination of workplace hazards, health incentives, such as monetary rewards or "well leave," subscribing to a Health Maintenance Organization (HMO) plan that can minimize excessive use of medical services, and provision of child-care facilities for single parents or families where both the husband and wife are employed outside the home.

4. **Sphere IV.** At the fourth level are health determinants beyond the immediate regulation of individuals or organizations except for voting and legal challenges. Among these determinants are economic conditions, legislation at the state and federal levels, as well as legal precedents and litigation. The most obviously visible areas are environmental protection measures and hazardous waste regulations, which have been suspended, postponed, or revised by the Environmental Protection Agency (EPA) under the Reagan Administration. In the "invisible" hazards area, M. Harvey Brenner of Johns Hopkins University presented evidence that a 1 percent jump in the unemployment rate is associated with an additional 36,887 deaths and related increases in hospital admissions, imprisonments, and violence in families (Brenner, 1979). Economic and political decisions have personal consequences.

Of equal significance is the crushing rise in population, where for more than 99 percent of all human history, the human population was under 7 million. Today, that many people are added every month. From both animal and human studies, overcrowding and competition for limited resources are major causes of disease and death. In actuality, these levels overlap and intersect and the individual is like a fish in an ocean of such influences, all of which must be changed if optimum health and longevity are to ensue. In addition, this model can be extended onward from the individual to the nervous system, to molecules and atoms, as well as outward to communities, subcultures, nations, and planetary ecology.

In this discussion, the first three "Spheres of Influence,"

emphasizing the individual within an adaptive organization, are the primary focus. From such efforts numerous benefits have already been reported for individual employees, such as the following: (1) improved job satisfaction (Manuso, 1979; Salvendy, 1980); (2) decreased blood pressure (Alderman, 1976; Alderman, Green & Flynn, 1979; Arnold, 1973; Bjurstrom & Alexiou, 1978); (3) improved morale (Blair et al., 1979; Kobasa, 1981); (4) better overall health (Jennings & Tager, 1981; Kreitner, 1976); and (5) economic awards and peer recognition from participation in health promotion programs (Kristein, 1977). For employers, the benefits have been as follows: (1) decreased absenteeism (Cox & Shepherd, 1979; Muchinsky, 1977); (2) improved productivity and a sharpened competitive edge (Kennedy, 1976; Martin, 1978; Nehrbass, 1976); (3) better employer and employee relations (*Business Week,* 1981; Kiefhaber, Weinberg, & Goldbeck, 1978); (4) decreased medical and disability claims (Corrigan, 1979; Larson, 1979; Love & Love, 1978; Lublin, 1978); and (5) decreased employee turnover (Patrinos, 1977). All of these benefits are mutually enhancing, and there are numerous other positive indications from various programs for both individuals and organizations.

Background of Health Promotion Program

It is widely recognized that health promotion is considered to be the most important approach to reduce illness and promote health. According to the most recent U.S. Surgeon General's Report, *Healthy People: Health Promotion and Disease Prevention* (DHEW, 1979):

> Further improvements in the health of the American people can and will be achieved—not alone through increased medical care and greater health expenditures—but through a renewed national commitment to efforts designed to prevent disease and to promote health.

Moving from a general statement of policy to suggest specific measures in a lengthy article in *Science,* Marvin M. Kristein, Chief Health Economist at the American Health Foundation, and his colleagues (Kristein, Arnold & Wynder; 1977) stated:

> It can be said unequivocally that a significant reduction in sedentary living and overnutrition, alcoholism, hypertension, and excessive cigarette smoking would save more lives in the age range 40 to 64 than the best current medical practices.

Most recently, Richard S. Schweiker, Secretary of the Department of Health and Human Services, stated:

> We are going to stress . . . preventive health care. Another word for preventive

health care is "wellness." For example, many companies are finding out that there is a big advantage to keeping their employees healthy. . . . We hope to encourage such programs by providing incentives for them . . . to give people incentives to stay well rather than get sick . . . to find out more about what people can do to live longer. (Cranton, 1981.)

Despite the prominence and consistency of these acknowledgments, health promotion programs have not advanced rapidly. This is quite inexplicable since a 1981 Task Force for the California Department of Health Services, under chief physician Kathleen H. Acree, has clearly stated a general consensus that, "In one sense the planning of the health promotion program is very straightforward. Only a limited number of health behaviors have been shown to at all improve health. Even the methodologies for influencing the behaviors, while not totally satisfactory, are well-identified" (Acree, 1981). If the need for individual and organizational health promotion has been unequivocally recognized and the means for implementation are readily available and such measures have been tested in limited but promising studies of cost-effectiveness, why are such programs not more evident? That is a highly consistent, well-founded, but "Catch-22" question, which essentially asks: Where are the data and cost-effective indications from programs and studies that have not been given adequate emphasis or funding for decades?

One dimension of this critical question is as simple as it is elusive, yet it has a direct bearing on every aspect of health promotion in the workplace. This is the dimension of *incentives* which reward either sickness-enhancing or health-promoting behaviors and is an issue that will recur in any health promotion program. There is a fundamental need to restore both individual involvement as well as the economic and health benefits for that participation, to any one person or group of people who voluntarily undertake health promotion efforts. Until such positive incentives are created to at least offset the innumerable negative incentives that are already present, any health promotion effort will not have enough inherent or immediate reward to sustain it over time, and it will fail. In an article entitled "The Health Promotion Organization," physician Lorenz K. Y. Ng, from the National Institute on Drug Abuse, and his colleagues (Ng, Davis, & Manderscheid, 1978), summarized the situation concerning sickness incentives as follows:

The health care system provides economic incentives for sickness rather than health, in that people receive financial rewards from most health care plans only when they are ill. Physicians are paid only for treating illnesses, and there is no incentive to focus on methods for promoting health. These negative incentives extend into other realms as well. . . . Such practices not only fail to reward those who are healthy or who make an effort to stay healthy, but also implicitly penalize them.

There is no other place, except in the public school system, where sickness incentives are more prominent than in the workplace where sick days can translate into time off, where disability claims can be tantamount to a paid vacation, and sickness care is seen as free or at least paid by a remote third party. To compound these tendencies, there are strong peer pressures toward alcohol abuse, macho images promoting cigarette smoking and high speed driving, as well as the ultimate incentive of time off from work. None of these behaviors implies deliberate manipulation or deception by the vast majority of working people; instead, we are simply noting that the incentives of the present system are conducive to certain behaviors and sickness incentives. At present, the scenario is one of infinitely escalating costs, diminishing returns, and negligible effects on the *health* of the population, within a system where supply and demand are controlled by the same organizations.

By contrast, health promotion programs have found a positive response to health incentives, such as courses in stress management, discounts on goods and services, memberships to recreational facilities and fitness clubs, economic incentives in the form of salary increases, and even the creation of planned "well days" in place of unused sick leave. Health incentives are a controversial realm with potential for manipulation and abuse. A fine line exists between an appropriate incentive or encouragement and an undue inducement or coercion in matters related to employee health. Based on his extensive work with more than 200 companies of the Fortune 500, Willis Goldbeck (1982), President of the Washington Business Group on Health, has clearly analyzed the need to be aware of multiple incentives and motivations in any health promotion program:

> One should never approach a potential corporate supporter with exaggerated claims of future savings. Cost savings are a very legitimate reason for employers to be interested, but that reason alone is not sufficient. Community groups need to work with employers along the whole spectrum of logical, health, financial, environmental and ethical motivations for developing a wellness agenda in every community.

With this background in mind, it is clear that any attempt to formulate an effective health care system, in essence, one that is designed to elicit and sustain optimum health and longevity for the population as a whole, must meet certain basic criteria. In a memorandum to Governor Jerry Brown, attorney Rick J. Carlson (1982), former chairman of the California Governor's Council on Wellness and Physical Fitness, noted three constraints on the development of effective health promotion programs:

> First the lack of disincentives to deter people from using unnecessary and inappropriate curative medical services and as a result, contain stampeding medical care costs. Second, the lack of sufficiently powerful incentives to encourage people to assume more responsibility for their health and therefore utilize less medical care also resulting in cost containment. And third, the availability of money to develop realistic health promotion options to aid them in assuming that responsibility.

These prerequisites apply to both individuals and organizations. Despite these restrictions, the movement to create the healthy workplace as an adjunct to and extension of individual lifestyle practices is becoming a reality in some innovative and prominent government agencies, school districts, major corporations, small businesses, several major universities, and numerous health promotion facilities.

Basic Principles of Workplace Health Promotion

Turning to the question of what constitutes an effective health promotion program, there are consistent, key prerequisites and assumptions that arise in any discussion, as well as in any individual or organizational program. First and foremost, there is one basic concept that requires emphasis since if it is not understood then all else fails, in essence, the factors and conditions promoting individual health and longevity are the same as those promoting the organizational health and longevity of the greater or "incorporated" body. It is necessary to achieve a balanced view of the interactions among individual, organizational, and environmental factors as the primary method for disease prevention and health promotion.

Among the more articulate analysts of such complexities in organizational, particularly corporate, programs is Richard L. Pyle, who is an instructor in Management at the University of Massachusetts. In his article in the *Training and Development Journal*, he noted: "Far from a singular concept, a 'fitness program' represents anything from insignificant financial support for physical activities to $1,000 per year expense per selected participant in exclusive corporate facilities" (Pyle, 1979). Fitness programs are usually highly limited and focus primarily on cardiovascular fitness through specific aerobic exercises, such as jogging or working out on a stationary exercycle. Even in the instances of a limited focus on "physical fitness," according to Pyle, "Considerable anxiety has been felt about corporate fitness programs. The anxiety results from the inability to justify the investment for a program and facilities to corporate stockholders" (Pyle, 1979). This is a common concern in any organization whether profits, tax dollars, or any budget allocations are at stake. Although this is a pervasive concern, it is possible to reconcile such issues within the demands of serving both the individual and the organization. There is increasingly widespread support for Pyle's claim that "Even strong proponents are recognizing the necessity of providing business with the logic and means of evaluation. . . . Fitness programs . . . must be viewed as functions concurrent with corporate objectives, and not simply as goodwill fringe operations" (Pyle, 1979). This is a challenge that can be met and should provide a more solid basis for the development of effective health promotion programs in the workplace as well as for the public as a whole. However, one important dimension to recognize is that a critical dimension of any health promotion activity will remain individual, personal, subjective fulfillment and that cannot be underestimated or dismissed. Rather than forcing individual or program evaluations into preconceived, scientific standards, it will be far more challenging to develop assessments that are sensitive and accurate enough to measure what people and programs are achieving. Of course, although it is possible to formulate a series of reasonable, attainable objectives over time, any program cannot meet all demands immediately or simultaneously.

For the present time, it is sufficient to note that virtually every organizational health promotion program, whether limited or holistic, has demonstrated itself to be consistent with individual and organizational objectives as well as cost-effective. Of even greater concern are the two major criticisms often cited regarding organizational health programs: "(1) the employer doesn't have a right to meddle in the personal lives of the employees; and (2) the testimonials are personal individual experiences which are not generally applicable to the overall population" (Pyle, 1979). By virtue of relocation, time demands, weekend travel, job termination, and many other factors, employers are inevitably influencing every employee's personal life and all too often in a negative fashion. It is the second area that is closely related to the first and addresses the area of the effectiveness of any health promotion effort in any setting. A major concern is that of both protecting and eliciting personal and organizational responsibility.

A highly emotional controversy is occurring in the courts, as well as in many workplaces, concerning the complexities of responsibility for the health of a person in the workplace. In six states, courts have extended workman's compensation coverage to include cases of illness caused by "gradual or cumulative injury from work stress," and nine other states have permitted compensations for anxiety, depression, and other disabling psychological disorders. This is a complex area as indicated by Alan McLean, eastern medical director for IBM, who has maintained that recent court decisions "have ignored preexisting emotional conditions as well as personality or family factors, all of which may underlie or contribute to an employee's mental illness" (Rice, 1981). Actually, IBM has long been noted for its humane employee policies but has drawn the line on psychological counseling and has encouraged the use of outside resources for its employees in this area. At one extreme is a misunderstood, but loudly extolled, model of the Japanese system of "lifetime security"; at the other are astonishingly arrogant pronouncements by a minority of organizations that health promotion programs are "coddling," or "knuckling under" (Crowley, 1981) to unrealistic and unfounded employee demands. Neither extreme is accurate. Closer to a middle ground is the observation made by psychiatrist Sidney Lecker, who is a consultant to several major corporations: "It is nonsense to say it's completely a matter of how the individual employee reacts to stress. The big factor is the overall quality of work life and the organizational climate" (Rice, 1981). Amid the controversy, it is clear that there is mutual responsibility in separate but related areas. At the conclusion of an insightful overview of workplace ethics, Berkely Rice (1981) concluded:

A more humane and, in the long run, perhaps more productive commitment might offer employees not only job security, but within reason, emotional security as well, particularly for those who have spent 20 years or more with the company . . . assuming responsibility for employee mental health may become a matter of legal obligation rather than an optional policy of management.

If individuals and organizations choose to accept a measure of responsibility, health promotion programs will be effective for individuals and organizations alike. If this imperative is ignored or overlooked, the results will be reflected in increased worker dissatisfaction, rampant absenteeism, greater disability and insurance costs, decreased productivity, higher sickness insurance premiums and disability costs for the organization, and premature disability and death for the individual. In a fundamental sense, when both the individual and the organization are sound, then it is an asset for all concerned.

While the medical care industry and most government bureaucracies have been predictably slow to react, private business has reacted swiftly and effectively. This is primarily due to the following observations contained in *Business Week* (1979):

Employers picked up the tab for about $40 billion of the nation's $183 billion in health care expenses in 1978. In fact, employee medical benefit costs have risen so rapidly that they account for as much as 10% of total compensation in some companies. . . . All share a common goal: to contain the runaway cost of medical benefits. If nothing else, such efforts serve notice on the health care delivery system that the business establishment has abandoned its unquestioning role.

These figures have grown considerably since 1978, and there is even greater support for the necessity of taking action with significant results ranging from "hospital utilization reviews," which indicated that patients admitted on Friday tended to stay longer in hospitals than did those admitted on Monday or Tuesday, to sophisticated and comprehensive health education and promotion programs. Despite these promising trends, they are still the exception rather than the rule. In fact, a 1981 survey of chief executives in 69 major companies by Harvey M. Sapolsky and his colleagues at the Massachusetts Institute of Technology concluded that measures such as "cost-sharing," "tightening claims control," "HMOs," and similar measures were not of great interest to corporations. Many of these measures were seen as coercive to employees and were unlikely to gain wide acceptance. However, on a more positive note, Sapolsky stated: "There is no doubt of the popularity of preventive health measures as additional employee benefits. And if the claims made for these prevention programs by their advocates even partially materialize, then the future medical costs of many corporations may decline" (Spolsky et al., 1981). At least, it is a note of optimism that organizations seem far more interested in genuine health promotion than in varieties of pathology management.

Health Promotion Programs

What are the components of a health promotion program? While the actual programs vary, one factor remains constant. Voluntary health promotion practices by aggregates of informed individuals is at the heart of any effective health promotion program. For this one essential requirement, no substitute exists. Closely related to this requirement is the question, Why should health promotion programs take place at the workplace, rather than elsewhere? This question is easier to address since many demonstrable advantages have been enumerated by physician Jonathan F. Fielding (1981), co-director of the Center for Health Enhancement (CHEER) at the UCLA School of Medicine:

> (1) access to people and time; (2) stability of the working population; (3) lower costs; (4) existence of management and organizational structure; (5) ability to conduct several interventions simultaneously; (6) strong social support networks; and (7) willingness of the working population to participate in occupationally sponsored programs.

Overall, this point is emphasized in Governor Jerry Brown's (1982) mandate to the Governor's Council on Wellness and Physical Fitness for the state of California: "More intensive, high cost treatment will do little to *prevent* chronic, debilitating and costly disease and the health care setting and the workplace are the ideal places to initiate health promotion programs". Since there is such an array of health promotion programs, it is useful to divide the majority of them into four major categories:

1. Toxic substance elimination and regulation;
2. Health promotion policies initiated usually by the employer and involving the controversial measures of "cost sharing" and "health incentives";
3. Limited health promotion programs;
4. Holistic health promotion programs.

Included under the first category of toxic substance regulation are the data gathering, policy setting, and regulatory activities of the National Institutes for Occupational Safety and Health (NIOSH). While exposure and toxic agents have received considerable attention and several instances of diseases caused by such agents are clearly evident, a recent article by Allport (1981) in *Medical Tribune* summarized the current status of this area as follows:

> Most occupational diseases are not so easily linked and are notoriously difficult to trace. Safe exposure limits have been set for only about 500 of the approximately 70,000 chemicals in industrial use today. . . . It has only recently become clear that in addition to such physical disorders as pneumoconioses,

cancers and skin diseases, mental health problems can be caused by many chemicals, such as carbon monoxide, carbon disulfide and trichlorethylene.

Lack of data does not minimize the magnitude of the problem, since the National Cancer Institute has conservatively estimated that 20 percent to 40 percent of all cancer is caused by occupational exposure. Moreover, 90 percent of the American companies employ fewer than 25 people. In these small companies, it has been estimated that one out of every four workers has an occupational disease. These are the effects of the known, visible hazards. For physicians and workers, a NIOSH publication entitled *A Guide to the Recognition of Occupational Diseases* (1981), which can be purchased from the U.S. Government Printing Office, is helpful for an individual becoming aware of toxic agents and indications of such exposure. Workers, managers, and physicians can also contact NIOSH's Health Hazard Evaluation Program in Cincinnati, which conducts workplace evaluations if toxic exposures are suspected. One fact that remains constant in any consideration of toxic exposure is that virtually every resulting disease is preventable.

In the second dimension of a health promotion program, the most important factor is health promotion policies initiated by employers. These policies are usually developed by management and affect issues relating to reducing medical utilization. While these are not inherently health promotion measures, they do indicate a significant change from traditionally accepted disease treatment. Numerous studies by Health Cost Containment Programs have indicated the following:

> People admitted on Fridays tend to stay longer for the same ailment than those admitted on Mondays and Tuesdays ... a quarter of the hospital days were logged by patients with two types of problems: nervous and mental disorders, and alcoholism—ailments often treated effectively on an outpatient basis. (*Business Week*, 1979)

These conditions are not only treatable outside of hospitals, they are also preventable. Out of economic necessity, employers have moved from a passive role in medical costs to one that is more active by undertaking significant steps. Among these measures are the following: (1) reduction of hospital care since 40 percent of disease care is spent for excessive use of hospitalization (*Business Week*, 1976; Cooper, Kehoe, & Murphy, 1978; Parkinson et al., 1982); (2) Blue Cross of California recently circulated a study of 700 simple procedures, such as most biopsies, which can be performed without hospitalization (Brown, 1982); (3) insurance coverages have been extended to include second and third opinions on elective surgery with large savings. This policy has raised the ire of many hospitals and physicians, although both the American Medical Association and the American Hospital Association have endorsed it. Second opinion programs have proven to be particularly effective cost containment measures (Adamson, 1981; *American Medical News*, 1981e; Goldbeck, 1982, 1974); (4) more reimbursement for health services at home (Berry, 1980; *Business Week,* 1979); (5) elimination of the costly annual physicals, which have little if any effect on overall health (Hackler, 1980; *Harvard Medical School Health Letter,* 1980); (6) greater use of preadmission testing and preauthorization for hospital admissions (Berry, 1980; *Business Week,* 1978; Fielding 1978, 1979, 1981); and (7) the enrollment of more employees in Health Maintenance Organizations (HMO), which provide virtually all medical coverages but have a vested interest in minimizing the use of high cost technologies (Green, 1979; Parkinson et al., 1982). Greater use of HMO facilities is definitely a major future direction, which is made even more viable with the addition of active health promotion programs, such as the "Total Health Care Project" currently underway at Kaiser Hospital in Oakland under the direction of the Kaiser system founder, Sidney R. Garfield and his colleagues. Each of these efforts has resulted in substantial savings with no loss of effectiveness. As these trends continue, it is important to note that, according to Joseph Califano, former Secretary of Health and Human Services, "Corporations could have saved up to $150 million last year (1979) if just 5% of the employees of the Fortune 500 companies belonged to Health Maintenance Organizations (HMOs)." These are not only savings to corporations but to individuals as well.

Perhaps the most significant and controversial innovations in the realm of organization policies involve "cost sharing" and "health incentives" by employers and employees. One invaluable aspect of savings based on this approach is that many cost containment strategies do save money for the organizations but the savings are seldom if ever passed on to the workers, even the middle management of the organizations involved. It is in the area of cost-sharing practices that both individuals and organizations can benefit greatly by working together. Cost sharing had been fought by labor unions, but unions also realized that medical costs were taking an increasingly larger share of funds that might be used for wage increases and other benefits. Far from a uniform practice, several different alternatives are being used or considered under cost-sharing strategies and benefits: (1) assume all or part of the cost of worksite programs, such as in the IBM, Johnson & Johnson, Xerox, and Scherer Brothers Lumber Company approaches; (2) pay the costs of interventions for high-risk employees, as done by Ford Motor Company; (3) pay only for interventions with clear evidence of effectiveness, as done by New York Telephone; (4) train employees to perform in-house medical self-care, such as monitoring blood pressure and cardiopulmonary resuscitation (CPR), as a basis in virtually all programs; (5) allow time off during work hours for health promotion activities, such as done by Pepsico, IBM, and Xerox; and (6) at the other extreme, suggestions that organizations completely drop all health benefits, raise workers' salaries, and let individuals create their own plans. While this last alternative may seem highly unlikely, many individuals would welcome the opportunity to choose between one wage plus health coverages versus a high wage without such coverages, since they would actually realize greater health and wealth over time. Ironically, this idea is extolled by both the "hard liners" who want to drop all health benefits in contempt of worker needs and the health promotion advocates in the name of personal responsibility, participation, and cooperation! It would be foolhardy for individuals or organizations to overlook the powerful implications of creating workplace environments that enhance individual responsibility and participation. Every major poll of all levels of workers is consistent with a 1979 Gallup poll conducted for the U.S. Chamber of Commerce, revealing that "the overwhelming majority believe that if they are more involved in making decisions that affect their job, they would work harder and do better ... nearly two-thirds would be happy to have their salaries linked to higher productivity" (Burk, 1981, p. 69), and this conclusion has recently been extended to other benefits as well. There are undeniable indications of a profound shift in individual and cultural values regarding work. Values are not abstract or intellectual constructs; they represent fundamental changes in attitudes and behavior that are challenging every aspect of work, the workplace, and life itself.

In the third dimension of health promotion, the focus is on the limitations in health promotion programs. They are "limited" in the sense that they focus on one particular aspect of health care and tend not to use multiple approaches or organizational change per se to attain employee health. This area reveals the clearest evidence of prepackaged programs, high-cost health technologies that are identical to the high-cost sickness technologies, and facile solutions by instant experts to complex considerations. At the outset, it is important to acknowledge that the actual materials to be used in any health promotion program are well-developed, well-defined, effective, inexpensive, and readily available. The challenge is not to package and market these materials like competing brands of canned vegetables, but to have an individual or organization be aware of the options available and to expend the time and effort to create an effective program suited to their individual requirements. According to a personnel manager for a national banks' stress management program, "The only people deriving clear benefits from packaged stress management programs are the companies marketing these programs." A growing health promotion industry promises, or perhaps threatens, to equal the pathology management industry in its technology and costs. By contrast, it is certainly possible to create a health promotion program with minimum cost and maximum effectiveness. To date, the limited health promotion programs have been defined by Jonathan F. Fielding as "the assessment of individual morbidity and mortality risk for chronic diseases (primarily heart disease, cancer, and stroke) through screening and a health risk appraisal tool" (Fielding, 1979). Although considerable merit and benefit to such early screening and detection programs probably exist, they are still within the disease model of the pathology management industry since the thrust is the earlier detection of disease, which is related to, but distinct from, health promotion. Medical assessments in these programs focus primarily on the risk factors associated with heart disease, stroke, and cancer.

This risk assessment is then followed by a risk-reduction program comprising one or more of the following components, depending on the individual and organizational level of commitment:

1. Accident and self-protective measures concerned with job safety and visible, environmental hazards of the workplace with an emphasis on occupational medicine (Ng, Davis, & Manderscheid, 1978, 1979).

2. Medical self-care programs including CPR training, basic first aid, blood pressure monitoring and other aspects of lay medical care (Piller, 1981; Stokes, 1979; Vickery & Fries, 1979).

3. Smoking cessation is an effective area since well-motivated participants can achieve initial abstinence rates of between 70 percent and 100 percent and long-term quit rates of up to 50 percent (*Business Week*, 1974; Danaher, 1979).

4. Alcohol and substance abuse programs, which are among the most long-standing and effective programs with success rates reported between 65 percent and 80 percent for long-term reduction in excessive drinking (Erfurt & Foote, 1977; Hilker, Asma, & Daghestani, 1975; Milbourn, 1981; Monagan, 1981; Rutherford, 1980).

5. Hypertension control, which would include early screening, detection, and follow-up, since the majority of people with high blood pressure do not exhibit symptoms (Foote & Erfurt, 1976; Koerner, 1973; Peters, Benson, & Porter, 1977; Stamler et al., 1978).

6. Weight reduction is an area in which business can assist employees by providing meeting times and places, as well as provision of low-calorie, low-fat foods and educational materials in the cafeteria (Clutterbuck, 1980; Melhuish, 1979; Pyle, 1980).

7. Exercise and physical fitness programs that are the most currently pervasive health promotion programs usually limited to certain employee levels and among the most expensive to implement (Garman, 1979; Levey, 1980; Mann, 1979; Meyer, 1980, 1981; *Newsweek*, 1976; Pelletier, 1979b).

8. Stress management programs, which are inundated by a tidal wave of "how to," prepackaged programs of limited effectiveness but certainly better than nothing at all. Currently, the area of stress management is the most frequent and prominent area of concern and has generated the greatest amount of activity (Lightbody, 1981; McGarrey, 1978; McLean, 1979; Odiorne, 1978; Pelletier, 1977a, 1977b, 1977c; Tubbs, 1980).

9. Nutrition and healthy dietary practices, which are relatively uncommon but growing rapidly, without a particular focus on weight control but on an overall, healthy diet (Calahese, 1980; Calundo, 1980; Foreyt, Scott, & Gotton, 1979; Pelletier, 1979b; Piller, 1981).

10. General health education approaches focused on information dissemination with no programs or incentives to encourage implementation (Chadwick, 1979; Sehnert & Tillotson, 1978).

As evident from this list, for virtually every practice, an individual can assume responsibility for a healthier lifestyle in his or her own personal life. One difference between personal and organizational health promotion is that the workplace must also change to serve as an inducement for these practices to generalize throughout a person's life.

Finally, health promotion programs per se address the remaining 85 percent to 90 percent of all employees who have no early indications of disease and are relatively healthy. Health promotion activities consist of those individual practices, organizational programs, and workplace environments and policies that elicit and sustain health in the population as a whole. No mystery exists concerning the components of such programs, since they usually consist of various mixes from the previous list of limited programs. Although the actual components are virtually identical, a holistic program is unique, because it is not isolated and compartmentalized while business continues as usual. One example of a limited program would be for an organization to conduct a stress management seminar, rather than seek the roots of the apparent stress in poor communications, disruptive noise levels, lack of health incentives, as well as instruction in stress management in a holistic model. While this is an extremely new area of work, research indicates that limited programs are often more expensive and less effective than comprehensive programs, which create more subtle but permanent changes. To address such programs, no definitive guidelines exist, but it is an area of great promise.

Clearly related to these observations are the findings of a 1980 publication of the Office of Health Information, *Health Promotion, Physical Fitness and Sports Medicine* of the U.S. Department of Health and Human Services. In *Toward a Healthy Community*, the authors identified six common shortcomings of health promotion programs. For individuals and organizations alike, these warrant careful attention in formulating any particular program:

1. "Fragmentation of effort" since programs are often implemented in a piecemeal fashion and this minimizes their impact.

2. "Overemphasis on initial motivation" since concerted advertising can induce momentary changes but programs fail to provide support and maintenance over extended periods of time.

3. "Appeal to individual heroics" indicates that group-based programs among aggregates of workers are more likely to be effective than victim-blaming approaches, which insist that people "pull themselves up by their own bootstraps."

4. "Overemphasis on activities as opposed to results," where sporadic programs such as periodic screening have minimal value if the subsequently recommended workstyle or health promotion practice are in fact not supported by the workplace environment or value system.

5. "Overemphasis on knowledge and information" is the greatest potential pitfall. With all good intentions, the plethora of information is usually well known and disseminated but not applied. It is in the realm of applying what is already known that lies the greatest challenge in health promotion.

6. A "We will do it for you" approach, rather than "Together we can do it ourselves" is a critical element. Individual workers need to feel a sense of participation and ownership of the program from the beginning as well as an ongoing responsibility for its continuation over time.

These potential problem areas can and have been remedied within a few programs and can serve as useful guidelines to ensure effective individual and organizational programs in the future. When all factors are considered, the one essential element remains the individual worker. In 1981, *Business Week* issued a cover story special report on "The New Industrial Relations," which reached some astonishing but widespread conclusions, such as: "The evidence is growing that rank-and-file workers for the most part want to be more deeply involved in their work," with the caution that workers also, "have an unerring ability to spot exploitive schemes." In the same article was the observation that, "If a work improvement aims only at improving productivity, it quickly loses worker support. But a program that has only a vague plan of making workers feel better about themselves is likely to collapse for lack of business perspective." What is a possible solution? A deceptively simple solution, which echoes an underlying theme throughout this book, was voiced by Michael Donduck of Digital Equipment: "Improved job satisfaction and improved productivity go hand in hand, and both are as important to workers as they are to managers," and this applies directly to the effective design and successful implementation of health promotion programs.

REFERENCES

Acree, K. H. (1981). *Worksite health promotion demonstration.* Draft submitted to California Governor's Council on Wellness and Physical Fitness.

Adamson, G. J. (1981). Health promotion and wellness: A marketing strategy. *Group Practice Journal,* 17–22.

Alderman, M. H. (1976). Detection and treatment of high blood pressure at the work place. *Proceedings of the National Conference on High Blood Pressure Control in the Worksetting.*

Alderman, M., Green, L. W., & Flynn, B. S. (1979). *Hypertension control programs in occupational settings.* National Conference on Health Promotion Programs in Occupational Settings, Washington, DC.

Allport, S. (1981, August 4). Sharpening doctor's awareness of "vague" occupational disease. *Medical Tribune,* p.4.

American Medical News. (1979, January 5). Many employers jump aboard employee fitness bandwagon, p.3.

American Medical News. (1980a, September 5). Brain cancer incidence found high at refineries, p.7.

American Medical News. (1980b, November 7). Lifestyle held exaggerated as cancer cause, p.14.

American Medical News. (1981a, February 6). Job-linked stress topic of book, p.17.

American Medical News. (1981b, March 27). Women's health on decline, report says, p.24.

American Medical News. (1981c, July 31). AMA urges caution on cancer-environment ruling, p.8.

American Medical News. (1981d, September 4). AFL-CIO, government plan cancer screening, p.22.

American Medical News. (1981e, November 6). New business attacks on costs seen, p.23

Arnold, W. B. (1973). Cardiovascular health program. *Recreation Management Magazine,* pp. 14–15.

Arnold, W. B. (1976, July). Before you start a fitness program. *The Journal of Employee Recreation, Health, and Education, 19,* 10.

Behavior Today. (1978, December 25). Pacific mutual: Behavioral research + maintenance = lower health costs, pp. 2–4.

Berry, C. A. (1980). *Good health for employees and reduced health care costs for industry.* Washington, DC: Health Insurance Institute.

Bjurstrom, L. A., & Alexious, N. G. (1978). A program of heart disease intervention for public employees—a five year report. *Journal of Occupational Medicine, 20*(8).

Blair, S. N., Pate, R. R., Howe, H. G., Blair, A. E., Rosenberg, N., & Parker, G. M. (1979). Leisure time physical activity and job satisfaction *Medicine and Science in Sport, 11*(1), 105. (Abstract.)

Brenner, M. H. (1979, March 24) Unemployment, economic growth, and mortality. *Lancet,* p.1043–1046.

Breslow, L. (1979). A positive strategy for the nation's health. *Journal of the American Medical Association, 242*(19), 2093–2095.

Brown, J. (1982). *Wellness in the workplace.* Sacramento: Office of the Governor.

Burk, C. G. (1981, June 15). Working smarter, *Fortune,* pp. 68–73.

Business Week. (1974, January 5). The new Rx for better health, pp. 91–96.

Business Week. (1976, May 17). The skyrocketing costs of health care, pp. 144–147.

Business Week. (1977, November 14). Using cancer's rates to track its cause, pp. 69–75.

Business Week. (1978, August 21). How companies cope with executive stress, pp. 107–108.

Business Week. (1979a, August 6). The corporate attack on rising medical costs, pp. 54–56.

Business Week. (1979b, April 30). Executive stress may not be all bad, pp. 96–103.

Business Week. (1980, March 3). Living with chronic illness, pp. 88–92.

Business Week. (1981a, May 11). The new industrial relations—special report, pp. 84–98.

Business Week. (1981b, September 21). Wellness programs paying dividends, pp. 1, 36, 37.

Calahese, E. J. (1980). Nutrition and environmental health: The influence of nutrition status on pollutant toxicity and carcinogenicity. In *The vitamins* (Vol. 2). New York: Wiley Interscience.

Carlson, R. J. (1982). Memorandum to Governor Edmund G. Brown and the California Governor's Council on Wellness and Physical Fitness.

Calundo, M. A. (1980). *Nutrition and preventive health care.* New York: MacMillan.

Chadwick, J. H. (1979). *Health behavior change at the worksite: A problem-oriented analysis.* National Conference on Health Promotion Programs in Occupational Settings. Washington DC: Department of Health, Education, and Welfare.

Clutterbuck, D. (1980). Executive fitness aids corporate health. *International Management, 35*(2), 19–22.

Cooper, P. D., Kehoe, W. J., & Murphy, P. E. (Eds.) (1978). *Marketing and preventive health care: Inter-disciplinary and interorganizational perspectives.* Chicago: American Marketing Association.

Corrigan, D. L., et al. (1979). *Effect of habitual exercise on total health as reflected by non-accidental insurance claims.* West Lafayette, IN: Purdue University.

Cox, M. H., & Shepard, R. J. (1979). Employee fitness, absenteeism, and job satisfaction. *Medicine and Science in Sport, 11*(1), 105. (Abstract.)

Cranton, E. M. (1981, November 1). Holistic medicine. *Newsletter of the AHMA.*

Crowley, D. (1981, Spring). Four top health care executives speak out. *World,* pp. 11–12.

Danaher, B. S. (1979). *Smoking cessation in occupational settings: State of the art report.* National Conference on Health Programs in Occupational Settings. Washington, DC: Department of Health, Education, and Welfare.

DuPont, R. L. (1979). *The control of alcohol and drug abuse in industry.* Office of Health Information and Health Promotion. Washington, DC: Department of Health, Education, and Welfare.

Erfurt, J. C., & Foote, A. (1977). *Occupational employee assistance programs for substance abuse and mental health problems.* Ann Arbor: University of Michigan.

Farquhar, J. (1978). *The American way of life need not be hazardous to your health.* New York: Norton.

Farquhar, J., Maccoby, N., Wood, Pl, Breitrose, H., Brown, B., Haskell, W., McAlister, A., Meyer, A., Nash, J., & Stern, M. (1977, June 4). Community education for cardiovascular health. *Lancet.*

Fielding, J. E. (1978). *Preventive medicine and the bottom line.* Boston: Massachusetts Department of Public Health.

Fielding, J. E. (1979). Preventive medicine and the bottom line. *Journal of Occupational Medicine, 21*(2), 79–88.

Fielding, J. E. (1981). Health promotion and disease prevention at the workplace. *Wellness Resource Bulletin—California Department of Mental Health, 1*(5), 1–12.

Foote, A., & Erfurt, J. (1976). A model system for high blood pressure control in the worksetting. *Proceedings of the National Conference on High Blood Pressure Control in the Worksetting.*

Forbes. (1974). Getting a move on. *113*(5).

Foreyt, J. P., Scott, L. W., & Gotton, A. M. (1979). *Weight control and nutrition education programs in occupational settings.* National Conference on Health Promotion Programs in Occupational Settings. Washington, DC: Department of Health, Education, and Welfare.

Garman, J. F. (1979). *Physical fitness, mental health, and job performance: A preliminary report.* Ann Arbor: University of Michigan. (AAFDBI Abstract).

Goldbeck, W. (1982a). *Evaluating corporate health programs.* Ongoing study of the Washington Business Group on Health. Washington, DC: Washington Business Group on Health.

Goldbeck, W. (1982b, March 2). Transcribed talk to the American Psychological Association Task Force on Promotion and Prevention.

Green, L. W. (1979). National policy in the promotion of health. *International Journal of Health Education, 22*(3), 161–168.

Hackler, T. (1980, June). The yearly physical: A costly indulgence? *Mainliner,* pp. 112–115.

Harvard Medical School Health Letter. (1980, July). Periodic health exams in perspective, pp. 1–4.

Hilker, R. J., Asma, F. E., Daghestani, A. N., & Ross, R. L. (1975). A drug abuse rehabilitation program. *Journal of Occupational Medicine, 17* 351–354.

Iglehart, J. K. (1982, January 14). Health care and American business. *New England Journal of Medicine,* pp. 120–124.

Jennings, C., & Tager, M. J. (1981, Summer). Good health is good business. *Medical Self-Care,* 14–18.

Kennedy, C. (1976). Maintaining top managers in good running order. *Director, 28*(11), 57.

Kiefhaber, A., Weinberg, A., & Goldbeck, W. (1978). *A survey of industry sponsored health promotion, prevention and education programs.* Washington, DC: Washington Business Group on Health.

Knowles, J. H. (1977, Winter). The responsibility of the individual. *Daedulus,* pp. 55–80. Adapted and reprinted in *Science,* 1977, December 16, *198* (4322).

Kobasa, S. C. (1981). The Hardy personality: Toward a social psychology of stress and health. In J. Suls & G. Sanders (Eds.), *Social psychology of health and illness.* Hillsdale, NJ: Erlbaum.

Koerner, D. R. (1973, September). Cardiovascular benefits from an industrial physical fitness program. *Journal of Occupational Medicine,* 700–707.

Kreitner, R. (1976). Employee physical fitness: Protecting an investment in human resources. *Personnel Journal, 55* (7), 334–344, 348.

Kristein, M. (1977). Economic issues in prevention. *Preventive Medicine,* 6.

Kristein, M. M., Arnold, C. B., & Wynder, E. L. (1977). Health economics and preventive care. *Science, 195,* 457–462.

Larson, K. (1979). How companies can rein in their health care costs. *Personnel Administrator, 24.*

Levey, R. (1980). Fitness fever: Everybody into the company gym. *Dun's Review, 116* (5), 115–118.

Lightbody, J. (1981). A safety valve for employee stress. *International Management, 36* (2), 17–18.

Love, J. R., & Love, L. B. (1978). Cost-benefit concepts in health: Examination of some prevention efforts. *Preventive Medicine, 7,* 414–423.

McGaffey, T. N. (1978). New horizons in organizational stress prevention approaches. *Personnel Administrator, 23* (11), 26–32.

McLean, A. A. (1979). *Work stress.* Reading, MA: Addison-Wesley.

Mann, G. V. (1979). A proposal for a trial of the efficiency of fitness training for employed persons. Nashville TW: Vanderbilt University. AAFDBI Abstract.

Manuso, J. (1978). Testimony to the President's Commission on Mental Health, Panel on Costs and Financing. *Report of the Presidents' Commission on Mental Health* (Vol. 2, Appendix). Washington, DC: US Superintendent of Documents.

Manuso, J. (1979). Corporate mental health programs and policies. In L. NG & D. Davis (Eds.), *Strategies for public health, promoting health and preventing disease.* New York: Van Nostrand.

Martin, J. (1978, March). Corporate health: A result of employee fitness. *Physician and Sports Medicine.*

Melhuish, A. (1979). Weighing the benefits of slimming. *International Management, 34* (9), 63.

Meyer, H. E. (1980). A fitness program for Canadian business. *Fortune, 101* (1), 94–98.

Milbourn, G., Jr. (1981). Alcohol and drugs: Poor remedies for stress. *Supervisory Management, 26* (3), 35–42.

Monagan, D. (1981, August 21). Auto industry seeks alcohol substitute. *Medical Tribune,* p. 12.

Muchinsky, P. M. (1977). Employee absenteeism—a review of the literature. *Journal of Vocational Behavior, 10* (3), 316–340.

Nehrbass, R. G. (1976). Physical fitness pays off in productivity. *Personnel Journal, 55* (12), 600.

Newsweek. (1976, April 19). Run for your life, p. 100.

Ng, L. K. Y., Davis, D. L., & Manderscheid, R. W. (1978). The health promotion organization: A practical intervention designed to promote healthy living. *Public Health Reports, 93* (5), 446–455.

Ng, L. K. Y., Davis, D. L., & Manderscheid, R. W. (1979). *Toward a conceptual formulation of health and well being.* Paper presented to AAAS, Houston, Texas.

Odiorne, G. S. (1978). Executives under siege: Strategies for survival. *Management Review, 67* (4), 7–12.

Parkinson, R. S., & Associates (Eds.). (1982). *Managing health promotion in the workplace: Guidelines for implementation and evaluation.* Palo Alto, CA: Mayfield.

Patrinos, D. (1977, April 19). Health pays industry dividends. *Milwaukee Sentinel.*

Pelletier, K. R. (1977a). Biofeedback. In *Collier's Encyclopedia.* New York:Macmillan, pp. 164–165.

Pelletier, K. R. (1977b). *Mind as healer, mind as slayer: A holistic approach to preventing stress disorders.* New York: Delacorte & Delta.

Pelletier, K. R. (1977c, February). Mind as healer, mind as slayer. *Psychology Today,* P. 3542. Reprinted in C. F. Wilson & D. L. Hall (Eds.), *Stress management for educators.* San Diego, CA: Department of Education.

Pelletier, K. R. (1979). Stress/unstress: A conversation with Kenneth R. Pelletier. *Medical Self-Care,* (5), 3–9.

Pelletier, K. R. (1980). The mind is health and disease. In A. Hastings, J. Fadiman & J. S. Gordon (Eds.), *Holistic medicine: An annotated bibliography.* Rockville, JD: National Institute of Mental Health. Reprinted as *Health care for the whole person: A comprehensive guide to holistic medicine.* Boulder, CO: Westview Press.

Pelletier, K. R. (1981). *Longevity: Fulfilling our biological potential.* New York: Delacorte & Delta.

Pelletier, K. R. (1983). *Corporate health promotion programs.* San Francisco: California NEXUS Foundation.

Peters, R. K., Benson, H., & Porter, D. (1977). Daily relaxation response breaks in a working population: I. Effects on self-reported measures of health, performance and well-being. II. Effects on blood pressure. *American Journal of Public Health, 67,* 946–953.

Piller, C. (1981, Summer). Staying healthy at work. *Medical Self-Care,* pp. 6–11.

Pyle, R. L. (1979). Performance measures for a corporate fitness program. *Training and Development Journal, 33*(7), 32–38.

Pyle, R. L. (1980). Trimming the corporate waist with fitness programs. *Business Horizons, 23*(2), 70–72.

Rice, B. (1981, June) Can companies kill? *Psychology Today,* pp. 6–11.

Rutherford, D. (1980). Alcoholic solutions. *The Accountant, 183*(5506), 309–310.

Salvendy, G. (1980). Effects of job pacing on job satisfaction, psychophysiological stress, and industrial productivity. *Proceedings of American Institute of Industrial Engineers,* Spring Annual Conference, pp. 433–442.

Sapolsky, H. M., Altman, D.,Greene, R., & Moore, J. D. (1981). Corporate attitudes toward health care costs. *Millbank Memorial Fund Quarterly, Health and Society, 59*(4), 561–585.

Shenert, K. W., & Tillotson, J. K. (1978). *A national health care strategy: How business can promote good health for employees and their families.* Washington, DC.: National Chamber Foundation.

Stacey, J. (1980, October 3). Business taking new look at care costs. *Amercian Medical News,* pp. 1–9.

Stamler, R., et al. (1978). A hypertension control program based on the workplace: A report on the Chicago center. *Journal of Occupational Medicine, 20,* 618–625.

Stokes, B. (1979). Self-care: A nation's best health insurance. *Science, 205*(4406), editorial page.

Tubbs, S. L. (1980). Effective stress management. *Proceedings of American Institute of Industrial Engineers,* Spring Annual Conference, pp. 227–229.

U.S. Department of Health, Education, & Welfare. (1979). *Healthy people: The Surgeon Generals report on health promotion and disease prevention.* Publication No. 79-55071. Washington, DC.: U.S. Government Printing Office.

U.S. News and World Report. (1980, January 28). As companies jump on fitness bandwagon, pp. 36–99.

Vichery, D. M., & Fries, J. F. (1979). *Take care of yourself: A consumer's guide to medical care.* Reading, MA: Addison-Wesley.

Will, G. F. (1978, August 7). A right to health? *Newsweek,* p. 88.

Yankelovich, D. (1978, May). The new job values. *Psychology Today,* pp. 46–50.

17

Evaluations, Results, and Problems of Worksite Health Promotion Programs

JONATHAN E. FIELDING

Employer-sponsored health promotion/risk reduction programs have proliferated among large- and medium-size employers. Descriptions of these programs suggest considerable variation in program configurations, modes of operation, educational techniques, costs, and rates of employee participation. Frequently left unanswered are·such central questions as the following: Are these programs generally effective? Which program type is likely to be the most effective for a particular employee population? How do program benefits compare with program costs?

Three characteristics of many written reports on these programs interfere with attempts to provide answers:

1. Program goals, objectives, and methods are not explicit or only described in a general fashion.
2. Results are not presented in ways that permit assessing whether goals and objectives have been met.
3. An inadequate evaluation design makes it difficult to determine if observed changes resulted from institution of the health promotion program.

Nonetheless, review of the growing body of reports on prominent health risks and risk reduction programs in clinical and worksite settings permits several general statements about reasonable expectations for worksite health improvement programs. Appropriate questions for such a review include the following:

1. Have the health risks, which are targets of these programs, been convincingly demonstrated to cause significant increases in morbidity and mortality?
2. Are the health risks frequently found in working populations?
3. Are the increased risks reversed by changes in habits or by other control measures?
4. Have health promotion activities been shown to change the target behavior and/or physiological or biochemical measures of risk?
5. Have careful analyses matched intervention program costs with benefits and/or degree of effect?
6. Is there a coherent body of literature on efforts to reduce the target health risk in the work environment?

To illustrate the range of data currently available to answer these questions, six risk reduction interventions will be considered: smoking cessation, hypertension control, physical fitness, weight management, stress management, and auto-accident injury prevention programs. In addition, multi-factoral interventions to reduce cardiovascular risk will be briefly examined.

HYPERTENSION

Elevated blood pressure (over 140/90) is a well-established risk factor, which leads to higher mortality and morbidity (VA, 1967, 1970, 1972). On average, hypertensives, when compared to normotensives, develop approximately three times as much coronary heart disease, six times as much congestive heart failure, and seven times as many strokes (Hypertension Detection F/U program, 1979; VA, 1972). Lowering blood pressure reduces the excess risk (VA, 1967, 1970, 1972). In a five year-clinical trial involving 11,000 hypertensive patients ages 30–69, those randomized for treatment using a standardized antihypertensive protocol experienced a significantly greater decline in both blood pressure and overall mortality than those referred to their private physician for care (VA, 1972). This differential derived from differences in degree of control achieved. After five years under protocol care, diastolic blood pressure decreased from 101.1 mm Hg to 84.1 mm Hg, compared to 101.2 mm Hg to 89.1 mm Hg under referred care (Hypertension Detection F/U Program, 1979). Based on the accumulated report of well-controlled studies, hypertension related morbidity and mortality can be reduced through adequate treatment over five years by 20 percent to 50 percent compared to similar untreated patient populations.

Hypertension is a prevalent problem in the work force, with increased frequency in older workers. Based on worksite and community blood pressure screening programs, hypertension (uncontrolled and controlled) is found in 15–25 percent of adult populations (NHLBI, 1981).

Company-sponsored hypertension screening programs have reported success in identifying hypertensive employees unaware of their condition and in increasing the percentage of hypertensives under adequate control. For example, during the first year of a voluntary onsite screening, referral, and follow-up program at the home office of Massachusetts Mutual Life Insurance Company the percentage of hypertensives under control increased from 36 to 82 percent (NHLBI, 1980).

A well-controlled hypertension management program of four methods at four different Ford Motor Company manufacturing plants found that at the end of three years, 56 to 62 percent of the hypertensive employees at the three sites whose interventions incorporated follow-up had blood pressures below 140/90 mm Hg, and 86 to 90 percent had readings below 160/95. Of employees receiving screening and referral but no follow-up, corresponding control rates were 21 to 47 percent. Systematic routine follow-up, including both patient education and support for maintenance of treatment, appeared to be essential to good outcomes (Foote & Erfurt, 1983).

In a three-site industrial hypertension screening, detection and follow-up program of 120 auto workers, 138 sanitation workers and 106 postal workers referred for high blood pressure (above 160/95), 92 percent saw their physician, 86 percent had treatment initiated, and 72 percent showed progress toward control (readings below 140/90 or reduced since screening and below 160/95) during an average follow-up period of 16 months (Foote & Erfurt, 1977). In a group of 218 Chicago area hypertensive employees attending a special high blood pressure control clinic near their workplace, diastolic blood pressure fell from an average of 102.6 at first screening and 98.8 at second screening to 83.1 at the end of the first year (Stamler et al., 1978).

One of the most important evaluation questions regarding employer-sponsored hypertension control efforts is the program effect on absenteeism. In one study of 8,467 members of three unions entering a screening and treatment program for hypertension, disability days for the hypertensives identified and treated on-site, or by their own physician, declined an average of 25.3 and 30.9 percent in the next two years, while disability for all employees increased an average of 9.2 percent in three years (Alderman & Davis, 1976). Days of hospitalization for cardiovascular diagnoses decreased for those treated on-site, but not for those treated by their private physician.

By contrast, absenteeism among male employees of Dominion Foundries and Steel (Canada) increased an average of 5.2 days per year (80 percent over baseline) after the diagnosis of hypertension was made. Increased absenteeism was unaffected by institution of antihypertensive drug treatment or degree of control achieved. The authors of the study speculate that informing employees that they had a health problem increased sick role behavior (Haynes, et al., 1978).

Contrasting results were found in the multicenter Hypertension Detection and Follow-up Program where labeling an individual as hypertensive was not associated with increased self-reported absenteeism. However, detection and treatment were independently associated with increased absenteeism among newly diagnosed hypertensives treated at their usual sources of medical care, while previously treated hypertensives referred to special hypertension control clinics demonstrated reduced absenteeism. These conflicting results from three well-controlled studies preclude conclusions on the effect of a clinically based or worksite-based hypertension control program on absenteeism (Polk et al., 1979).

Available data suggest that worksite treatment for high blood pressure may be more cost-effective than community care. For example, in one study 457 hypertensive workers ages 18 to 69 in metropolitan Toronto were randomly assigned to worksite or community (regular) care (Logan et al., 1981). While average yearly cost of care did not significantly differ between the two groups (worksite $243, community $211), the incremental cost-effectiveness ratio of the worksite-treated groups ($5.63 per mm Hg reduction) was well below the base cost-effectiveness ratio for the regular care groups ($32.51 per mm Hg reduction). In the same study, specially trained nurses at the worksite were found to be more effective in achieving blood pressure goals with employees at the worksite (48.5 percent) in the first six months than was the patient's family doctor (27.5 percent), despite random assignment of employees to the two sources of care (Logan et al., 1981). Another workplace hypertension control program reported average patient costs in 1978 of $194.77, of which drugs accounted for 23.9 percent (Ruchlin & Alderman, 1980).

Models have been developed to estimate the overall cost to benefit ratios for worksite hypertension programs. While reasonable estimates exist for many key terms in these analyses, several critical variables still defy reliable estimation, in particular, the effects of hypertension and its treatment on absenteeism rates and the fraction of possible benefits actually realized from any particular program. Despite these important limitations, modeling of costs and benefits for hypertension control is further advanced than for any other health risk reduction program in the industrial setting. Although some preliminary runs of one model suggested a significant dollar benefit in excess of costs under several sets of assumptions, the sensitivity of the net figure to small changes in critical variables, such as treatment effects on absenteeism and the proportion of hypertensives already under effective control makes it difficult to consider the results as proof of a positive benefit-to-cost ratio (Hannan & Graham, 1978).

One limitation of modeling costs and benefits of a program from only the employer perspective is that personal and societal costs and benefits are ignored, including reduced suffering for the individual and family members, improved quality of life, and changes in tax collections and social security benefits.

The life insurance industry has provided financial incentives for effective blood pressure control by reducing or eliminating the extra premiums levied on hypertensives if their blood pressure is under effective control. At least one major reinsurer, Lincoln National Life, accepts treated blood pressure readings as the basis for calculating premiums when the insured has been under continued treatment for more than five years and blood pressure is successfully lowered (below 150/96 or 160/94) (NIH, 1975).

SMOKING

Smoking is strongly associated with excess incidence of heart disease, stroke, chronic obstructive lung disease, and many cancers (especially lung, oesophagus, oropharnyx, and bladder) (Cowell & Hirst, 1979; DHEW, 1979; Hammond & Seidman, 1980; Fielding, 1985; National Center for Health Statistics, 1984). Coronary heart disease death rates for two-plus packs a day male smokers are 335 percent of that of nonsmokers (standardized for age) in the 45–54 decade, and 213 percent of nonsmoking rates for the 55–64 age range. The stroke death rate of women smokers is 211 percent of non-smokers in the 45–54 age range.

A large, 15-year actuarial study of life insurees of one major carrier found the following mortality ratios (smokers/nonsmokers): respiratory cancer, 15.0; pneumonia and influenza, 14.7; arteriosclerosis and degenerative heart disease and myocardial insufficiency, 2.7; hypertensive heart disease and hypertension, 8.1; digestive diseases, 8.1; and motorcycle accidents, 2.2 (Cowell & Hirst, 1979).

Recent data from the National Health Interview Survey show that the percentage of current smokers was 35 percent in 1983 for all U.S. males over age 20. For females of the same age group, the corresponding figure was 29.1 percent in 1983. A comparison with 1965 prevalences of 52.4 percent for males and 34.1 percent for females underscores that for almost two decades there has been a pronounced decline in smoking, particularly among males (National Center for Health Statistics, 1984).

Smoking-attributable risk for most serious diseases declines after cessation, although at different rates for different diseases. Post-myocardial infarction patients who discontinue smoking halve their risk of subsequent myocardial infarction and cardiovascular disease death (Salonen, 1980). On average, cessation reduces the excess coronary heart disease mortality risk from 100 percent to 50 percent during the first year. After 5 to 10 years the post-cessation mortality rate is virtually equivalent to the nonsmoking rate (DHEW, 1979). Lung cancer risk declines more slowly after smoking cessation, with smoking-attributable risk declining by approximately two-thirds after about the first five years and ex-smoker risk approaching that of a nonsmoker after 10 to 15 years (Salonen, 1980).

Although different smoking-cessation programs have been used in occupational settings, reports of careful studies of worksite programs are sparse. Campbell Soup estimated a 20 percent quit rate in 70 volunteers and a cost per quit of $500 (mostly time off to attend cessation classes) (Wear, personal communication).

The most innovative and promising worksite smoking-cessation program involved competition and monetary incentives. A seven-month program involved 20 one-hour meetings and required cessation on the first day of the second month. Employee and employer contributions went toward successful individuals, with bonuses for the team with the best nonsmoking record. Preliminary results, with six-month abstinence corroborated by observation of peers at the workplace and contact with family and friends of each participant, are indicated in Table 17.1. Both abstinence rates and the high percentage of smokers who enrolled in the smoking-cessation program are impressive. If the results are reproducible in other worksites, the approach will constitute a large step forward in workplace smoking control (Stachnik & Stappelmays, 1983).

Table 17.1. Summary Results of Three Pilot Studies

Program (%)	Percentage of Smokers at Worksite Enrolled in Program	Percentage of Participants Abstinent after 6 Months	Net Reduction in Worksite Smokers as a Result of Program (%)
Worksite I	70	91	65
Worksite II	47	80	37
Worksite III	54	85	46

Source: Stachnick and Stoffelmayr (1983).

Strategies for employers to maximize cost-effectiveness of smoking-cessation classes for employees include the following: (1) paying for cessation classes, preferably on-site to minimize barriers to participation and maximize employee convenience, (2) only contracting for cessation programs with organizations that can validate a sustained quit rate in a reasonable range, and (3) requiring at least a token copayment from participating employees (which can be returned to successful quitters) to improve commitment. Most important is to combine the opportunity to participate in organized cessation activities with a clearly stated and publicized smoking policy that discourages smoking at the worksite and makes clear that the social norm of the company is not to smoke. Although evidence is lacking on the degree of synergy between these strategies, both theoretical considerations and results from community studies support the enhanced effect of this tripartite approach.

Reports from clinical settings suggest that 33–80 percent of participants receiving intensive clinical-educational assistance are able to quit initially, while a smaller percentage (in the range of 10–25 percent) achieve cessation only if provided with educational materials, brief instructions, and encouragement (Danaher, 1980). Long-term abstinence, even after the most intensive program, is more difficult to achieve. By 6 to 12 months after the intervention, from 15–50 percent of cessation-program participants are non-smokers (Leventhal & Cleary, 1980). Recent clinical studies of nicotine gum suggest that its inclusion in a behaviorally oriented smoking-cessation program can improve both initial cessation rates and long-term abstinence, the latter by up to 50 percent (Schneider et al., 1983).

Large-scale community risk reduction efforts are another useful source of information on the effectiveness of smoking reduction efforts. In the Belgian Heart Disease Prevention Project, the percentage of smokers in an industrial high-risk group receiving semi-annual individual health counseling, complemented by a worksite health education campaign, declined in two years by 6.5 percent more than the controls (Kornitzer et al., 1980). In the Stanford Three Community Study, the smoking prevalence in the high-risk group who received both a mass media campaign on heart disease prevention and face-to-face counseling declined 37 percent in two years, compared to controls. The group exposed only to mass media had a decline of 21 percent, compared to a control group (Farquahar et al., 1977; Meyer et al., 1976). In a community-based cardiovascular disease prevention program in North Karelia, Finland, mass media, training of health professionals, and environmental supports were associated with a 13 percent greater decline in the percentage of smokers among men and 8 percent greater decline among women in the intervention group than in the reference group in a five-year period (Salonen et al., 1981).

The importance of environmental influences is reinforced by one economist concluding that demand for cigarettes in the United States in 1975 was 20 percent to 30 percent lower than anticipated in the absence of the Surgeon General's initial 1964 Report and the cumulative effect of persistent publicity and other public policies to discourage smoking. Based on both community and econometric studies, it is probable that aggressive health education activities and policies to discourage smoking at the workplace can act synergistically with sponsorship of specific smoking-cessation activities (Warner, 1977).

Smoking is costly to employers. Smoking employees, on average, use more health care benefits than do nonsmokers. Although estimating these costs is fraught with methodological problems, at least one analysis projected $190 per year (during the entire lifespan) in excess medical costs for smokers (Luce & Schweitzer, 1978; Warner, 1977). Several studies have shown that adult heavy smokers use the health care system at least 50 percent more than do nonsmokers, with a higher prevalence of acute and chronic conditions and disability days (Kristein, 1977). Control Data Corporation reported that smokers have 25 percent higher health care costs and that their hospital stays averaged 114 percent longer than those of nonsmokers. However, these figures should be interpreted with considerable caution since these observed differences were not adjusted for age, sex, or other health-habit differences between smokers and nonsmokers (Naditch, 1984).

Some studies suggest that smokers experience a significant increment in work accidents (perhaps as much as twofold), and one economist has estimated an added $40 per smoking employee per year for worker's compensation insurance (Kristein, 1980).

Smokers average two to three more days of absenteeism per year than do nonsmokers. While the marginal cost of this extra time away from the job on work output is difficult to determine, the figure must be significant (Kristein, 1980).

Other potential increased dollar and human costs attributable to smoking are fire and life insurance, energy (extra air filtration to clear the smoke), family suffering, increased turnover due to excess morbidity and mortality, and possible effects of secondhand smoke inhalation. One overall estimate of the annual cost to business for each smoker is $400 (Kristein, 1980), and a reasonable range of estimates in 1985 is $350–$750.

FITNESS PROGRAMS

Two national surveys conducted in 1978 and 1979 found 37 percent and 36 percent of adults receiving regular exercise (General Mills, 1979; Pacific Mutual Life Ins., 1979). A large number of epidemiologic studies, both in the United States and in Europe, support the hypothesis that lack of physical activity is an independent risk factor for higher age-specific rates of cardio-vascular events and associated deaths, with a dose-response relationship (Kannel & Sorlie, 1979; Paffenbarger et al., 1978). For example, Harvard alumni reporting the expenditure of fewer than 2,000 kilocalories per week at work and play had a 64 percent higher risk of heart attack than the more active (Paffenbarger et al., 1978).

Industrial fitness programs have been widely established, particularly in larger employers, with the expectation of reducing known risk factors, ill-ness, reimbursible insurance claims, and absenteeism. Equally important to employers is the anticipation that such programs can improve productivity, morale, and attitudes toward job and employer, and both the recruitment and retention of employees. While those who have participated in industrial fitness programs tend to have lower than average risk characteristics, including lower cigarette consumption and lower age-adjusted blood pressure and cholesterol levels, those *entering* fitness programs share these advantages (Barnard & Anthony, 1980; Durbeck et al., 1972; Yarvote et al., 1974). Unresolved to date is whether participation causes a reduction in those risk indicators and what percentage of workers with poor physical fitness can be expected to become regular program participants. A six month study of 3,231 white-collar workers (divided into four job categories—management, professional, clerical, and other) offered a corporate fitness program found that a strong positive association existed between current job performance and high adherence to exercise. In addition, at each level of adherence no differences in pre- versus post-work performance were observed (Bernacki & Baun, 1984). Unfortunately, most published reports on the effects of employer-sponsored fitness programs involve studies without adequate controls or insufficient information to assess adequacy of methodology.

One of the most productive but least representative types of studies involves occupations where fitness can be required as a condition of work. A five-year compulsory fitness program for Los Angeles City Firefighters was associated with a significant decline in diastolic (although not systolic) blood pressure and a significant reduction in serum cholesterol but no change in body weight. None of these variables changed among a control sample of Los Angeles Police Officers. The percentage of firefighters in the program with poor or very poor fitness scores on the Kasch Step Test decreased from 38.6 percent to 12.7 percent in four years and to 9.1 percent in seven years (Barnard & Anthony, 1980).

In voluntary employee exercise programs, regular participation has been associated with significant reduction in weight, fitness (recovery pulse, performance on standardized graded exercise tests, etc.), systolic and diastolic blood pressures, and skin fold thickness (Barnard & Anthony, 1980; Bjurstrom & Alexious, 1978; Durbeck et al., 1972; Horne, 1975; Yarvote et al., 1974). The degree of change is directly related to frequency and intensity of exercise with participation often reversing age-related increases in these values (Horne, 1975). However, reductions for any single-risk variable other than fitness are not seen consistently among studies of exercise programs for employed populations. Studies reported from North America, Western Europe, and Russia all support reduced absenteeism in exercise program participants, although study design limitations often preclude definitive conclusions (Donoghue, 1977; Haskell & Blair, 1980; Linden, 1969; Yarvote, 1974). Short-term reductions in absenteeism are probably a function of attitudinal and morale changes rather than reductions in illness. Net reductions in absenteeism compared to baseline and control groups have ranged from about 0.5 to 2 days per year. Illustrative is a pilot fitness program at Metropolitan Life, which compared 100 participants with a control group of 100 employees stratified for age and sex. In two years the absenteeism rate for the exercise group decreased to 4.9 days/year, while climbing in the control group to 7.0 days/year. Complicating interpretation of these results is that the control group initially had, on the average, lower job groupings, higher cholesterol, higher percentage of cigarette smokers (56 versus 38 percent) and lower baseline exercise levels (R.D. Garson, unpublished data, January 1977).

A controlled study of a six-month worksite aerobic fitness program in a Toronto insurance company with 1,281 employees found that the high adherent group (two or more exercise classes per week) decreased 42 percent in average monthly absenteeism during the postintervention period, compared to a 20 percent decline in both the test company overall and in a control Toronto insurance company (Cox et al., 1981). Both the pre- to post-changes for the high adherent group and this group's relative change versus other groups were highly significant ($p < 0.01$ and $p < 0.001$) using simple chi-square analysis, but were not statistically significant by analysis of variance using each subject as his or her own control. Both low adherents (sporadic, inconsistent attendance) and high adherents had significantly less turnover ($p < 0.05$) than nonparticipants or dropouts during a 10-month period, which started four to five months before the program was initiated. The fitness participants' 10-month turnover rate was 1.5 percent versus 15 percent for the other company employees. Major limitations include lack of information collected on the individuals comprising the various participation and dropout groups prior to program initiation and lack of comparisons between turnover rates in the test and control companies.

Although the same study reported significant reductions in overall hospital use in the test company, compared to the controls during the program, methodological problems of the study raise serious questions whether the fitness program can be considered the cause of the findings (Shephard et al., 1982). The authors claimed health care savings due to the program. These savings are based on the hospital use changes and an almost 50 percent year to year increase in control company costs, while there was no appreciable change in test company costs, with $p = .07$. There is a strong need for controlled studies with good baseline data and use of individuals as their own controls as well as pre-post group differences and multiyear follow-ups.

Current information does not permit valid estimates of dollar benefits from the effect of worksite physical fitness programs on frequency of illness (as distinct from absenteeism), or use and cost of health care services, workers' compensation, and disability insurance.

Physical exercise has been shown to decrease perceived stress and feelings of depression, anger, and anxiety, but how these effects are seen in terms of job performance is not known. Direct measurements of increased productivity secondary to worksite exercise programs have not been reported in the North American literature. Also lacking are controlled studies on the effect of such programs on employee turnover, despite the hope that programs will lead to decreased turnover. However, self-reported improvements in energy level, attitude toward job and toward company, overall morale, and work performance have been widely observed and related to the level of program adherence (Durbeck et al., 1972; Heinzelmann & Bagley, 1970).

In one illustrative study, it was found that men from a variety of occupations randomly assigned to an exercise program reported improvements in work performance 60 percent of the time, compared to 3 percent for the nonparticipant group (Durbeck et al., 1972; Heinzelmann & Bagley, 1970). One inherent problem with this type of study is trying to convince people to enroll in the exercise program without citing its potential benefits—in essence, feeling better increased energy, and greater work satisfaction since such promotion can influence posttest responses to questions on these variables.

Positive effects of physical activity programs require participation and adherence. These two variables are in turn influenced by socioeconomic status, age, health practices, gender, proximity to work stations, flexibility of job scheduling, variety of activities offered, hours per day of program operation, support from participation of top management and immediate supervisor, availability of program supervision, seasonal factors, criteria for entry, and degree and type of recruitment and maintenance efforts (Bjurstrom & Alexious, 1978; Durbeck et al., 1972; Haskell & Blair, 1980; Horne, 1975; Oja et al., 1974; Yarvote et al., 1974). On the average, 20–40 percent of employees at a worksite with an on-site exercise program complete preexercise screening and at least initiate an exercise program. Corresponding participation rates in offsite company sponsored programs (e.g., YMCA membership) are 10–25 percent.

High attrition rates are universal program concerns, with the rate directly correlated with the frequency and intensity of the recommended or self-defined regimen. Among the best reported adherence rates are those from a heart disease prevention program for public employees. Sixty-one percent of the initial participants remained active in the exercise program at the end of one year, 42 percent at the end of three years, and 25 percent at the end of five years (Bjurstrom & Alexious, 1978).

More typical results derive from a supervised on-site circuit training program for Exxon executives. After one year, 9 percent of 110 "active" participants were exercising less than one-half session per week, 24.5 percent used the facility an average of once per week, 53 percent twice per week, and 9 percent three times per week (Yarvote et al., 1974). In a variety of NASA-sponsored exercise programs, 47.5 percent of 237 initial participants exercised an average of one or more days per average week and 38.4 percent exercised two or more days per average week. Factors reported as interfering with adherence included workload (28 percent), travel (20 percent), scheduling conflict (15 percent), physical problems (13 percent), and operational aspects of the program (11 percent) (Durbeck et al., 1972). In other studies and presentations on worksite programs, other important factors cited as increasing dropout rates are lack of personal motivation and negative spousal attitude toward the program (Linden, 1969).

Different definitions of "active" participation and adherence make program comparisons difficult. However, what is clear is that convincing a high percentage of the work force to participate, even in an on-site program, is difficult. Even an initial participation rate of 30 percent, combined with an unusually high one-year adherence rate of 50 percent, results in only 15 percent of employees benefiting from the program. In addition, employees who enter and remain in company-sponsored programs tend to have below-average health risk factors and already established exercise habits, with underrepresentation of those in the upper quartile of risk who could derive the greatest program benefit (Cox et al., 1981). Special efforts to recruit those with the highest levels of risk into the program could increase its potential effectiveness in forestalling preventable illness.

Good analyses of costs and cost-effectiveness of corporate fitness programs are needed. In such analyses, it will be important to define clearly the sponsorship and the nature of the program. At one end of the spectrum, a limited program may seek only to educate employees about the benefits of regular physical activity and possibly recommend appropriate exercise routines to be followed outside of work. By contrast, some programs include elaborate fitness facilities (e.g., gymnasium, Nautilus and other equipment, aerobic exercise classes, etc.) with full-time staffing.

In general, employees are not charged for use of programs and facilities at the work setting, although programs sponsored by employee clubs or recreation groups may involve a small fee. Private health clubs or fitness centers for which employers underwrite some cost are more apt to involve employee cost-sharing. For example, at one fitness center in Los Angeles, employers typically pay the initial evaluation fee of $100 per participant and one-half the monthly program charges ($24–$45/month based on access times).

The costs for an on-site physical fitness facility are both for construction and operating costs. Construction costs vary widely according to geographic location and desired interior finishes. At construction costs of $80 per square foot and equipment costs of $30,000, the costs for a 3,000 square foot facility would be $720,000, plus land costs, planning time, fees, and so forth. More elaborate facilities, with such features as an indoor track, swimming pools, and racketball courts, can cost in the millions of dollars.

Operating costs are heavily influenced not only by features of the physical facility but also by the level of support staff. For those companies choosing to employ only maintenance personnel, costs have been estimated at around $150 per participant per year. Companies that staff their facility with exercise physiologists 10 to 14 hours per day can expect to incur costs of between $500 and $1,000 per participant per year. The range is probably even greater at the high end if active company participation is defined as averaging at least two visits per week.

WEIGHT REDUCTION

Actuarial and epidemiologic studies relating body build, weight, and mortality have consistently demonstrated a direct relationship between above-average weight (for given height) and specific chance of death (Society of Actuaries, 1959; Sorlie et al., 1980). The influence of obesity is primarily mediated (reversibly) through increases in blood pressure, higher total cholesterol, and blood glucose (Ashley & Kannel, 1974). Obesity, when defined as greater than 120 percent of ideal body weight, affects 25 to 45 percent of Americans over 30 years old. Of workers, 13.6 percent of men and 21.0 percent of women are overweight by this definition (National Heart, Lung, and Blood Institute, 1981). Prevalence of obesity differs by socio-economic status and by cultural and ethnic group (Srole et al., 1962).

While control of obesity can reduce risk indicators for cardiovascular disease and diabetes mellitus (Type II), prolonged weight control in obese individuals has been difficult to achieve. Only in the last 15 years has significant weight reduction been achieved in obese populations without fasting or enforcing consumption of nutrient forms that do not occur in nature, such as "liquid protein" formulas. In general, the most successful programs have used various behavioral therapies.

One major behavioral approach used to help individuals gain more control over their eating habits is recordkeeping, with individuals maintaining a diary of what foods they eat, the quantities, and circumstances. This strategy provides a more objective view of what is consumed, while simultaneously underscoring that the individual can be effective in self-monitoring. Other approaches include altering the environmental stimuli to eating, eliminating or reducing the automatic triggers to overeating, and teaching ways to slow down the eating process. The consequences of eating may also be structured to reduce intake by self-contracting or contracting with someone else to reward good eating habits promptly (Bellack, 1975; Stunkard, 1979).

The first major trial of these behavioral techniques by Stuart led to weight reduction in one year of 26 to 47 pounds in eight obese women (Stuart, 1967). However, most studies of behavioral therapy have shown more limited success. A careful review of studies incorporating sound behavioral approaches found that average weight losses were only 11 pounds (Jeffrey et al., 1976). One of the more promising studies found that 29 subjects who trained with their cooperative spouse in a variety of behavioral techniques in 10 weekly 1½ hour sessions lost considerably more weight (average about 30 pounds) in 8½ months than those individuals who were similarly trained alone (Brownell et al., 1978).

By 1975, the largest organized weight-reduction program, Weight Watchers, had served approximately 400,000 people in the United States and in 25 foreign countries (Stunkard, 1977). Behavioral weight management components have been integrated into their program, which usually involves a one-hour weekly meeting. A number of other commercial programs and nonprofit self-help groups have been established using a combination of a recommended diet, behavioral techniques, suggestions for increasing physical activity, and sometimes pharmacotherapy. Systematic studies of the effectiveness of most of these programs have not been reported. One major problem in trying to compare results of alternative programs is determining how to treat dropouts in reporting weight loss. A recent study of an unnamed commercial weight reduction program highlighted this problem by finding a 50 percent dropout rate in the first six weeks and a 70 percent dropout rate within 12 weeks (Volkman et al., 1981).

Critical questions still to be answered about most available programs include the nature of the population served, dropout rates, average and median weight loss during the period of treatment, and average weight change during the maintenance phase. Answers to these questions would permit relating the cost of each program to its weight-reduction result and associated reductions in risk factors.

One controlled clinical trial comparing behavioral therapy, pharmaco-therapy, and their combination found six-month weight losses of 24, 32, and 34 pounds respectively. However, at a one-year follow-up, behavioral therapy alone appeared most effective, with an average loss of 20 pounds from baseline compared with a loss of 14 pounds for pharmacotherapy and a loss of 10 pounds for their combination (Craighead et al., 1981).

Recidivism remains the most serious program in weight-control efforts. During the first year posttreatment, it is difficult for even the initially most effective programs to increase or maintain the loss. Of nine studies of behavioral programs, few reported clinically significant changes from base-line considering the initial percentages overweight (20–78 percent), either after 8 to 16 weeks of initial treatment or at one-year follow-up (Stunkard & Penick, 1981). Of 26 groups in the nine studies, 14 showed weight gains during the 12 months following treatment. In the other 12, although weight loss continued, it averaged only an additional 3.5 pounds.

In what was reported as the first trial comparing worksite treatment for obesity to medical site treatment, 40 women volunteers averaging 57.1 percent overweight were assigned from stratified blocks based on percentage overweight to one of four experimental settings: (1) professional therapist, worksite, weekly treatment; (2) professional therapist, medical site, weekly treatment; (3) lay therapist, worksite, four treatments per week; and (4) lay therapist, worksite, weekly treatment. At the end of a 16-week behavioral program, including self-monitoring, stimulus control, self-reinforcement, exercise management and calorie counting, attrition rates were lowest in the groups led by lay therapists (31 percent four times per week and 50 percent weekly) than by professional therapists (75 percent medical setting and 82 percent work setting). Unfortunately, the high attrition rates in the professional therapist groups precluded an adequate assessment of the relative efficacy of worksite versus medical site treatment. Participants adhering to the program through the sixteenth week achieved an average weight loss of 7.9 ± 2.2 pounds, with no significant differences in weight loss between groups. Six months later, the average weight loss for program completers had declined to 2.6 ± 2.0 pounds. No appreciable differences in rates of absenteeism were found between program groups (Stunkard & Brownell, 1980).

Encouraging results have recently been reported using competition as the principal technique to motivate weight reduction in worksite populations. In one study, a team of employees from each of three small banks in the same market area competed against each other in a three-month weight-loss program. Teams competed for a pool of money contributed primarily by team members and awarded to the winning team. The major program activities were efforts to solicit city-wide publicity about the contest and weekly weigh-ins, at which time team members were given chapters from a weight-loss manual. Average weight loss was 13.2 pounds with a remarkably low attrition rate of 0.05 percent (attrition defined as individuals attending fewer than eight weekly weigh-ins and not completing the final questionnaire). Based on the cost and weight losses of this and two similar competitions with a company and a union group, the average cost per 1 percent reduction in weight was $2.93, which compared favorably to the cost per 1 percent reduction reported by other clinical and worksite programs (Brownell et al., 1984).

Despite limited data from worksite studies, available data suggest that significant reductions in weight may be obtained by both traditional behavioral techniques and by harnessing the competitive feelings associated with business. Similar to clinical programs, the greatest problem in worksite weight-loss programs is recidivism. However, particular characteristics of some worksites might contribute to long-term program effectiveness. The availability of treatment and especially maintenance groups in which individuals are already spending most of their waking hours, as well as holding groups at convenient hours (before or after work, during lunch breaks), could increase maintenance of altered eating habits. Availability of co-worker friends who could be trained in a buddy system to help with food selection and eating habits during the workday could improve the amount of weight loss and possibly enhance maintenance. Follow-up and reinforcement by trained personnel from an employer's medical office might also enhance adherence in a similar manner to the effect noted in hypertension detection and follow-up programs.

Costs of organized weight-control programs vary widely. In general, behaviorally oriented classes without medical examination or treatment require a one-time membership or entry fee. Weight-control clinics may prescribe or dispense appetite suppressants, vitamins and/or food substitutes such as "liquid protein," in addition to the standard counseling on diets and on modifying eating habits. These are run under physician supervision and may cost $50–$200 for an initial evaluation and $10–$30 per week. The fee sometimes includes medication or nutritional supplements. While the most active phase of weight management activities varies considerably in duration, in most programs it is at least 8 to 12 weeks. Many have follow-up reinforcement sessions and some suggest long-term continuing program involvement.

The impressive, although preliminary, results from reported worksite weight-loss competitions suggest providing financial incentives for sustained weight loss by organized work units (e.g., a division, a branch, a store, a union local) (Stachnik & Stoffelmayr, 1983). The worksite environment can be made more supportive of employee efforts to manage their weight by improving food choices in worksite cafeterias, making available nutritious snacks in vending machines, and installing scales to encourage self-monitoring of weight. Employer sponsorship, either on or offsite, of regular exercise programs can also be a major adjunct to weight control. It can increase the amount of energy (calories) used, increase the percentage of lean body tissue, and reduce perceived stress that can contribute to excessive eating.

STRESS MANAGEMENT

Specific health conditions associated with stress include such diverse problems as peptic ulcer, coronary heart disease, and migraine headaches. The mechanisms through which stress increases susceptibility to a broad range of health conditions are not well established, although deleterious effects on the immunological system have been demonstrated. Although the relationship between stress at work and illness has received limited attention, quality of work life is recognized as having a major impact on employee health status. Only a few systematic studies have been reported that reveal direct effects of occupational stress management on level of stress, other risk factors for illness, or use of health benefits or other employer-sponsored health services. Assessment of efficiency and effectiveness of occupational stress management programs is complicated by (1) lack of objective measures of stress, (2) a large number of work and outside factors that can affect perceptions of stress, and (3) the heterogeneity of individual, group, and environmental interventions, which are all called "stress management."

Many stress management programs sponsored by employers have been targeted at either those identified as being at highest stress, identified through a stress-level questionnaire, an employee assistance program, the medical department or supervisor referral, or being a member of a particular job group, such as middle manager. Other programs are oriented to the entire employee population.

A review of 13 occupational stress management programs found that interventions varied from one instructional session to 15, with one study involving home self-instruction of relaxation techniques (Murphy, 1983). All the techniques generally used in worksite stress control programs, including muscle relaxation, biofeedback, meditation, and cognitive restructuring/behavioral skills training, have been shown to reduce arousal level and psychological signs of stress (Murphy, 1983; Pomerleau & Brady, 1979). Most studies have demonstrated benefits on one or more of the following: self-reported stress measures (anxiety, stress, mood, coping, sleep) (Seamands, 1982); attitudes toward work (Peters et al., 1977); job satisfaction, job performance, directly measured EMG, systolic and/or diastolic blood pressure, (Peters et al., 1980) and level of urinary catecholamines excretion. Six-week to six-month follow-ups reveal continued benefit based on self-report, but contradictory results when reductions in arousal level are measured (Murphy, 1983; Peterson, 1981). To cite a program example, at the Converse Rubber Company 126 volunteers were randomly divided into a group taught the relaxation response (Group 1), a group instructed to sit quietly (Group 2), and a group that received no instruction (Group 3). The first two groups were asked to take two daily 15-minute relaxation breaks. After eight weeks, Group 1 reported significantly fewer symptoms, less illness, greater job performance, and better levels of satisfaction and sociability than did Group 3, with Group 2 reporting intermediate values. Group 1 also exhibited significantly greater decline in both mean systolic and mean diastolic blood pressure than did Groups 2 and 3. A study of New York Telephone employees found that those taught how to practice stress management techniques showed clinical improvement in reported decreases in stress symptoms compared to waiting list controls. These effects were seen regardless of the frequency that subjects practiced their techniques (Carrington et al., 1980).

While the majority of studies use controls, few directly compare alternative stress management techniques (Peterson, 1981). Nonspecific effects that simulate program benefits are almost always observed in the controls. The major reason is the increased attention given the controls. Control groups frequently are not included in the follow-up, limiting the effects that can be attributed to the program.

ACCIDENT PREVENTION

Use of lap and shoulder belts can prevent about 60 percent of fatalities from automobile accidents, but where use is voluntary only 10 percent to 14 percent of drivers and a small percentage of passengers use them. On average, traumatic deaths are associated with a loss of more than 20 years of working life, much longer than heart disease deaths (2 years) and cancer deaths (slightly longer than 5 years) (Beck, 1982). In 1978, approximately one-third of all work-related fatalities were caused by motor vehicle crashes, with an average employer cost for each death of $120,000 (Brennan, 1983). At least 19 states require *state* employees to use safety belts in vehicles used on the job (DHHS, 1983), and as of August 1985, 14 states required use of safety belts by all automobile occupants.

At least 20 private and public employers, among them DuPont, General Motors, Ford, the Air Force, Exxon, several Blue Cross–Blue Shield plans, Southwestern Bell and Teletype Corporation, have conducted incentive programs to increase safety belt use by employees (Richman, 1983). While results vary considerably by site and program, preprogram use based on direct observation was almost always in the range of 10–25 percent, with notable exceptions in an Air Force base (40–50 percent) that had stressed safety belt usage prior to the program. Most incentive programs about double usage rates during the period when incentives are offered (Block, 1983). In employer-sponsored programs, drivers using safety belts at entry to the worksite parking areas have been given immediate rewards on a random basis or tickets good for drawings of large prizes. Common prizes have included gift certificates, vacations, cars, lottery tickets, cash, and food coupons (Provato, 1983). In general, blue-collar workers have lower pre-program safety belt use but sometimes show a greater percentage increase in safety belt use than do white-collar employees. In those programs that have assessed compliance, postprogram use has significantly declined from program levels, but has remained higher than before the intervention was introduced.

MULTIPLE RISK INDICATOR STUDIES

Most large risk factor reduction studies have targeted cardiovascular disease. Despite the high prevalence of cardiovascular diseases, considerable time and large study populations are required to assess the impact of risk reduction programs on morbidity and mortality. Most heart disease prevention projects instead use a predictive equation that combines the quantitative relationships between known cardiovascular risk indicators (e.g., smoking, high blood pressure, total serum cholesterol) and cardiovascular problems (e.g., heart attack) and mortality. A reduction in the age-adjusted risk based on this equation translates into reduced risk of a cardiovascular incident and/or dying from cardiovascular disease during a defined period, usually 10 years.

The largest reported heart disease prevention project at the worksite was a controlled multifactorial trial involving 19,300 men aged 40–59 employed by 30 Belgian industries (Kornitzer, De Backer, Pramaix, et al., 1980). Intervention company employees whose risk profiles were in the top quintile (20 percent) received a brief individual couseling session from a physician twice a year. A cardiovascular health education campaign was also organized in each intervention factory. After two years, the risk equation product of the intervention high-risk group had decreased by 20 percent, while it had increased by 12.5 percent in the high-risk control group. Thus the risk difference for cardiovascular disease between these two high-risk groups was more than 30 percent, based on a low-intensity intervention. Younger workers in general experienced the greatest changes in risk variables.

The closest comparable American disease prevention program, the Stanford Three Communities Study, used community-wide, rather than worksite only, intervention approaches (Meyer et al., 1976). Stanford studied both sexes and had a slightly broader age range, 35–59. The cardiovascular high-risk group received face-to-face counseling in addition to mass media risk reduction health education approaches. At the end of two years, the risk equation difference between intervention and control groups was 28 percent, strikingly similar to the Belgian study.

Both studies showed a significant effect of health education/communication campaigns alone on random sample groups of the entire population, including both high- and low-risk individuals. Differences in cardiovascular-risk-equation-change-score between intervention and control groups were 27.3 percent and 20.6 percent, respectively, in the Belgian and Stanford Studies, with the intervention groups demonstrating the greatest changes in both studies.

DISCUSSION

As illustrated by the examples above, health practices that increase individual risk cannot be considered homogeneous. Some habits are more refractory on the average than others. Overeating appears more difficult to curb permanently than cigarette smoking. Preventing the adverse effects of hypertension may involve taking only one or two pills per day, requiring minimal effort, but the development of a new habit such as exercise requires an ongoing struggle to find time and to reorder priorities. Stopping smoking requires a major effort initially but the impulses to smoke diminish in frequency and intensity in time. Modifying eating habits entails making difficult decisions at least three times daily every day, adapting to different tastes or at least excluding some favorite foods, and effectively coping with social norms that lead to the serving of copious portions. Although the struggle may become easier in time, it is rarely eliminated entirely. Vigorous exercise does not require such difficult choices but does involve a high degree of continuing motivation. It imposes strong demands for time in an already crowded schedule. An exercise routine is boring for many, at least at first. Exercising can also lead to orthopedic injuries, which can make developing a continuing exercise habit even more difficult. Satisfactions of improved body shape, reduced stress, and the "exerciser's high" take some time to develop. Stress management techniques require practice and frequent use to be maximally effective. Seat belt use is a simple habit to adopt, but to become established, requires a volitional effort each time a car is entered.

In the cases of exercise and reduced calorie consumption, the rewards can be seen and felt in a matter of weeks. Some smokers, but not all, may find cessation soon leads to reduced cough and improved respiration. But much of the payoff for smokers, virtually all of it for hypertensives, and some for exercisers, includes reducing longer term risks—for heart attacks, lung cancer, emphysema, strokes, and other morbid events whose chances of appearance can only be described as probabilities, not certainties.

Complicating estimation of the benefits of risk reduction efforts is that reducing the variables that confer risk has not in every case been shown to lower risk. While smoking cessation and seat belt use and hypertension control do lower risk, physical exercise has not yet prospectively been demonstrated to reduce cardiovascular incidents. Elimination of obesity will reduce cardiovascular risk if it causes a lowering of high blood pressure, will probably reduce risk if weight loss mediates a reduction in total cholesterol, and probably would not substantially affect risk in a person with low cholesterol and blood pressure unless he or she is 25 or more percent overweight.

For only a few risk reduction efforts can benefits be accurately quantified, and in fewer instances still can costs and benefits be weighed against each other. Even in the most propitious circumstances, such as smoking cessation and hypertension control, the variation in program costs and effectiveness and lack of valid information on timing, degree of benefits, and costs of problems averted limit conclusions. A cost-effectiveness analysis, which relates program costs to benefits stated in nonmonetary terms, may be a better way to view the desirability of program investment. What is the value to a company of a program that can gradually reduce in five years the number of heart attacks per 1,000 male employees annually from 10 to 6? What is it worth to have the percentage of employees who say they are happy with their jobs increase from 30 to 60 percent after institution of a company-sponsored exercise program? In most cases, employers' interest in the health of their employees justifies a significant investment to achieve these types of results, even if the return-on-investment in dollar terms cannot be accurately predicted.

REFERENCES

Alderman, M. H., & Davis, T. K., (1976, December). Hypertension control at the work site. *Journal of Occupational Medicine, 18,* 793–796.

Ashley, F. W., & Kannel W. B. (1974). Relation of weight change to changes in atherogenic traits: The Framingham Study. *Journal of Chronic Diseases, 27,* 203–214.

Barnard, R. J., & Anthony, D. F. (1980, October) Effects of health maintenance in Los Angeles City Firefighters, *Journal of Occupational Medicine, 22,* 667–669.

Beck, R. N. (1982). *Health promotion and health protection: The role of industry.* Paper presented at the 2nd Annual Lester Breslow Distinguished Lectureship, UCLA, School of Public Health, Los Angeles, CA.

Bellack, A. S. (1975). Behavior therapy for weight reduction. *Additive Behaviors, 1,* 73–82.

Bernacki, E. J. & Baun, W. B. (1984, July). The relationship of job performance to exercise adherence in a corporate fitness program. *Journal of Occupational Medicine, 26*(17).

Bjurstrom, L. A., & Alexious, N.J. (1978). A program of heart disease intervention for public employees. *Journal of Occupational Medicine, 20,* 521–31.

Block, D. L. (1983, April). *Executive management and medical leadership roles in health care cost containment: A medical director's perspective.* Paper presented at American Occupational Medical Association Annual Meeting, Washington, DC.

Brennan, A. (1983). *Management Review,* 41–47.

Brownell, K. D., Heckeiman, C. L., Westlake, R. J., Hayes, S. C., & Monte, P.M. (1978). The effect of couples training and partner cooperativeness in the bahavioral treatment of obesity. *Behavioral Research and Therapy, 16,* 323–333.

Brownell, K. D., Yopp-Cohen, R., Stunkard, A. J., Felix, M., & Cosley, W. B. (1984). *Weight loss competition at the worksite: Impact on weight, morale, and cost-effectiveness.* ASPH.

Cantlon, A. (1983, April 25). *Southern New England Telephone's Health Promotion Program.* Lecture presented at Bell System Nurses and Health Educator's Conference, Washington, DC.

Carrington, P., et al (1980). The use of meditation—relaxation techniques for the management of stress in a working population. *Journal of Medicine, 22,* 221–231.

Cowell, M. J., & Hirst, B. L. (1979, October). *Mortality differences between smokers and nonsmokers.* State Mutual Life Assurance Company of America.

Cox, M., Shephard, R.J., & Corey, P. (1981). Influence of an employee fitness programme upon fitness productivity and absenteeism. *Ergonomics, 24,* 795–806.

Craighead, L. W., Stunkard, A. J. & O'Brien, R. M. (1981, July). Behavior therapy and pharmacotherapy for obesity. *Archives of General Psychiatry, 38,* 763–768.

Danaher, B. (1980). Smoking cessation in occupational settings. *Public Health Reports, 95,* 119–126.

Donoghue, S. (1977, May–June). Correlation between physical fitness, absenteeism and work performance. *Canadian Journal of Public Health, 68,* 201–203.

Durbeck, D.C., Heinzelmann, F., Schaeter, J., Haskell, W. L., Payne, G. H., Moxley, R. T., Nemiroff, M., Limoncelli, D. O., Arnoldi, L. B., & Fox, S. M. (1972, November 20). The National Aeronautics and Space Administration–U.S. Public Health Service Evaluation and Enhancement Program. *American Journal of Cardiology, 30,* 784–790.

Farquhar, J., Maccoby, N., Wood, P. D., et al. (1977). Community intervention for cardiovascular health. *Lancet,* 1192–1195.

Fielding, J. E. (1985). Smoking: Health effects and control. *Public Health and Preventive Medicine. Maxcy-Rosenau.*

Fielding, J. E., & Breslow, L. E. (1983). Health promotion programs sponsored by California employers. *American Journal of Public Health, 73,* 538–542.

Foote, A., & Erfurt, S. C. (1983). Hypertension control at the worksite. *New England Journal of Medicine, 308,* 809–813.

Foote, A., & Erfurt, S. C. (1977). Controlling hypertension, a cost-effective model. *Preventive Medicine, 6,* 319–343.

Hammond, E. C., & Seidman, H. (1980, March). Smoking and cancer in the United States. *Preventive Medicine, 9,* 169–173.

Hannan, E. L., & Graham, J. K. (1978, December). A cost-benefit study of a hypertension screening and treatment program at the work setting. *Inquiry, 15,* 345–358.

Haskell, W. L., & Blair, S. N. (1980, March–April). The physical activity component of health promotion in occupational settings. *Public Health Reports, 95,* 109–118.

General Mills. (1979). *Family health in an era of stress: The General Mills American Family Report, 1978–1979.* Minneapolis, MN: General Mills.

Haynes, R. B., Sackett, D. L., Taylor, D. W., Gibson, E. S., & Johnson, A. L. (1978, October 5). Increased absenteeism for work after detection and labeling of hypertensive patients. *New England Journal of Medicine, 299,* 741–744.

Heinzelmann, F., & Bagley, R. W. (1970). Response to physical activity programs and their effects on health and behavior. *Public Health Reports, 85,* 905–911.

Horne, W. M. (1975, March). Effects of a physical activity program on middle-aged sedentary corporation executives. *American Industrial Hygiene Association Journal,* 241–246.

"Hypertension Detection and Follow-up Program Cooperative Study." (1979, December). Five-year findings of the Hypertension Detection and Follow-up Program. 1. Reductions in mortality of persons with high blood pressure, including mild hypertension. *Journal of American Health Association, 242,* 2562–2577.

Jeffrey, R. W., Wing, R. R., & Stunkard, A. J. (1976). Behavioral treatment of obesity: The state of the art. *Behavioral Therapy, 9,* 189–199.

Kannel, W. B., & Sorlie, P. (1979, August). Some health benefits of physical activity: The Framingham Study. *Archives of Internal Medicine, 139,* 857–862.

Kornitzer, M., De Backer, G., Pramaix, M. et al. (1980). The Belgian heart disease prevention project, *Circulation, 51,* 18–25.

Kristein, M. M. (1977). Economic issues in prevention. *Preventive Medicine, 6,* 252–264.

Kristein, M. M. (1980, January 9). *How much can business expect to earn from smoking cessation?* Presentation at the National Interagency Council on Smoking and Health's National Conference: "Smoking and the Workplace," Chicago, IL.

Leventhal, H., & Cleary, P. D. (1980). The smoking problem: A review of the research and theory in behavioral risk modification. *Psychological Bulletin, 88,* 370–405.

Linden, V. (1969). Absence from work and physical fitness. *British Journal of Industrial Medicine, 26,* 47–53.

Logan, A., Achber, C., Milne, B., Campbell, W., & Haynes, R. B. (1979, December). Site treatment of hypertension by specially trained nurses. *Lancet, 2,* 1175–1178.

Logan, A., Milne, B., Achber, C., Campbell, W., & Haynes, R. B. (1981, March–April). Cost-effectiveness of worksite hypertension treatment program. *Hypertension 3,* 211–219.

Luce, B. R., & Schweitzer, S. O. (1978, March 9) Smoking and alcohol abuse: A comparison of their economic consequences. *New England Journal of Medicine, 298,* 569–571.

Malotte, K., Fielding, J. E., & Danaher, B. G. (1981, August). Description and evaluation of the smoking cessation component of a multiple risk intervention program. *American Journal of Public Health, 71,* 884–847.

Meyer, A. J., Macalister, A., Nash, J., Maccoby, N., & Farquhar, J. W. (1976). Maintenance of cardiovascular risk reduction results in high risk subjects. *Circulation, 54* (Supplement 2), 211–226.

Murphy, L. R. (1983, January). *Occupational stress management: A review and appraisal.* NIOSH, Centers for Disease control. (unpublished).

Naditch, M. P. (1984). The Staywell Program. In J. D. Matarazzo, N. E. Miller, S. M. Weiss, J. A. Herd, & S. M. Weiss. (Eds.), *Behavioral health: A handbook of health enhancement and disease prevention.* New York: Wiley.

National Center for Health Statistics, Division of Health Interview Statistics. (1984, May). Data from the *National Health Interview Survey. Washington, DC.*

National Heart Institute. (1975). *The underwriting significance of hypertension for the life insurance industry.* DHEW Publication No. (NIH) 75-426. Washington, DC: U.S. Government Printing Office.

National Heart, Lung, and Blood Institute. (1981, January). *Cardiovascular primer for the workplace.* DHEW Publication No. (NIH) 81-2210. Washington, DC: U.S. Government Printing Office.

National Heart, Lung, and Blood Institute. (1980, Winter). *National high blood pressure education program at Mass. Mutual: Off-site care and good monitoring reduce medical costs, re: high blood pressure control in the worksetting.* Washington, DC: U.S. Government Printing Office.

National Heart, Lung, and Blood Institute. (1979). *Smoking and health.* A Report of the Surgeon General Office on Smoking and Health. DHEW Publication No. (PHS) 79-50066. Washington, DC: U.S. Government Printing Office.

Oja, P., Tersalina, P., Partanen, T., & Karava, R. (1974). Feasibility of an 18-month physical training program for middle-aged men and its effects on physical fitness. *American Journal of Public Health, 64,* 459–464.

Pacific Mutual Life Insurance Co. (1979). *Health maintenance.* Newport Beach, CA: Pacific Mutual Life Insurance Co.

Paffenbarger, R. S., Wing, A. L., & Hyde, R. T. (1978). Physical activity as an index of heart attack risk in college alumni. *American Journal of Epidemiology, 108,* 161–175.

Peters, R. K., Benson, J., & Porter, D. (1977). Daily relaxation response breaks in a working population: I. effects on self-reported measure of health, performance, and well-being. *American Journal of Public Health, 67,* 946–953.

Peters, R. K., Benson, J., & Peters, J. M. (1980). Daily relaxation techniques for the management of stress in a working population. *Journal of Occupational Medicine, 22,* 221–231.

Peterson, P. (1981). *Comparison of relaxation training, cognitive restructuring behavioral training, and multimodal stress management training seminars in an occupational setting.* Dissertation submitted to Fuller Theological Seminary, Los Angeles, CA.

Polk, F. B., Harlan, L. C., Pozner-Cooper, S., Stromer, M., Ignatus, J., Mull, H., & Blaszowski, T. P. Disability days associated with detection and treatment in a hypertension control program. *American Journal of Epidemiology, 119*(1) 44–53.

Pomerleau, O. F., & Brady. J. P. (1979). *Behavioral medicine, theory and practice.* Baltimore, MD: Williams & Wilkins.

Provato, F. L. (1983, April 25). *Corporate health care costs: A medical economic problem at Southern New England Telephone.* (unpublished).

Richman, L. K. (1983). *Fortune,* 95–110.

Ruchlin, J. S., & Alderman, M. H. (1980, December). Cost of hypertension control at the workplace. *Journal of Occupational Medicine, 22,* 795–800.

Salonen, J. T. (1980). Stopping smoking and long-term mortality after acute myocardial infarction. *British Heart Journal, 43,* 463–469

Salonen, J. T., Maccoby, N., Wood, P. D., et al. (1977). Community intervention for cardiovascular health. *Lancet, 1,* 1192–1195.

Schneider, N. G., Javerik, M. E., Forsythe, A. B., Read, L. L., Elliot, M. L., & Schweiger, A. (1983). Nicotine gum in smoking cessation: A placebo–controlled double blind trial. *Addictive Behaviors, 8,* 253–261.

Seamands, B. C. (1982). Stress factors and their effect on absenteeism in a corporate employee group. *Journal of Occupational Medicine, 24,* 393–397.

Shephard, R. J., Corey, P., Renzland, P., & Cox, M. H. (1982, January). Fitness program reduces health care costs. *Dimensions,* 1–15.

Sorlie, M. S., Gordon, T., & Kannel, W. B. (1980, May 9). Body build and mortality. *Journal of American Medical Association, 243,* 1828–1831.

Srole, L., Langer, T. S., Michael, S. T., et al. (1962). *Mental health in the metropolis: The Midtown Manhattan Study.* New York: McGraw-Hill

Stachnik, T., & Stoffelmayr, B. (1983, December). Worksite smoking cessation programs: A potential for national impact. *American Journal of Public Health, 73 (12),* 1395–1396.

Stamler, R., Gosch, F. C., Stamler, J., Lindberg, H., & Hilker, R. (1979, September). A hypertension control program based on the workplace. *Journal of Occupational Medicine, 20,* 618–625.

Stuart, R. B. (1967). Behavioral control of overeating. *Behavior Research and Therapy, 5,* 357–365.

Society of Actuaries. (1959). *Build and blood pressure study* (Vol. 1), pp. 1–268. Chicago, IL: Society of Actuaries.

Stunkard, A. J. (1977, November 30). Obesity and social environment: Current status, future prospects. *Annals of the New York Academy of Sciences, 300,* 298–320.

Stunkard, A. J. (1979). Behavioral medicine and beyond: The example of obesity. In O. F. Pomerleau & J. P. Brady (Eds.) *Behavioral Medicine: Theory and Practice.* Baltimore, MD: Williams & Wilkins.

Stunkard, A. J., & Brownell, K. D. (1980). Worksite treatment for obesity. *American Journal of Psychiatry, 137,* 252–253.

Stunkard, A. J. & Penick, S. B. (1981, July). Behavior modification in the treatment of obesity. *Archives of General Psychiatry, 38,* 763–768.

U.S. Department of Health, Education, and Welfare. (1979) Smoking and health. Report of the Surgeon General Office on Smoking and Health (DHEW Publication No. PHS 79-50066). Washington, DC: U.S. Government Printing Office.

U.S. Department of Health and Human Services. (1983, July). Phone conversation with Willis Goldbeck, Washington Business Group on Health.

Veterans Administration Cooperative Study on Antihypertensive Agents. (1967, December 1). Effects on morbidity in hypertension: Results in patients with diastolic blood pressures averaging 115 through 129 mm Hg. *Journal of American Medical Association, 202,* 1028–1034.

Veterans Administration Cooperative Study Group on Antihypertensive Agents. (1970, August 17). Effects on morbidity in hypertension, II. Results in patients with diastolic blood pressures averaging 90 through 114 mm Hg. *Journal of American Medical Association, 213,* 1143–1152.

Veterans Administration Cooperative Study Group on Antihypertensive Agents. (1972, May). Effects of treatment on morbidity in hypertension, IV. Influence of age, diastolic pressure, and prior cardiovascular disease: Further analysis of side effects. *Circulation, 15,* 991–1004.

Volkmar, F. W., Stunkard, A. J., & Bailey, R. A. (1981, March). High attrition rates in commercial weight reduction programs. *Archives of Internal Medicine, 141,* 426–428.

Warner, K. E. (1977, July). The effect of the anti-smoking campaign on cigarette consumption. *American Journal of Public Health, 67,* 645–650.

Yarvote, P. M., McDonagh, T. S., Goldman, M. E., & Zuckerman, J. (1974, September). Organization and evaluation of a physical fitness program in industry. *Journal of Occupational Medicine, 16,* 589–598.

FOUR

Policy Issues

18

Preventive Medicine and the Corporate Environment: Challenge to Behavioral Medicine

MICHAEL F. CATALDO, LAWRENCE W. GREEN,
J. ALAN HERD, REBECCA S. PARKINSON,
and WILLIS B. GOLDBECK

The American people enjoy the most advanced health care system in the world and are healthier than at any previous time in history. From this position, we continue to attack new problems, to improve health care procedures and systems, and to anticipate the future. At times, these efforts occur in an atmosphere of cooperation, and at other times, amid controversy. As we anticipate the future, increasingly we can foresee significant problems for health care, significant enough to lead to a decline in the extent, quality, and benefits of the current care system we enjoy. This chapter will discuss four different perspectives relevant to how predicted problems related to health and health care may be addressed increasingly through industry.

Several factors indicate that the demands on our health care system will greatly increase, perhaps to crisis proportions, as we enter the twenty-first century (Cataldo, 1983):

1. **The Changing Nature of Health Care Problems.** During the twentieth century, this country's major health problems, in terms of

This chapter is based on a symposium held at the Fourth Annual Scientific Meeting of the Society of Behavioral Medicine in Baltimore, MD, March 1983. The chapter was a collaborative effort, and the order of authorship does not indicate ranking of contribution but has been based on the conceptual order of each topic in the chapter. The topic authors are "Health Education", L.W. Green; "Physician", J.A. Herd; "Management", R.S. Parkinson; "Business", W.B. Goldbeck; "Organization, Introduction, and Conclusion", M.F. Cataldo.

cause of death, have changed from acute to chronic diseases. By definition, chronic disorders such as cancer and cardiovascular disease are long-term problems, often with associated disability, in which onset may begin in the first three decades of life and remain subclinical in manifestation until the fourth, fifth, and sixth decades of life.

2. **An Increase in Chronic Disorders Will Result in More Americans Needing Health Care for a Longer Period.** As chronic disorders with clinical onset in mid-life become more prevalent, an increasing number of people with activity limitations will require longer term medical treatment, workers' compensation, and job retraining.

3. **The Impact of the Post-World War II Baby Boom: The Health Time Bomb.** The rise in the birthrate, beginning in the 1940s and lasting through the early 1960s, has caused various social, economic, and political problems as a larger number of individuals have entered each age category during the past 30 years. During the next 20 years, this predominant proportion of our population will reach the age when the most prevalent chronic diseases begin to be clinically manifested.

4. **The Escalating Cost of Health Care.** Health care costs have risen exponentially in the last half of the century: from $3.6 billion in 1929 to $355 billion in 1983. Since 1929, the percentage of our gross national product (GNP) devoted to health care has risen from 3.5 percent to more than 10 percent. This represents an increase of more than 2,000 percent since 1950 and more than 8,000 percent since 1929.

5. **The Problem of the Cumulated Federal Deficit.** For more than two decades—with two exceptions, 1960 and 1969—as a nation we have been deficit spending such that the cumulative deficit is now approximately $1.5 trillion. If we began to pay off this debt at the rate of $1 million per day, beginning today and with no further deficit spending, then the current cumulated deficit would be paid off by the year 6093.

Unless some other national priorities are greatly cut back, or we choose to go further into debt, health care costs per capita must be reduced soon. Consider, for example, the following statistics: within the next three to four decades, 25 percent of our population will be over 65 years of age, and 20 percent will be under 18; this will mean that for the first time in our history, 55 percent of the population will be attempting to support 45 percent of the population.

As we confront the current and anticipated problems, we do so in an inter-disciplinary fashion, now including not only the various disciplines concerned with health care (e.g., medicine, public health, nursing, psychology, sociology, economics), but also factions in industry (e.g., labor and management). This phenomenon adds a potential area of conflict, an added problem, namely, working together across (and despite) disciplinary boundaries. One way to address this potential conflict area is to determine the basic assumptions, the unique viewpoints that each area brings to such an interdisciplinary roundtable. This chapter provides perspectives from public health (health education), medicine, management, and business about behavioral medicine approaches to health in industry. Each section is authored by a leader in one of these fields and is structured around (1) basic assumptions, (2) priorities, and (3) present and future barriers.

HEALTH EDUCATION PERSPECTIVE

Why the Worksite?

The first consideration concerns the demographics of public health target groups. Approximately 70 percent of the adult population is employed. Through no other single channel can we reach such a large proportion of the adult population. More than 85 percent of the adult male population is employed. This population is hard to reach through other traditional medical care and public health channels, because men do not use medical care services to the same extent as women; and public health services are not as specialized for men as these services are for maternal and child health programs. In addition, males are at higher risk for premature death and disability. On the other hand, females, in addition to having higher medical user rates, represent an increasing proportion of the work force, and the proportion of working women should increase to more than 60 percent by 1990.

The second consideration concerns the logistical, scientific, and professional advantages of the worksite for promoting health. These advantages include lower costs of medical services, greater access to people and time, stability of the working population as a "captive" population, willingness of workers to participate in employer-sponsored programs, existence of supportive management and organizational structures, and, most important, the opportunity to improve the competence of our overall health programs. With regard to this last point, worksite programs may increase our ability to mount more comprehensive or holistic health programs than is possible in traditional medical care and public health institutions. Worksite programs allow for strong social support networks. In some special cases, such as alcohol and other substance abuse, employers can use the worker's job as leverage to encourage him or her to accept treatment. Within this context, industry can provide the full spectrum of services from early detection, to referral, to treatment—dimensions difficult to provide through traditional medical care and public health institutions. The National Heart, Lung, and Blood Institute (NHLBI) lists other additional reasons for focusing on worksite for high blood pressure programs, including access to data, tracking facilities and mechanisms, and existing medical care and health personnel to support the programs.

Priorities

What about the priorities and advantages to industry? In which programs should they invest? Priorities for industry must include some combination of considerations about the timeframe for the realization of benefits, whether short-term or long-term; whether the program will benefit or serve the interest of all employees or only a selected few high-risk individuals and groups; and whether the program has demonstrated some cost-savings in some worksites, or at least some cost-savings in other settings from which we can extrapolate the potential cost-savings across sites. Based on these criteria, the following represent logical priorities for industry disease prevention and health promotion programs.

1. **Employee Assistance Programs.** This priority is based heavily on the track record of acceptance for employee assistance programs already established in industry. Within industry a momentum has developed for such programs because of the great magnitude and scope of the problems that must be addressed. Paramount among these problems is that of alcohol abuse. It is estimated that 57 percent of all industrial accidents are associated with alcohol. To the cost of this must be added other alcohol but non-job related accidents. For example, more than 50 percent of all motor vehicle accidents are alcohol related; and according to the National Highway Traffic and Safety Administration, industry pays an estimated $120,000 for every accident incurred by an employee, even if the accident occurs while the employee was not on the job.

2. **Hypertension Detection and Treatment or Referral Programs.** Based on 1972 estimates of workdays lost and 1976 data on earnings, the NHLBI has concluded that more than $1 billion in lost earnings per year are associated with cardiovascular diseases related to high blood pressure.

3. **Anti-Smoking Programs (i.e., cessation and prevention).** This is a difficult case because industry will not necessarily realize an early return on their investment with highly mobile work forces. Nonetheless, for the truly health conscious company, anti-smoking programs must have high priority in light of the overwhelming, incontrovertible evidence of the association between smoking and serious chronic disease. Less compelling but financially important to industry are the variety of other savings that are associated with smoking cessation and prevention, including smoking rituals, extra cleanup costs, extra

damage to furniture and equipment, inefficiencies and errors due to factors such as high carbon monoxide levels and eye irritation, and sickness while on the job (which can lead to absenteeism and lost productivity). Other reasons for corporate programs are legal challenges and state/local regulations (e.g., Minnesota, San Francisco, Los Angeles).

4. **Exercise and Fitness Programs.** The benefits to employees of exercise and fitness programs are projections from short-term gains, mostly based on self-report. The President's Council on Physical Fitness and Sports continues to cite the NASA Study, which was neither experimental in design nor objective in its source of measures, but which seemed to have yielded palpable benefits that are of interest to management regardless of the validity of the data. The President's Council appeals to industry's bottom line with an enigmatic reference to the estimate that regular exercise can reduce absenteeism by three to five days per person per year, a savings that would soon pay for the installation and maintenance of a corporate fitness program. Furthermore, the enthusiasm and momentum of the advocates and devotees are irrepressible, including that of many CEOs themselves. Fitness programs offer a competitive edge in recruiting and holding some of the most vigorous corporate talent and seem to be an opening wedge for people's involvement in other areas of health promotion, especially weight control, which would otherwise gain little support.

In addition to the priorities mentioned above, other priorities would be the use of seat belts, self-care education, obesity and weight management, and stress management.

Barriers and Pitfalls to Avoid

We need not adopt the above set of priorities simply because the case can be made to industry from the research standpoint. Indeed, from a research standpoint we ought to reverse this list of priorities because the list is based primarily on what we know. Research priorities should also be based on what we do not know.

We must be careful in attributing blame for illness and responsibility for health. We must avoid political rhetoric about self-help and personal responsibility that is akin to the practice of victim-blaming and ultimately results in forcing workers to fend for themselves (Allegrante & Green, 1981). Unfortunately, this inappropriate context for health education and health promotion has been implanted in most of the federal documents that have been produced during the transition from the Carter to the Reagan administration.

Finally, one other social and ethical trap to avoid is *health fascism*. One manifestation of such fascism is defining for other people what is a proper lifestyle. A second example of health fascism, which we fall into with the trade secret mentality of the private sector, is the trap of hoarding and secreting knowledge. This second example would be most unfortunate for those of us who, as scientists and health educators, are committed to the dissemination and use of knowledge. In either example, if we fall into the trap of health fascism, we would undermine the purposes that drew us to this enterprise in the first place.

Summary

The health education perspective on health promotion in the corporate environment derives from a public health perspective, but recognizes that the acceptance of programs by corporations depends on more than epidemiological evidence. A hierarchy of priorities for corporate adoption of programs, considers costs and benefits accruing to the firm, as well as public relations and considerations of employee demand and enthusiasm. These criteria place highest priority on employee-assistance programs, anti-smoking programs, exercise and fitness programs, in that order. While bowing to corporate priorities, professionals must resist pressures to bow to some corporate practices that would compromise professional ethics and obligations, such as the commitment of health educators and behavioral scientists to the freedom of information and knowledge acquisition.

PHYSICIAN PERSPECTIVE

Why a Physician Perspective?

A point of departure for the physician perspective is to ask why this should be considered an important perspective. Physicians can, and sometimes do, disrupt programs. There is nothing more deadly to a preventive medicine effort or health program than to find that the physicians in the community (for whatever reason) have turned their backs or actively campaigned against such programs. Yet we must accept the fact that physicians can have a powerful influence, that they do influence their patients, and that their patients are often decision makers, opinion makers, and participants (or not) in health and prevention programs.

Another important reason for attending to the perspective of physicians is that, like it or not, they are the gatekeepers for reimbursement. In the past, all the policies that have been established concerning what should be paid for—both in medical care and now in this burgeoning area of preventive medicine—have been controlled by physicians. Although this may eventually prove to be outmoded, dated, and conservative, these are the facts about how reimbursement is decided and will likely continue to be the case for some time. Currently, we are also interested in behavioral medicine. We study behavior and consider health from a behavioral medicine. We study behavior and consider health from a behavioral viewpoint. We may be well advised to do a behavior analysis of the viewpoint of the physician, and *use* that knowledge as a means of designing our approach toward the preventive medicine programs.

Priorities: Categories of Preventive Medicine Activities

Categories of preventive medicine often include health promotion, fitness, nutrition, and so forth. These categories are generally prescribed by physicians in the spirit that at least they won't hurt and probably will do some good. From the physician's perspective, the categories are primary, secondary, and tertiary. By definition, *primary* approaches would include the identification and alteration of risk factors; *secondary* refers to the early detection and treatment of disease; and *tertiary* is the comprehensive rehabilitation of symptomatic cases. Secondary and tertiary approaches to preventive medicine occupy far more of the physician's attention. They offer more powerful means of demonstrating benefits in terms of cost-effectiveness. That is not to imply that these categories are more cost-effective, but rather that they offer a better *means of demonstrating* cost-effective benefits in light of the current methods for reimbursement and time-frames for analysis of outcome.

Therefore one is often led to the conclusion that prevention is expensive and inefficient. This is not because prevention really is or has to be, but because it appears to be expensive and inefficient. At best, this is because of practical matters and limitations. For example, adherence to lifestyle recommendations is often a problem—seconded only by the problem of maintenance of lifestyle change. People simply do what they do for reasons that we have yet to understand sufficiently. Another practical problem is that many industry-based prevention programs are directed toward young people who change jobs frequently. Thus many do not stay with a particular corporation long enough to develop the complications of hypertension or some other form of heart disease, cancer, or stroke. In that instance, it is extremely difficult to demonstrate that a preventive program actually has some health benefits to the employees, let alone finanacial benefits to the company. Should such programs exist and survive, they often do so for corporate public relations purposes, for employee morale and recruitment, or as a labor/management package, rather than for their demonstrated health benefits.

Barriers to Cooperation

Problems for the Physician in Preventive Medicine. The problems for physicians in preventive medicine are many. First of all, there is a "magic bullet" expectation. Patients believe that just doing something will result in a remarkable outcome. A passive versus an active role has evolved in medicine, and the realistic expectations in medicine are that, in fact, people do not change their behavior quickly and, furthermore, will most probably relapse. This attitude has led to considerable discrepancies in expert opinion about what we think patients are willing to do, what they know, and what their expectations are. Unfortunately, our knowledge base in this area is incomplete.

Problems for Corporations in Preventive Medicine. An especially important problem for corporations in preventive medicine is that the cost-benefit data are poor, nor is it likely that in the next 10 years we will be able to show that preventive medicine is cost-effective in reducing health care costs. For example, treating somebody for hypertension costs at least $200 a year, and industry would be lucky to get back, on the average (considering turn-over rates, age, and sex), $10 a year. For the foreseeable future, it will be more expensive for industry to treat a disorder such as hypertension than to let the workers with hypertension go untreated, waiting for them to change jobs eventually, and have their stroke while working for another employer. There-fore, there must be some other reason for introducing a preventive medicine program. This cost-benefit problem is compounded by other factors. The reimbursement mechanism is difficult, despite the fact that there is consider-able interest in changing ways in which people pay for health care. Resources are limited, specialized services are hard to obtain and expensive, and co-ordination with the community resources is exceedingly complex.

Potential Conflicts between Physicians and Corporations in Preventive Medicine Programs.   First, the *physician–patient relationship* is, in part, some-thing that the physician uses as a means to persuade patients to do various things such as take medication or undergo a diagnostic or therapeutic precedure. It is an important part of the practice of medicine, even though we would all agree that something more intellectual, more active, or more inter-active might also be appropriate. Nonetheless, this physician–patient relationship is a strong tradition that physicians consider important and will fight to retain. Second, there is the potential conflict surrounding *attitudes and expectations*. Physicians do not have high expectations that preventive measures will have major effects. Because of this, the *advice and encouragement* that they might give to their patients in private practice are somewhat different than are heard in the corporate setting or in the preventive medicine/health promotion program. This discrepancy is a source of concern to physicians and leads to the next potential area of conflict—the physician's perception of threat by the current climate wherein corporations are taking an increasingly positive, active role in medicine. While in the long run such a role by industry must come, it is nonetheless a present threat to physicians. Furthermore, the *threat* of the health promotion/preventive medicine program may represent something that is far more ominous than any particular program itself.

Another potential area of conflict is positive test results that are false—a reality of medical diagnostics that is often magnified when viewed in isolation. When a positive results shows up on a worksite screening, the worker is often referred to a physician for treatment. The worker comes to the physician with concerns for his or her health problem and expects treatment. If the screening resulted in a false-positive result, the physician must explain the error; alternatively, if identified in the course of a regular diagnostic procedure, it could have been obtained in proper context, without unduly alarming the worker/patient.

The last area for potential conflict concerns how popular trends and folk wisdom influence attitudes and expectations. Physicians insist on a medical-scientific basis for the recommendations that are made, and no amount of explaining, hand waving, and enthusiasm will convince physicians to do something that has not been supported by scientific investigation.

Summary

In summary, the bias of the physician is important because of this group's high leverage in the health care system and, for that matter, opinion-making systems and processes. As long as our current health care system exists, the physician must be attended to in a direct and positive manner.

MANAGEMENT PERSPECTIVE

Why Industry Supports Health Promotion Programs

Health care costs of employee benefits are rising at an alarming rate. In most companies the health care benefit packages provide for the treatment of disease and its associated disability. But these benefits often do not include payment for preventive services. In recent years, however, corporate leaders have begun to adopt comprehensive health care strategies that include preventive approaches such as employee health promotion. These health promotion programs are emerging as part of a viable strategy for managing health care costs, improving employee health and morale, and offering employees preventive health care benefits.

Employee health promotion specifically addresses the behavioral changes associated with individual lifestyles and corporate cultures. Health and business norms, which are deeply rooted in corporate cultures, either support or inhibit employee health promotion activities. These norms affect the way business managers allocate scarce resources and their expectation of health promotion outcomes. From the perspective of behavioral medicine, the effort to effect change in corporate health norms and practices raises a set of issues that have considerable value to both the health professional and the business manager. These issues form a critical decision path that can be used to guide the establishment of employee health promotion activities.

Determining Priorities: Needs Assessment

Why do managers and health practitioners want to help employees adopt healthier lifestyles? Why is this partnership important to behavioral medicine? In many respects an organization, such as a corporation, is a living organism. The word corporation derives from the word *corpus*, which means body. If one looks at a corporation as composed of people and, therefore, their collective behaviors, one can think about a corporation as having a culture that must undergo a lifestyle change, just as an individual would undergo such a change. An understanding of behavioral science and workplace employee health promotion thus can be obtained through a brief tour of some of the organizational behaviors that managers in corporations experience in designing and implementing health promotion programs.

The objective in health promotion is change. While behavioral scientists usually are interested in changing the individual and his or her lifestyle behavior, an organization is interested in changing its organizational behavior or culture. The individuals that compose a corporate culture practice a myriad of lifestyles that influence both their health and ultimately the company's health care costs. To create a healthy corporate culture, the norms of the culture must change to support healthy lifestyles. This change must occur at both the executive and the individual level to facilitate health promotion.

It is of interest, and perhaps importance, to note that changing an organization's culture follows a process that is similar to an individual's experience in changing his or her lifestyle. The process includes identifying and perceiving the need for change, setting the change objective, learning and practicing the change, and obtaining feedback or evaluation data on the degree of success achieved. Such a parallel between individual lifestyle change and organizational change is a reasonable comparison from a behavioral science perspective because organizations are composed of and directed by individuals.

The first step in considering a change in the corporate culture is to assess the *need* for such a change. This assessment should be conducted at all levels of the organization, but especially at the executive level where the creation of corporate norms occurs. Personal interviews, focus groups, and questionnaires are three useful assessment tools. These tools reveal the predominant norms of the culture (written and unwritten), norms that act as either facilitators or barriers to change, and the need for new norms.

Identified norms may pertain *directly* to health, such as policies on absenteeism, the type of medical services offered, and the benefits provided, or *indirectly*, such as structured work hours that influence flextime or company time being used for health promotion activities. Other norms pertain to the culture as a whole. For example, highly bureaucratic companies exhibit a slower rate of cultural change than do high-risk companies whose cultures are more receptive to the creation of new norms (Deal & Kennedy, 1982). Companies striving to maintain the public image of a leader in improving the quality of life for employees will adopt health promotion policies more rapidly than will companies choosing to be followers. The financial position of the company and its ability to relate health promotion to the bottom line also affect the speed at which an economic risk is taken for a new venture such as health promotion.

A sound cultural assessment will reveal the business rationale that can be used to support health promotion. Most commonly, this rationale is based on an economic case for health care cost-savings and the improvement in employee morale. Other aspects of the rationale often include improving employee health and the quality of life, aiding in employee recruitment, promoting good community relations through the use of local health organizations, and fostering corporate social responsibility.

In addition to health and business norms, a corporate needs assessment should provide baseline data on the health, health risk, and lifestyles of employees. Until recently, most corporate databases were built solely or primarily on treatment or health care use data. The health risk appraisal adds a vital element to this database by providing preventive and risk data as well.

The most modern and efficient tool for such an assessment is the health risk appraisal (HRA). This self-administered questionnaire provides feedback in two areas. Individuals receive a personal health risk profile that helps to motivate them to participate in lifestyle change programs. Management receives a group risk profile, which, when combined with claims data, can provide a sound strategy for estimating and managing future health care costs. The HRA provides an essential element for establishing a computerized database from which health promotion programs and cost-benefit analyses can be generated.

Barriers and How to Avoid Them

Part of the assessment process is to examine what acts as a barrier or reinforcer to potential change. One of the more disconcerting findings from a cultural assessment will be the norms that negatively affect or serve as barriers to the acceptance of health promotion programs. Often one of the first of these norms to be encountered pertains to management's perception of the role of medical services in the company. Such services are often viewed as important, but not necessary to the business operations. This perception ignores the major economic contribution medical departments make in cost avoidance, through the early prevention and detection of disease, monitoring disability, and controlling the quality and cost of care.

Business managers want to be convinced of a return on their investment. The most successful medical directors or health managers are those who demonstrate that the health promotion investment fits into a business plan for cost containment. This means aligning the goals and functions of their profession with their business goals. It may require a behavioral change on their part—reshaping their perspective of their own profession. But the evidence of a direct relationship between health promotion and the bottom line will qualify health promotion to be more than of mere social value to the corporation. It will integrate the medical manager into the business operations (Bernstein, 1984; Wilbur, 1983).

Part of the business case for health promotion includes an increased understanding by medical and business managers about the relationship of lifestyle change to cost-savings from reductions in health risk and the associated disability, hospitalization, and premature mortality. A significant part of this understanding pertains to the presentation of realistic outcomes from health education and lifestyle change programs. Traditional health education, or the provision of information through brochures and audio-visuals, has been found to be necessary, but not sufficient, to change health behavior. However, the techniques of *health promotion* combine this traditional approach with more modern health education techniques, such as behavior modification, social support skills, and environmental changes to create lifestyles and corporate cultures supportive of such change. For example, recent research has shown that new health education techniques used to control blood pressure and to stop smoking not only are successful but also reduce the set of risk factors known to contribute to cardiovascular, respiratory, and neoplastic diseases (Alderman & Davis, 1976; Levy, 1981). Increasing evidence supports the positive health and economic effects of other lifestyle changes that result from exercise, cholesterol reduction, and weight-loss programs (Parkinson & Fielding, in press).

Finally, cultural assessments often reveal norms that pertain to employee perceptions about "new" programs. On the positive side, employees perceive new health benefits as an extension of the company's maternalistic culture. On the negative side, health promotion can be perceived as a hidden way to increase productivity without sufficient personal reward. In most companies the case is not black and white—both objectives are present. The key question is: Will the personal value of an employee health promotion program be perceived by employees as being of equal importance to the company as the resulting cost-savings? If the answer to this question is yes, employee health promotion has a solid foundation on which to start. If the answer is no, the program may be viewed with suspicion and not accrue the level of participation needed to incur changes in lifestyle, disability, and associated health care costs.

Setting Objectives.   The second critical step in effecting change is to set objectives. For individuals, setting achievable, personal objectives is an important part of attaining the desired behavioral or lifestyle change; like-wise for an organization. Creating a corporate culture that is supportive of health promotion requires setting achievable health promotion objectives. They should specifically state the business outcomes for which the program will be held accountable. These objectives should be linked directly to business measures, such as disability, medical care costs, and absenteeism, or to other measures, such as employee health and morale.

Evaluation. Individuals need feedback and reinforcement concerning the achievement of their lifestyle change objectives. Corporations build in feedback loops in management operations to assess progress being made toward increased earnings and cost-savings. The company's business objectives for the health promotion program will determine the type of evaluation conducted and the financial resources allocated. Companies wanting to improve morale or to offer health promotion as a benefit or executive perk will most often measure participation rates or attitude changes. Companies wanting cost-benefit results may design long-term epidemiological studies to measure change over time. If marketing health promotion to internal divisions or outside buyers is the objective, a company will usually monitor behavior change, health care costs, and morale data (Dickerson & Mandelbilt, 1983; McCauley, 1984; Wilbur, 1983).

For those companies wanting to measure changes in health, health risk, morale, medical care, and benefit use, two types of evaluation databases are helpful: one for measuring changes in health care treatment costs, and one for measuring changes in costs associated with preventive lifestyles. The treatment database is built on benefit and insurance claims providing disability, hospitalization, and other health care use data. The preventive database is built on a health risk appraisal (or some other type of combined health history and health risk questionnaire) and the behavioral change data resulting from health promotion programs. The combination of these computerized databases brings discrete data into an aggregate file that reflects changes in group health, health risk, and resulting health care use. Initially, health risk, disability, and medical care costs payments are the three essential data sets for managers because these can be specifically tied to business costs. Later, other databases can be built on this foundation.

Program Implementation. Individuals endeavoring to change lifestyles enroll in one or more lifestyle change programs. An organization trying to change its corporate culture to support these lifestyles will need numerous program resources. There are four available options: in-house staff, community resources, employee volunteers, or a combination of these. It is important to note that the level of resource commitment demonstrated by an organization is often directly related to the strength of the business case that was built for such preventive action.

A health promotion program directed at changing lifestyles in a corporate culture requires both personal *and* organizational commitment. It requires total involvement of the employee population. In short, every employee and department must "own" part of the program. The ownership is achieved through employees enrolling in lifestyle-change programs or even serving as faculty members. Through their work activities, employees can participate by developing materials, analyzing data, or serving as members of health promotion leadership committees. Likewise, executives need to be active participants in the lifestyle programs and demonstrate their support through letters to the employee body, interviews, and other visible endorsements. The program cannot belong to management, to medical, or to a few dedicated healthy individuals. It must become part of "the way things are done around here," part of the corporate culture (Deal & Kennedy, 1982).

Summary

In summary, employee health promotion programs involve change—change directed at the creation of healthy lifestyles, new company norms, and new corporate cultures. The potential of such a change for society is significant. Industry has at hand a vital opportunity to impact the health and health care costs of the 1990s. The question is; How will industry respond to the challenge?

BUSINESS PERSPECTIVE

Why the Problem?

In a time of economic downturn, more than a thousand major corporations in the United States decided to invest in corporate health promotion and wellness programs. While a few have reduced the intensity of their programs for practical reasons, virtually none has stopped. Compeling economic pressures for initiating and continuing these programs include the recognition, finally, of the severe limitations of medicine and the increasing pressure from changing values on the part of the work force itself.

These health promotion programs pose the threat that physicians will be neglected. But it has not been medicine alone that has led to a healthier public. Public health measures claim considerable credit for improving the health of us all. That physicians have been the controllers of reimbursement is a reflection of the neglect of the rest of us. That will be changing and changing rapidly. For example, consider the degree to which the physicians have been fighting congressional changes.

Also important to recognize is that health care costs in the United States are not the only issues related to the GNP. We spend as much of the GNP methodically making ourselves sick as we do, less methodically, making ourselves somewhat better off. Yet we do not question that percentage of the GNP. The problem is one of waste, of poor quality care, and of an imbalance in the resource allocation. In essence, a growing share of the population is recognizing that we are receiving a poor return on our health care investment in a time when there is greater competition for these resources from areas of our society outside health and medicine entirely.

An Evolution in Wellness Programs

Health promotion or wellness programs in the worksite have experienced a series of generational advances. The first programs had virtually nothing to do with health and were more product related. The second generation was single-shot programs—those that focused on executives of a particular type, or in one location, disease entity, or procedural approach. The third generation, which is where the bulk of the leaders are today, is beginning for the first time to handle comprehensive programs and to consider behavioral science principles.

The fourth generation is where the excitement is—where wellness really begins to reflect a change that is changing attitudes and philosophy—not just a program. Fourth generation wellness programs are those that integrate wellness with the total corporate health care strategy, those that reflect the understanding of corporate culture. Clearly, we are not talking about the average company in this fourth generation. We are talking about companies that understand that stress management programs are not given without doing something about the causal agents of stress within the work setting itself. These fourth generation programs reflect the changing nature of organizations and society itself.

The fourth generation is a recognition that the person, not the machine, is the key to corporate success—to that bottom line. It is a shift from cost containment to cost management, a shift in the philosophy that benefits are something that are given away and therefore neglected to the philosophy that benefits are an asset to be managed by both the company and the individual participants, the employees and dependents. In addition, the fourth generation begins finally and aggressively to include dependents and retirees in the wellness programs.

Serious corporate wellness programs are based on a set of principles, a few of which stand out in relation to behavioral medicine's particular interests. An actual benefit is a mix of wellness, curative, and chronic care. An actual wellness benefit recognizes that quality of health and medicine can be increased by better management, and by better knowledge and skills given to the user. Wellness is for all ages and for all levels of employees. Wellness is never to be used to hide health hazards of the work setting. Wellness should be used in the corporate setting to assist self-responsibility, not to induce victim-blaming. We should reject the concept that a company should avoid doing a prevention program because the ultimate cost benefit might accrue to some other employer down the road. That is anathema to the concept of wellness itself.

Avoiding the Barriers: Aggressive Approach to Cost Factors and Innovations in Corporate Thinking

Companies are becoming increasingly aggressive on the payoff issue. Particularly corporate chairpersons and other senior executives understand the rather basic logic that healthier employees are better to have than sick employees. For example, there is a rather large epidemiological database that demonstrates smoking is not healthful. Corporate leaders can understand that their employees are not really inherently different from those who form this database. Therefore, if the risk of cardiovascular disease for men over 45 who smoke two packs of cigarettes a day is 335 percent greater than it is for those who do not smoke, every company in the United States does not have to do a separate cost-benefit study of their 45-year-old men to know that their employees are better off not smoking. We also know that a wide variety of corporate measures exist apart from those traditionally viewed as medical or health that justify the investment. The investment in wellness is often a small one. It does not require a large amount of money to conduct these programs and to make these changes. Using the smoking example, health care and medical cost-reduction aside, there are cost-savings on life insurance, fire insurance, janitorial costs, productivity—a whole array of areas that can add up to more than the cost of most smoking-cessation programs.

We have better data on the effectiveness of keeping people healthy than we have on the effectiveness of many of the high-tech medical programs and procedures that we pay for throughout the United States, which an examination of physician practice patterns will demonstrate. For example, we know that, on the average, less than 10 percent of all medical tests done in the hospital are ever used to make a clinical decision. In fact, it is because of these practice patterns that fourth generation wellness programs must begin to integrate wellness with the basic design of the corporate benefit structure itself.

We know that in the United States we average more than 30,000 people a year dying from iatrogenic infections, diseases caught in the hospital, not those that were the purpose of the hospital admission. Forty thousand people a year are injured in New York hospitals alone. According to studies done by the California Hospital Association, approximately 20 percent of all the people that go to the hospital experience a medically compensable event. We know that throughout the United States 55 percent of the hospitals that do open-heart surgery do not do enough volume to achieve optimal medical outcome. And the response is not to discontinue such surgery. The response is to try to demonstrate that more people need open-heart surgery.

Reimbursement is not a problem for wellness. We would be better off if we kept wellness out of the insurance mode to begin with. Not that there cannot be some incentives to insurance. For example, there is a proposed tax cap on the amount of employee benefits for health insurance; currently, this does not include wellness programs. Wellness programs would be outside the cap because they are an uninsured benefit. If these programs were inside the cap, then they would probably be one of the first programs to be eliminated by many large companies, who could not possibly reduce their hospital surgical and medical costs quickly enough to fall under the cap.

Whereas the prevention movement and the worksite wellness programs have a patina of orientation toward the physical (e.g., cardiovascular risk reduction, physical fitness, smoking cessation, etc.), an exciting avenue of innovation is the growing recognition of the importance of our emotional side (Barrie, Smirnow, Webber, Kiefhaber, & Goldbeck, 1977). For example, with regard to accident statistics, troubled employees average two to four times the amount of absenteeism; they account for twice the rate of medical insurance use and more than 40 percent of all disability costs for American corporations result from mental disorders. Add to this an increase in court decisions that expand the definition of mental disability and the corporate responsibility, and these numbers increase rapidly. On the other hand, we have observed the situation of one Blue Cross Plan proudly announcing on television the reduction of its mental health benefit because "the people were using it for emotional problems," as though this represented a failure to understand what the benefit was about. Fortunately, some people recognize that investment in emotional support programs is a wise business decision.

Summary

In conclusion, one challenge that behavioral medicine and industry will face in the next five to ten years is the ability to remain open to the new and the innovative. What are commonplace, standard practices today were totally rejected only a few decades ago. The future has the potential for considerable innovation.

AREAS OF CONCURRENCE AND CONFLICT BETWEEN PERSPECTIVES

Assumptions

All the perspectives agree on the appropriateness of approaching preventive medicine programs through the mechanism of industry and at the worksite. All also concur on the importance of designing these programs so that they integrate with the fiscal priorities and corporate structures of industry.

A major area for conflict appears to be with the role of the medical profession in meeting a health or wellness agenda through the use of work-site programs; in fact, the physician perspective specifically identifies the importance of this issue. Where, then, is the conflict if the matter is commonly recognized? Primarily, it would appear in the way the matter is addressed. If the historical and financial approach of medicine has been to undervalue and underemphasize prevention, then, as industry makes prevention an increasingly preeminent objective, traditional medicine will either become part of the solution or become increasingly part of the problem. The physician perspective urges a behavioral analysis of the current role of medicine and consideration of how this role can be modified yet continue as an integral part of industry's health agenda. In contrast, management and business perspectives imply that change in financial support for health programs and the widespread use of these programs in industry will occur, perhaps despite the posture of traditional medicine about the matter. While the methods by which workers' health is improved and sustained are behavioral (lifestyle) in nature, the desired outcome pertains to biological functioning; and as such, it would be difficult to see how interface with, at least, some areas of medicine would not be inevitable. This area of conflict has the potential for diverting attention and priorities about the nature of the primary of concern. That is, if not efficiently resolved, the health and industry agenda may become a "turf" problem rather than one that addresses the true focus of concern: the health of the nation.

A second area of conflict, only casually alluded to but dramatic in its implications, is what we can conclude about the efficacy and health impact of various programs for workers' health. The fact that an employer has instituted a particular program does not necessarily mean that employees will be healthier and more productive if they participate. For example, exercise programs are popular, yet the health impact in terms of such important variables as morbidity, mortality, and quality of life has yet to be demonstrated in well-controlled studies. For centuries, medicine has witnessed the folly of falsely concluding that procedures are effective when they are not. Scientists are even more sensitive to the importance of basing conclusions on careful data collection and study. Both the public health and physician perspectives address this concern. In contrast, management and business perspectives' enthusiasm and support appear to exceed what we can conclude with confidence from research to date. The advantage of moving forward in many areas with health programs is that if these programs are truly beneficial, the health of the nation will be improved sooner, and we will have more data to prove that fact. However, if some approaches are not beneficial, the risk includes not only the cost but also the dissatisfaction and disillusionment of industry with the general health promotion approach, including those programs that could benefit workers.

Priorities

None of the four perspectives approaches the matter of priorities in the same way. The public health priorities were based on disease and risk factor areas (e.g., employee assistance, hypertension, smoking, exercise and fitness), their prevalence, and the knowledge base that exists for intervention on the areas. The physician perspective stressed the historical and current priorities in medicine on secondary and tertiary approaches and the need to somehow shift activity (and finances) to prevention or primary approaches.

Industry's approach to priorities appears to emphasize ways of determining, on an individual industry basis, the needs or interests of industry. This needs assessment approach would certainly be an important step to take in implementing health programs in industry; it should not obviate consideration of priorities based on what we know how to change; that is, discrepancies between the results of needs assessments in industry and research studies on program effectiveness are important areas for concurrence or conflict. When the expressed needs of industry identify programs that have been shown to be effective through well-controlled studies, there is concurrence, and these programs should be implemented in that industrial setting. When the needs of industry are for programs not documented to be effective in previous research, such a discrepancy or conflict should stimulate more research and not necessarily the immediate implementation of such programs in that industry before such study.

The business priority is most exciting. The notion of a fourth generation of approaches to wellness that result in a general and pervasive change in societal attitudes about health would have great and sustained impact. How to cause such a change and measure its impact are challenges of some magnitude. This priority is also exciting in its breadth of extending beyond the bounds of industry-based programs and activities, extending wellness activities to retirees and dependents, and so forth.

Despite the difference in the stated priorities across perspectives, these differences do not appear to be major areas for conflict. In each section, support can be found for concepts espoused by other perspectives about priorities. Thus, by-and-large, all the perspectives agree on each other's priorities, tending merely to have different emphases—with perhaps one exception: whether the cost-benefit factor as a priority is a matter of conflict. All agree it should be considered and measured, but whether health programs should immediately, or even eventually, pay for themselves is an area of disagreement. The business perspective is the most divergent on this point in the consideration that such programs are almost a given, despite cost factors, for any particular industry. This is a divergent opinion but not necessarily an inappropriate one if even some of the anticipated events discussed in the beginning of the chapter occur.

Barriers

Again, the different perspectives approach the question of barriers with different emphases, but not necessarily with much conflict resulting. All emphasize attitudes, knowing what attitudes are now and determining how to change them. This is approached and discussed differently but is a general theme in all perspectives—public health stressing worker versus industry attribution for responsibility about health; physician perspective emphasizing attitudes of physicians; management concerned about assessing its own attitudes; and business concerned with general societal attitudes and effects on industry.

While not mentioned by all perspectives, a barrier of concern is differentiating between research and intervention activities. The public health and physician perspectives are most sensitive to this issue, but management and business do not appear to be unsympathetic. If anything, this is a potential area of conflict in terms of basic assumptions and knowledge about what research is and how it is conducted. Perhaps as the science community learns more about the intricacies and priorities of industry, industry can obtain a better understanding of the workings of science. Fortunately, all perspectives agree on the need for objectivity and data on program effectiveness and accountability—which is not a bad beginning.

CONCLUSION

If these brief samples of four different perspectives are representative of considerations about preventive medicine (health promotion, wellness) programs in industry, then we can anticipate agreement and cooperation about (1) the appropriateness of such programs, (2) the fact that these programs should be integrated into fiscal parameters and corporate structures of industry, and (3) the need for changes in attitudes about such programs by physicians, management, labor, and so forth. Similarly, we can also anticipate disagreement and a distinct challenge to success about (1) the role of the physician in such programs, (2) the efficacy of various types of programs, (3) the priorities for program implementation, and (4) the need for programs to be cost-effective.

The need to base industry health programs on research findings cannot be overemphasized. This chapter departs from such a strategy. It considers divergent or concurrent opinions; and the n is small—one from each of four groups. Nevertheless, it would be difficult to deny that understanding the entering repertoires of these four essential ingredients for industry-based health programs is critical. If research is the result of observation and experiment, then this chapter can be considered a preliminary "observation session" in the study of translating research into practice.

REFERENCES

Alderman, M. H., & Davis, T. K. (1976). Hypertension control in the workplace. *Journal of Occupational Medicine, 18,* 793–796.

Allegrante, J. P., & Green, L. W. (1981). When health policy becomes victim-blaming. New England Journal of Medicine, 305(25), 1528–1529.

Barrie, K., Smirnow, B., Webber, A., Kiefhaber, A., & Goldbeck, W. B. (1977). Mental distress as a problem for industry. In R. H. Egdahl & D. C. Walsh (Eds.), *Mental wellness programs for employees.* New York: Springer-Verlag.

Bernstein, J. E. (1984, February). *Health risk management in the corporate setting.* Paper presented at Organization Resource Counselors Occupational Health and Safety Group, Washington, D.C.

Cataldo, M. F. (1983). *Health Armageddon of the twenty-first century: Priorities for the experimental analysis of bio-behavioral interaction.* Paper presented at 91st Annual Convention of the American Psychological Association, Anaheim, CA.

Deal, T. E., & Kennedy, A. A. (1982). Corporate cultures. Menlo Park, CA: Addison-Wesley.

Dickerson, O. B.,& Mandelbilt, C. (1983). A new mode for employer-provided health education programs. *Journal of Occupational Medicine, 25,* 471–474.

Levy, R. I. (1981). The decline in cardiovascular disease. *Annual Review of Public Health, 2,* 49–70

McCauley, M. (1984). *AT&T communications, Total Life Concept Program.* New York: AT&T. (Unpublished data).

Parkinson, R., & Fielding, J. (in press). Health promotion in the workplace. In L. J. Cralley & L. V. Cralley (Eds.), *Pattys industrial hygiene and toxicology* (Vol. 2, 2nd ed.). New York: Wiley.

Wilbur, C. (1983). The Johnson & Johnson Program. *Preventive Medicine, 12,* 672–681.

19

Behavioral Medicine in Industry: A Labor Perspective

BARRY R. SNOW, DAVID LeGRANDE, JUDY BEREK, JUNE McMAHON, ROD WILFORD, and JEANNE M. STELLMAN

During the past decade, numerous pressures have affected the American worker. These pressures include automation, the threat of unemployment, increased demands for productivity, exposure to toxic or hazardous working conditions, and a generally depressed economic climate. These factors have had a profound impact on the lifestyle of workers and their health and well-being. As such, these issues are a concern for corporate management, organized labor, and the general society.

Yet historically the viewpoints of workers on these and other issues have not been studied by the scientific and professional community (Huszczo, Wiggins, & Currie, 1984). In some cases, the labor viewpoint was assumed to be the opposite of management's and, therefore, not requiring further detailed study. In other cases, the labor viewpoint was assumed to be similar to management's and, therefore, unnecessarily redundant. The concept that labor may have a unique perspective or, more likely, a variety of viewpoints regarding many issues was not recognized. This lack of communication between the scientific and labor communities made it difficult for assessment and intervention projects conducted in the workplace to be fully sensitive to the needs of the workers and their organizational milieu.

This chapter will present an overview of the fundamental issues that behavioral medicine professionals must confront when designing workplace-based surveys and intervention projects. These issues were identified in collaboration with union leaders (See the Appendix for a brief description of the unions whose representatives contributed to these remarks.) This chapter will describe specific concerns of workers and their representatives that must be handled at all phases of the contact between the scientific and workplace community. Recommendations for addressing these issues and thus facilitating the communication process between behavioral medicine professionals and the industrial, service, craft, and office-based labor force will be made.

PREPARATORY PHASE

One important step that the professional can take to enhance the quality of a project occurs before initial contact is made with representatives of a specific workplace. This step involves familiarizing oneself with the organization of labor bodies (Pryor, 1981). Many clinicians and researchers are not familiar with the structure of unions or the influence they have on their members. The role of union health and safety committees, grievance procedures, shop stewards, and other institutional structures that may affect the proposed survey or intervention project may thus be unclear. In some cases, the health professional may not even be aware of the existence of such bodies. This lack of awareness may cause the professional to ignore certain factors that both workers and management accept as an intrinsic part of their environment. These factors may play an important role in determining the outcome of a proposed intervention program. An example of such influence in the smoking-cessation area will clarify this point.

Various investigators (Graham & Gibson, 1977; Skewchuk, 1976) have suggested that the long-term maintenance of smoking cessation may be facilitated when "buddy" systems are used. These buddies can serve as resource people who offer mutual support as well as monitor the behavior of their coworkers. Such networks can be more effectively established whenever the preliminary survey determines that union representatives hold positions of authority, which make them powerful norm-setting group leaders. Furthermore, many unions provide an outlet for social activities among workers, whose job responsibilities restrict their daily interaction. This contact is especially important among craftsworkers who may not have a steady work location. Union leaders can help identify the patterns of friendship that have emerged in these settings. Finally, union representatives can provide important clues regarding the places and conditions under which individuals smoke, including such environmental factors as perceived job stress and peer group norms on the use of cigarettes. Health professionals who evaluate the union presence in a particular worksite may identify important channels for obtaining information on the natural history of smoking behavior in a particular setting. Such information has been deemed essential for the construction of more psychologically based intervention programs (Leventhal & Cleary, 1980).

The review of the labor viewpoint by health professionals may also alert them to key fallacies in their perspective of the union-management relationship. Although much collective bargaining is performed on a cooperative basis, newspaper reports frequently highlight the dramatic confrontations that may occur between union and management representatives. These isolated incidents give an inaccurate view of the process of determining workplace conditions. Such inaccurate views may cause the researcher to ignore labor-management negotiations, which typically precede workplace changes in such areas as staffing ratios and work schedules. These negotiations may curb worker objections to specific job changes. The behavioral medicine professional developing a study of the impact of these changes on worker health and well-being may benefit from becoming familiar with these issues in the preparatory phase of his or her work.

Even health professionals who are sufficiently aware of workplace organization and functioning may have orientations that are not broad enough for the union professional. Union officials represent workers in environments where a variety of hazards are present. These hazards stem from the physical, chemical, ergonomic, and psychosocial aspects of the work situation. Strong disciplinary boundaries, however, exist, which typically restrict the perspective one brings to the study of the occupational environment. Psychologists, for example, typically focus on the psychosocial features of the workplace and how they affect employees. They receive little or no formal training in the assessment and treatment of the physical hazards that affect a worker. These concerns demand the attention of the industrial hygienist and safety specialist who typically has minimal psychological training. Even among psychologists themselves there typically exists a fragmented approach to the assessment and treatment of workplace concerns. Industrial/organizational psychologists typically focus on the impact of the workplace on such organizational outcomes as absenteeism, turnover, and productivity, while clinical psychologists look for mental health outcomes relating to work. Physicians and epidemiologists, on the other hand, typically look for signs of physical illness. These approaches may initially appear too fragmented for union professionals who handle a wide range of problems, from traditional issues of wages, hours, and job security to complex questions of job safety and health.

The behavioral medicine professional may discover during the preparatory phase of the project that part of his or her professional responsibilities will involve educating workplace representatives regarding the limitations on research and intervention projects that exist within a particular work setting. Such educational activities may, for example, inform union leaders that an evaluative study of lifting practices among nurses may not fully evaluate the contributing role of occupational stress to orthopaedic injuries. Library research, which identifies workplace projects that successfully obtained their objectives using a similarly restricted focus, will help the professional obtain the cooperation of the union in executing this study.

INITIAL CONTACT

Once the health professional is familiar with the labor-management community that is to be studied, the initial contact with representatives of the workplace can be made. As previously indicated, this contact has frequently been made with management alone in recognition of management's entrepreneurial responsibility for the activity of its employees. This chapter suggests that the routine adoption of such a limited contact may be detrimental to the long-term success of a project. Conversations with union officials may for example, identify specific workplace factors that may either enhance or hinder the execution of a particular project. These factors may not be immediately obvious to the professional from the outside. Discussion with union representatives may also help the behavioral medicine professional to identify those management groups that are readily accessible and those that are more difficult to approach. Such examples typically occur where management in one area or division of the company is more reluctant to interact with outside professionals than other branches would be. Finally, most unions contain the organizational infrastructure of stewards and representatives, which can stimulate worker interest and thus ensure a high rate of participation in the proposed project. Such cooperation may enhance the quality of projects that depend on an adequate sampling of the population for meaningful results to be obtained. This support in data collecting is especially important for projects that enjoy only limited financial support from federal and other sources. An example of the role of unions in the initial contact and implementation phases derives from a recently completed survey of office work in which the first author was involved.

Stellman, Gordon, Snow, & Klitzman (1984) conducted a cross-sectional survey of office workers performing a variety of different types of work. A major objective of this study was to evaluate the effects of different physical and psychosocial conditions on workers' reports of mental and physical health and well-being. To initiate this study, direct contact was made with representatives of the unions at the different worksites. This contact was helpful in convincing managers at specific sites of the willingness of their workers to cooperate in such a study. Such expression of worker interest is important to management groups who seek to demonstrate their ongoing sensitivity to labor demands. The labor contact was also important in promoting acceptance of the study among workers who were being asked to reveal their attitudes toward their job responsibilities and supervisory staff. An overall response rate of 86 percent was, in fact, obtained at the various worksites that were surveyed. Such a high rate of questionnaire completion was necessary to ensure the meaningfulness of the study results. Union channels were also most effective in educating members to respond to the questionnaire in an honest manner, rather than use the survey as an opportunity to highlight personal complaints against management. Such nonobjective responses might have been obtained if the workers had been given a questionnaire that appeared to be solely a product of management influence. This outcome would have detracted from the significance of the study.

Union contacts were also helpful in clarifying the issues around which subsequent discussions were to be held. Union officials, for example, assisted in identifying key areas in which "site specific" questions should be developed. These questions typically concerned such issues as type of physical environment, specific work schedule, and other factors unique to a particular worksite. Inclusion of these items helped clarify the meaning of the more general items that were included in the surveys administered at all the worksites. The familiarity of the union leaders with workplace characteristics was also important in the design of the sampling technique. To evaluate adequately the impact of different work conditions, the office workers' study required a sample of workers who differed sufficiently on job characteristics at each of the specific worksites. Union officals were able to indicate that, although particular job titles may appear to be similar, significant differences existed in actual job responsibilities to warrant the inclusion of these groups in the sample being studied. Such detailed cooperation can immeasurably add to the discriminative ability of a study.

DEVELOPMENT PHASE

Once the preliminary steps of preparation and initial contact have been taken, the investigator must clarify the nature of the study that is being proposed. Some worksites may be more amenable to epidemiological surveys, while others may be more interested in targeted intervention projects. Once again, discussions with union officials may help the investigator develop his or her project in a way that facilitates its successful completion. These discussions can also alert the investigator to one of the most troublesome issues that may arise in the execution of the study—namely, the focus of a proposed intervention project.

Most health hazard interventions can be conducted in one of two broad ways (Snow, 1984). The first way, which focuses on environmental change, is for the clinician to intervene on the organizational level by either directly altering the environmental hazards or by strengthening the available resources that face the worker. Examples of the first type of intervention strategy are the introduction of engineering designs, which modify workplace exposure to toxic chemicals, and the establishment of nursing staffing ratios, which regulate the amount of work that each nurse performs. Potential forms of environmental restructuring to strengthen environmental resources include the enhancing of worker control over work pace and methods and the strengthening of support systems at the workplace. These methods require considerable coordination with management and union officials so that interventions consistent with company policy, as well as amenable to precise quantification and follow-up,can be introduced.

A second type of intervention program may, however, be mounted. This type of project does not focus on environmental change, but rather emphasizes training that alters either the person's assessment of the workplace or his or her response to hazardous working conditions. Examples of this type of intervention strategy include programs that instruct workers in preventive health practices and the reduction of such maladaptive coping responses as smoking and substance abuse. Comprehensive training in relaxation techniques and the provision of counseling to handle worker problems can also be made available.

The clinician proposing a study of the impact of specific environmental changes on worker health and well-being has the satisfaction of implementing a protocol that will have clear ramifications for the specific occupational hazards that workers on that site confront. As such, he or she is likely to receive the strong cooperation of labor in the design and execution phases. This support will be consistent with the union view of the workplace as a site where applied intervention projects to reduce environmental hazards and enhance group support should be undertaken. This approach to worksite-based intervention programs, however, may not be amenable to management. Structural workplace changes may be too expensive, time consuming, or at odds with corporate policy. Management may prefer to focus intervention efforts on changing the person's work behavior or health habits rather than on changing the organization. These projects may, however, be unacceptable to union officials. These leaders may stridently object to the view of the workplace as a catchment area for the delivery of individualized health services.

The behavioral medicine professional may be surprised by the vehemence of the labor-management controversy that may arise in this area. He or she may, for example, wonder what motive a union may have for not enthusiastically supporting a smoking-cessation program when the benefits are clearly directed towards helping their members. The training of the behavioral medicine professional may, moreover, make him or her suspicious of any group that does not want to avail themselves of sound, scientific intervention techniques for altering health behavior. Indeed, such a response may be viewed as justifying the tendency of many professional investigators to forego consultation with union officials.

In resolving this potentially strong barrier between the scientific and labor community, the preparatory educational step outlined earlier becomes especially important. Health professionals studying the role of unions may, for example, discover that workers who smoke were denied compensation benefits even though there was sufficient evidence of the role of environmental hazards in contributing to the disease process. Alternately, the researcher may uncover case histories in which the existence of stress reduction groups among nurses delayed the introduction of staffing changes that could have reduced workload demands (Berman, 1978; Mancuso, 1976; Weiner et al., 1983). These readings should alert the behavioral medicine professional to the need to be flexible in both the content and design of the intervention proposal. Once again, the professional may find that part of his or her role lies in educating management and labor groups in the virtues of the proposed project. Such an educational process may forge a compromise between the opposing views of workplace groups.

Clinicians and researchers should also be aware of certain high priority items that have been identified by different labor unions as worthy of interest. These items have attained their importance either because of the large numbers of workers affected by these hazards or by the nature of the adverse outcome that these hazards produce. Identification of these issues can help scientists find common ground with union representatives in their joint discussions about the feasibility of a particular project. Unions may also have specific hypotheses of findings that they have generated through discussions with their members. This information may be implicitly understood or explicitly contained in educational brochures or classes the union gives. The next section of this chapter reviews several important projects that have enjoyed the support of different labor organizations.

PRIORITY PROJECTS

Recent estimates by the U.S. Department of Labor indicate that the vast majority of adults will spend at least one-third of their working life in office settings (Stellman & Henifin, 1983). This factor is of special concern in view of recent changes in the design and organization of office settings. Many of these changes focus on the introduction of automated information-processing units such as VDTs (Visual Display Terminals). Many workers cite incidents where VDTs were introduced into the work setting after only minimal employee consultation. Labor is concerned by this management move because of reports of the adverse impact of terminals on health and well-being. These reports include complaints of eyestrain, blurred vision, and difficulty in focusing after extended use of VDTs. Concerns about the radiation leakage from these terminals, the use of workstations not designed for VDT viewing and the possible deskilling of their positions as their job becomes more routinized are also present.

Office workers are also faced with hazards associated with the physical plant in which their work must be conducted. Such factors as noise, unpleasant appearance of the workplace, lack of privacy, and poor air quality may have an adverse impact on worker health. The psychosocial environment of the job, which includes such factors as uninteresting work, poor supervisor support, hostility between workers and their clients, and the threat of unemployment, is also significant. Unions welcome studies that can collect the scientific information to evaluate worker complaints and test proposed intervention projects directed toward these risk factors. These studies acquire special significance when one realizes that office hazards have typically been ignored in the belief that these environments are relatively safe in comparison to the health hazards found in heavy industrial settings. The newness of many of these environmental changes also makes it unlikely that unions and management would have a ready-made database that would allow them to evaluate objectively the health impact of these environments.

Another large at-risk group from the perspective of union officials are hospital workers. These individuals, which include nurses, nurses' aides, and orderlies, work in a rapidly changing, high-tension environment. Nurses, for example, frequently work in settings where staffing has been reduced as a cost-control measure. They may be required to work variable shifts, where the hours of work change from one week to the next. Furthermore, the professional training of a nurse, which trains him or her to exercise professional judgement, may seem wasted in a work setting where his or her duties primarily consist of following doctors' orders and charting patient behavior. In addition, nurses may work in settings where they are exposed to such chemical hazards as ethylene oxide (used for equipment sterilization), anesthetic gases, radiation, and potentially carcinogenic cancer treatment medications. These hazards have rarely been studied in the comprehensive way that is necessary for sound occupational planning to be conducted. Even when studied, the population at risk has typically been restricted to higher-level professionals and has excluded the technicians, orderlies, and aides who may also be exposed to these hazards.

Unions have also been concerned with the hazards faced by individuals who do not work at fixed worksites. For example, painters and other individuals in related crafts may work at temporary building construction sites and, therefore, change their workplace throughout the year. These individuals may be exposed to solvents and other potentially hazardous agents that place them at risk for developing upper-respiratory, neurological, and reproductive defects, as well as cancer. Because of their frequent job changes, however, these workers do not typically benefit from design changes in their worksite, frequent industrial hygiene surveys, and changes in the methods of work organization and distribution that help workers in fixed worksites. The behavioral medicine professional interested in working with these groups faces a significant challenge for using his or her research and clinical skills in adapting a proposed project to the needs of this mobile group. The next section of this chapter reviews a number of projects that have been conducted in both the fixed and more mobile settings.

SPECIFIC PROJECTS

Various research projects, focusing on evaluation and intervention of the previously discussed worker groups, have been conducted during the past few years. The Communications Workers of America (CWA), for example, worked with representatives of Columbia University's School of Public Health and management representatives in the design and execution of the previously mentioned cross-sectional survey of office workers. This study has identified many of the risk factors associated with office work, which have been discussed earlier in this chapter. Detailed reports summarizing findings of these studies are currently being prepared (Stellman et al., 1984).

Studies of the impact of new technology represented by the introduction of VDTs to directory assistance operators working for a telephone company have also been conducted by the CWA with the assistance of the National Institute of Occupational Safety and Health (NIOSH) and the Department of Preventive Medicine of the University of Wisconsin. These surveys have helped identify significant issues that arise when VDTs are introduced into a workplace. Based on this work, several recommendations have been made, which are currently being implemented through collective bargaining between the union and management (Arndt, 1983).

Important progress has also been made in training workers to recognize and intervene with occupational hazards. The International Brotherhood of Painters and Allied Trades (IBPAT), for example, has worked with the Occupational Safety and Health Administration (OSHA) to develop interactive videotape modules and educational materials that teach the specific skills of job site monitoring and industrial hygiene sampling techniques. Special methods and forms for recording workplace ventilations and respiratory practice at the beginning and end of the training program have also been introduced. These procedures represent the direct outgrowth of various projects conducted with researchers at such diverse institutions as Mt. Sinai School of Medicine in New York (Selikoff, 1975), Stanford University, and other organizations that have worked on cooperative projects with union representatives,

Other projects have been proposed by various unions and still remain to be conducted. The Service Employees International Union (SEIU), which represents a large proportion of health care workers and office workers, has, for example, noted the scarcity of detailed information regarding the short and long-term impact of many standard work practices. These practices include long periods of night work for hospital workers and the tendency to automate as many jobs as can be done. The National Union of Health Care and Hospital Workers is similarly interested in supporting targeted studies that more clearly identify the specific hazards to which their workers are exposed. These studies should ideally be developed in a framework that allows for a second phase of intervention where the risk factors identified in the assessment portion of the study are then manipulated in a way that should result in measurable improvement in worker health and well-being.

Yet another model of union involvement in industrial projects is for union officials to contribute to the long-term development of clinical and research centers that focus on problems affecting the worker. These centers can act as educational clearinghouses, outreach centers, conference organizers, and centers of continuing education and training. They can also be instrumental in providing the organizational and technical expertise to mount clinical and research projects in the workplace. Union involvement in these activities can take the form of assisting in the development of training and educational materials, active service on advisory bodies, facilitators of liaison between the workplace and the academic and clinical communities, and providers of financial support for specific projects.

The first author is involved in two programs that provide models for the integration of labor and management expertise for the prevention and treatment of work-related orthopaedic problems. The first of these programs, which functions as an interdisciplinary clinical and research center in industrial orthopaedics and bioengineering, is the Occupational and Industrial Orthopaedic Center of the Hospital for Joint Diseases Orthopaedic Institute in New York City. This organization has been involved in sponsoring conferences, promoting studies of workplace hazards that promote injury to the musculoskeletal system, and developing primary and secondary prevention projects. A close working relationship has been established with specific management, labor, and academic bodies for the execution of evaluation, prevention, and educational projects. An affiliate body of this center, the Orthopaedic–Arthritis Pain Center, is similarly studying muscular dysfunction among musicians. This project uses key contact individuals from professional and fraternal organizations to involve musicians whose freelance type of work makes them inaccessible through traditional workplace channels. This venture points to yet another way in which labor input can be solicited in the design and implementation of workplace projects on a long-term basis.

IMPLEMENTATION AND ANALYSIS PHASE

After a suitable worksite has been chosen, appropriate management and labor contacts initiated, and the project suitably defined, the crucial phases of project implementation and data analysis remain. During these phases, the behavioral medicine professional will be helped by maintaining open lines of communication with these two groups. He or she should, for example, be prepared to make copies available of the specific research protocol and survey instrument for comment. The researcher should, however, be aware that worker and management groups may have different expectations. One group, for example, may simply request periodic reports of the study progress from the investigator, while another group may want to play a more active role in studies that directly affect issues of concern. Clear statements by the investigator regarding limitations that may exist on the conduct of the study and the sharing of confidential data that the study produces can promote an atmosphere of cooperation, which facilitates the execution of the study.

The behavioral medicine professional faces another set of challenges in the data analysis phase. Because the data that are being collected may have clear implications for the adequacy of current management and union practices, the investigator may face unanticipated pressures to complete the analysis and issue the results before the study is completed. These pressures can arise when the deadline for contract negotiations approaches and the data that have been collected may be relevant to an issue that could be raised in collective bargaining. Unexpected publicity that is being generated by the study may also pressure the investigator to make a statement on the study implications. These pressures are especially threatening when the questions that develop include those that the investigators feel cannot be adequately answered from the data that have been gathered.

The investigator that has worked on a union-sponsored project should also be prepared for the results of the study to be translated into language that is suitable for worker educational programs and informational brochures. As earlier indicated, union-sponsored occupational safety and health programs are interested in developing instructional materials that increase the level of awareness of labor and management on how to identify and eliminate occupational hazards. Both survey and intervention studies can produce recommendations that are suitable for such educational ventures. Unions may also use some of these recommendations to draft model contract language that provides health and safety protection for workers. Examples of these outcomes include collective bargaining agreements, which limit the amount of time spent in performing boring VDT work, provisions for providing career ladder opportunities for office workers, and the development of staffing ratio requirements to prevent overload on the nursing unit. The clinician and researcher may be unfamiliar with these practices, which directly translate empirical findings into applied recommendations.

Finally, the investigator should be aware of specific methodological issues that affect the execution of the study. Because the study is being conducted in a naturalistic setting, it may be difficult to introduce the rigorous experimental controls that researchers favor. A suitable quasi-experimental design may not account for the various contingencies that may arise as specific intervention projects are carried out. Labor for example, may successfully conclude discussions with management to alter a particular feature of the workplace (e.g., staffing ratio) that is currently being investigated. Personnel shifts in both management and labor groups may alter key aspects of the workplace that are being studied. Furthermore, such uncontrollable factors as strikes and union elections may delay the execution and analysis of the study. These factors must be accepted by the investigator as one of the inevitable costs of developing projects that have potential for immediate impact because of their specific focus on the environment. Close collaboration with union officials involved in these changes can work to minimize the effect of these changes when they occur.

CONCLUSION

Most adults spend from one to two-thirds of their waking life at work. Despite the profound influence that the work environment may, therefore, have on their health and well-being, the concerns of the labor unions representing these workers have only infrequently been recognized by the professional community. This chapter reviews the specific health and safety concerns of representatives from the industrial, service, craft, and office settings. Specific concerns of workers, as well as recommendations for research and modes of interventions, are made. The impact of these recommendations and their relevance to the different steps of the intervention process are addressed. Methodological issues affecting collaborative research with the union are also discussed.

APPENDIX

Communications Workers of America (CWA). The CWA represents more than 675,000 U.S. workers in the public and private sector. Members are employed in such fields as public service, general manufacturing, construction, the telephone industry, nontelephone communications and interconnect industries, sound and electronics, cable television, gas and electric utilities, and the media. These members are organized into more than 900 local unions who live and work in some 10,000 different communities across the United States.

Service Employees International Union (SEIU). SEIU represents more than 750,000 workers in the United States and Canada. Members include professional and non professional health care workers in hospitals and nursing homes; workers who provide a wide range of building services; state, local, and federal government employees; and workers in such miscellaneous industries as race tracks, cemetaries, utilities, and jewelry. Efforts to organize clerical workers are currently underway.

International   Brotherhood   of   Painters   and   Allied   Trades (IBPAT). IBPAT is a former AFL crafts union, which represents more than 675,000 workers involved in the many phases of the manufacture and application of paint and allied products. Approximately 180,000 of these workers are directly involved in paint manufacture at specific industrial sites. Other workers include manufacturing applicators, who apply paint in specific manufacturing settings, and construction applicators, who work at a variety of construction sites.

District 1199, National Union of Hospital and Health Care Employees. District 1199 is the Metropolitan N.Y. unit of the National Union of Hospital and Health Care Employees organization. This unit represents more than 80,000 members in approximately 100 health care facilities in the New York City and Long Island area. This grouping represents a concentrated force of workers who provide health services in one of the nation's largest metropolitan areas.

REFERENCES

Arndt, R. (1983). Working posture and musculoskeletal problems of video display terminal operators–Review and reappraisal. *American Industrial Hygiene Association Journal, 44,* 437–446.

Berman, D. (1978). *Death on the job.* New York: Monthly Review Press.

Graham, S., & Gibson, R. W. (1977). Cessation of patterned behavior: Withdrawal from smoking. *Social Science and Medicine, 5,* 319–337.

Huszczo, G. E., Wiggins, J. G., & Currie, J. S. (1984). The relationship between psychology and organized labor: Past, present and future. *American Psychologist, 39,* 432–440.

Leventhal, H., & Cleary, P. D. (1980). The smoking problem: A review of the research and theory in behavioral risk modification. *Psychological Bulletin, 88,* 370–405.

Mancuso, T. F. (1976). Medical aspects. In U.S. Department of Labor (Ed.), *Interdepartmental Workers' Compensation Task Force Conference on Occupational Diseases and Workers' Compensation.* Washington, DC.: U.S. Department of Labor.

Pryor, J. (1981). Occupational safety and health: An introduction to the literature and resources. *Labor Studies Journal, 6,* 133–139.

Selikoff, I.J. (1975). *Investigations of health hazards in the painting trades.* Report to the National Institute of Occupational Safety and Health (CDC 99-74-91).

Skewchuk, L. A. (1976). Special report: Smoking cessation programs of the American Health Foundation. *Preventive Medicine, 5,* 454–474.

Snow, B. R. (1984). *Occupational stress: Treatment and assessment issues.* (Manuscript submitted for publication).

Stellman, J., Gordon, G. C., Snow, B. R., & Klitzman, S. (1984). *Workers' health and well-being survey: A summary of findings on prevalence of physical and psychological health and working conditions for 2074 office workers.* (Manuscript submitted for publication).

Stellman, J., & Henifen, M.S. (1983). Office work can be dangerous to your health. New York: Pantheon.

Weiner, M. F., Caldwell, T., & Tyson, J. (1983). Stresses and coping in ICU nursing: Why support groups fail. *General Hospital Psychiatry, 5,* 179–183.

20

Evaluation of Industrial Health Promotion Programs: Return-on-Investment and Survival of the Fittest

NANCY MARWICK DeMUTH, JONATHAN E. FIELDING,
ALBERT J. STUNKARD, and ROBERTA B. HOLLANDER

Probably the only thing more difficult than trying to evaluate the effectiveness and costs of worksite health promotion programs is trying to draw conclusions from what has been published on the subject (Fielding, 1983). Problems in quantifying effectiveness variables and inconsistencies in definitions and methods render it difficult, and often misleading, to compare reported findings within or across categorical programs. Similarly, although to a lesser degree, quantifying economic or cost variables is not without controversy (Cohodes, 1982; Stoddart, 1982). Compelling evidence is still lacking that even the most studied worksite interventions, although effective in improving health, provide particular rates of return-on-investment (e.g., Fielding, 1979, 1982; Green & Lewis, 1981; Parkinson et al., 1982). Hence, companies with health promotion programs in place still do not have sufficient, scientifically rigorous data to justify their programs from a strictly economic viewpoint. Arguments for worksite health promotion are still based more on the inherent value of a healthier workforce and employee relations than on the magnitude of guaranteed cost-savings attainable by employers (Chadwick, 1982; Wilbur, 1983).

Health promotion programs typically involve multiple outcomes that are difficult, but increasingly important, to substantiate scientifically and

We thank Russell Leupker, Murray Naditch, Maggie Peterson, William Stason, Joseph Stokes, Beverly Ware, and Curtis Wilbur for their contributions to earlier drafts of this chapter.

practically (see Table 20.1). Behavioral scientists in industry must respond to this challenge by simultaneously studying variables relevant both for management decision making and scientific advancement. Further complicating evaluation efforts, even well-planned attempts to analyze health promotion outcomes may be thwarted or at least biased because variables of interest—for example, health premium costs, absenteeism, turnover, competitive recruiting edge—are also influenced by other unrelated and frequently unmeasurable factors such as sick leave rules, early retirement policies, or management styles (Fielding & Breslow, 1983). Inherent difficulties related to generating meaningful data for most variables in Table 20.1 are thus unavoidable. However, some specific and attainable ways that the evaluation science of worksite health promotion programs can and should improve are too often overlooked or ignored.

Table 20.1. Criteria for Evaluating Worksite Health Promotion Programs (WHPPs) and Activities

Participation rates in programs

Employee satisfaction with programs

Management satisfaction with programs meeting stated goals and objectives

Productivity as measured by quality of work, absenteeism, turnover, etc.

Health status as measured by individual and aggregated data such as changes in health education knowledge, behaviors, long term disability days, premature deaths, job related accidents, or reductions in particular risk factors

Costs of programs, health insurance, etc.

Less tangible effects: Employee self-image, job satisfaction, employee morale, company loyalty, competitive recruiting edge, hoped-for long run economic returns, community image

The purpose of this chapter is to draw debate about evaluation of health promotion programs to three fundamental levels—concepts, methods, and policies—and thereby highlight opportunities for innovation and progress in the field. We assume that problems in concepts, methods, and policies currently impede interdisciplinary cooperation, confuse meanings of findings to date, and limit the data on which to base employer decisions about the value of sponsoring health promotion programs. If advances are to be made in evaluating worksite health promotion outcomes, more discussion and clarity are needed regarding (1) terms and approaches used to describe programs and their evaluation (concepts), (2) how processes and effects of programs are to be measured and reported (methods), and (3) possibilities for creating environments and incentives to reinforce health-related choices as well as consideration of related societal and ethical issues (policies).

No attempt is made to review systematically the effectiveness or cost-effectiveness of categorical health promotion programs or of prevention in general since these purposes have been addressed elsewhere in the literature (e.g., Banta & Luce, 1983; Fielding, 1982; Kristein, 1982, 1983) and in this book. Guidelines for implementing particular evaluative methodologies and for integrating the health promotion and evaluation literatures have likewise been discussed by others (Flay & Best, 1982; Fuchs, 1980; Hollander, Lengermann, & DeMuth, 1985; Shepard & Thompson, 1979; Stoddart, 1982). Only recently has attention focused on possible relationships among economic incentives, other organizational incentives, and the practice of healthful behaviors (Moser, Rafter, & Gajewski, 1984; Stokes, 1983; Warner & Murt, 1984). As a result, this chapter explores some emerging needs and opportunities for improvement of health promotion evaluation "science" at conceptual, methodological, and policy levels, and for enhancement of manager and consultant/researcher communications within and across these levels.

UNTANGLING CONCEPTS AND APPROACHES

Amid the recent proliferation of new or planned worksite health promotion efforts (Fielding & Breslow, 1983), confusion abounds regarding what constitutes activities (e.g., company policies related to "no smoking"; appropriate linkages with medical or personnel departments) versus programs, that is, particular categories of risk reduction education for hypertension control, smoking cessation, exercise maintenance, and so forth. For purposes of this chapter, evaluation of worksite health promotion refers to analyzing impacts (as well as processes) of both programs and activities.

Sharpening Focus on Terminology, Concepts, and Requisite Skills

Considerable misunderstanding exists among behavioral consultants/ researchers and providers—and to a lesser degree among economists and business people—about the meaning and calculation of terms or methods, such as return-on-investment (ROI), cost-effectiveness (CE), and cost-benefit (CB) (e.g., Banta & Luce, 1983; Weinstein & Stason, 1977). At least two caveats are appropriate.

First, confusion in terminology and concepts is primarily due to divergent values held by participants entering the health promotion arena. At least nine perspectives for viewing worksite programs are described in Table 20.2. Depending on the particular analytic perspective taken, perceptions and definitions of cost-effectiveness or other analyses vary. Employees, for example, may intuitively view participation in sports and exercise programs as most desirable, fun, or valued. Companies, by contrast, may view other programs related to cigarette smoking, hypertension, or alcohol abuse as more attractive because they are more likely to offer immediate or higher returns-on-investment (e.g., Fielding, 1983; Kristein, 1982, 1983). These same programs, however, are often less popular with employees. Hence, what is most appealing or cost-effective from employee and corporate viewpoints may differ.

Table 20.2.   Multiple Perspectives for Evaluating Private Sector WHPPs (Short Run and Long Run)

1. Chief executive officers/management endorsing and funding WHPP
2. Researchers/independent consultants hired to evaluate the scientific rigor of WHPP
3. Providers/staff delivering WHPP services
4. Employees (or employee family members) participating in WHPP
5. Private third-party payers, external to the company or internal to the company if self-insured
6. Public third-party payers
7. Federal and state regulatory agencies
8. Societal context and trends (distinguished from programmatic effects)
9. The entire health sector

Second, worksite behavioral consultants/researchers must develop competencies in assimilating and applying evaluative techniques traditionally more familiar to business and economics (e.g., ROI, CE) than to behavioral medicine. Fortunately, a growing literature (Coates & DeMuth, 1984; DeMuth & Yates, 1984; DeMuth, Yates, & Coates, 1984; McDonagh, 1982) addresses this assimilation issue and makes it increasingly possible for behavioral researchers to access managerial training opportunities. Practical dilemmas and growth challenges are also created by requirements for versatility and sensitivity in (1) moving across business and academic worlds and (2) understanding the differing languages, organizational structures, priorities, and rewards of both systems.

Table 20.3.   Manager Dilemmas: Issues Faced by Corporate Decision Makers

Moving competently within the business world, including risk taking for long-run returns

Determining programs (rigorous, low-keyed, none) that meet diverse management and employee needs/budget constraints

Specifying how program effects and costs to company and employees are to be measured and valued

Identifying minimal information needed for optimal business decisions

Confronting low-level interest in funding long-term epidemiological research but needing "hard data"

Making final decisions about release/retention of short-term and negative results

Categorizing where worksite health promotion programs fit into broader company policies/plans

Additional manager and researcher dilemmas are elaborated in Tables 20.3 and 20.4, respectively. It is not within the scope of this section to discuss the full range of topics suggested in these tables. Their review, however, gives

Table 20.4.   Researcher Dilemmas: Issues Faced by Behavioral Medicine Investigators/Consultants

Moving competently within and across business and academic worlds

Creating measures/programs important to management, science, and employees within budget constraints

Testing theories to expand scientific (e.g. epidemiological) data and improve on-site programs

Identifying and pursuing new research areas applicable to WHPP (e.g., motivational research)

Considering how "long-term" is long-term enough for evaluating worksite interventions

Sharing results with the larger research community within guidelines endorsed by management

Preserving confidentiality of clients and at the same time providing management with requested data

Expanding health promotion's purview beyond exclusive focus on individual behavior

a context for understanding the array of divergent values and priorities that may serve to escalate both semantic confusion and strains in communication among evaluators, managers, and employees.

Cost and Outcome Concepts.   Evaluative approaches for combining cost and outcome information (e.g., ROI, CE, CB) are typically presented as ratios. Varying viewpoints described in Table 20.2, of course, change the numbers which are counted and included as costs and outcomes in these ratios. As a result, definitions of numerators and denominators in ratios vary across studies, making meaningful comparisons difficult. Because of this variability, less significance should be attached to numerator and denominator labels than to clear descriptions of particular variables included as components for any particular analysis. Likewise, separating programmatic changes from societal changes that contribute to worksite health promotion effects is essential (Weinstein & Stason, 1977), but introduces yet another complication. This refinement in analysis, as a result, is frequently overlooked. Overall, specifying the meaning of cost and outcome terms and their relationships is a complex matter and not one of simple accounting procedures (Hollander, Lengermann, & DeMuth, 1985).

Threshold Spending, Booster Spending, and Health Benefit Curve Concepts.   Green (1977) suggested that these concepts, related to the level of corporate expenditures for health promotion programs, need to be expanded and used in conjunction with cost-outcome concepts. *Threshold spending* refers to the minimum level of investment required to obtain desired effects or behavioral changes. *Booster spending* involves the determination of optimal magnitude, timing, and sequencing of additional expenditures for maintenance, expansion, and/or further evaluation of existing programs. The *health benefit curve* concept attempts to specify relationships between particular health promotion interventions and their longitudinal effects.

It may not be true that some health promotion interventions are better than none. If the worksite effort proves insufficient to achieve anticipated corporate effects, minimal expenditures may be wasted, or even worse, may place health promotion efforts in desrepute. Lost opportunities, as well as poor returns-on-investment, may also be created by investing heavily in one or two categorical programs to the exclusion of more diversified approaches. Little, however, is known about (1) optimal mixes of programs for particular types and categories of industries, employees, or geographic areas, (2) appropriate times for additional or booster spending to enhance the continuing impact of existing programs, or (3) the shape of the health benefit curves for particular risk factors and interventions.

Problems related to conceptualizing health benefit curves may severly interfere with practical needs for outcome evaluations. Epidemiologic evidence, for example, suggests that lack of exercise increases cardiovascular health risk and regular exercise decreases risk. It is unclear, however, how long it takes to achieve and maintain this effect and the related shape of the health benefit curve for different types of exercise is likewise unknown. While the current state-of-the-art does not provide answers to these questions, knowledge of such curves and underlying assumptions about their behavior and their value are important for evaluating outcomes and benefits in terms of health status and costs (Fielding, 1983),

Untested Theory, Practical Decisions, and Inadvertent Policy-making. In the absence of data and tested theory regarding cost-effectiveness of worksite health promotion, a number of questions remain for managers (whether to fund programs; at what levels and how long; what type of end products are to be expected) and for the research community (what type of longitudinal studies to fund; whether the general population or high-risk groups are the best candidates for cost-effective long-term treatments; what follow-up periods are "long-term" enough). Of course, practical decisions in all these areas inadvertently create policies. Such policies necessarily rely on incompletely supported assumptions about whether worksite health promotion benefits are expected to accrue rapidly or slowly, temporarily or permanently, in the general population or high-risk groups, and in what relationship to the economy (Green, 1977),

Importance of Evaluation in Worksite Health Promotion Programs

It is well known that worksite health promotion program evaluation, at its best, involves complex epidemiological and economic models, substantial financial commitment, computerized-information systems, and is executed by only a few companies (Chadwick, 1982). Two model worksite programs, Control Data's Staywell and Johnson and Johnson's LIVE FOR LIFE, still have not published sufficient data to determine the long-run successes of particular interventions or composite efforts. Published outcome results are available for some worksite components of the CHIP program (e.g., Brownell, Cohen, Stunkard, Felix, & Cooley 1984; Brownell, Stunkard, & McKeon, 1985). All three programs are based on steps considered essential parts of a rigorous health promotion model: (1) needs assessment, including a survey of employee interests; (2) program development based on assessed needs and surveyed interests; and (3) evaluation of results involving long-term epidemiological, cost and management information data, as well as feedback to corporate goals and objectives. In part, the newness of many worksite health promotion programs limits analysis of longitudinal data. Even more important, however, multiple other factors—many of which are detailed in this chapter as problems in concepts, methods, or policy—may sabotage or at least detract from the collection of "good" long-term follow-up data and, hence, from the fulfillment of the third step of a rigorous evaluation model.

Dickerson and Mandelblit (1983) describe a tension that exists between this most rigorous health promotion evaluation paradigm and a more low-keyed approach where resources are used to implement programs based on results from other settings. These authors state that if a rigorous health promotion evaluation paradigm is held as the ideal, this approach may become an effective deterrent to implementing more modest but still practical programs. One popular alternative, of course, is to offer programs such as exercise, weight control, and nutrition education with the hope that they will have a positive effect on both employee health and employee relations, without devoting resources to evaluation. Wilbur (1983) warns, however, that although a low-key approach may minimize program costs, it could also permanently handicap efforts to maximize optimal program impact. This argument corroborates Green's (1981) warning about the "cycle of poverty" in which low-level investments are made in programs which ultimately yield modest results. As a result, subsequent funding is minimal and the "cycle of poverty" is further perpetuated by inadequate tests of health education or health promotion impacts.

Different Types of Evaluation Data

While researchers tend to be worried about experimental designs and rigorous evaluations (Flay & Best, 1982), intuitive or subjective feedback may be more relevant to corporate decision makers than are cost-effectiveness data. For example, Dr. C.C. Wright (1982), Director of Health Services for the Xerox Corporation, notes that while formal studies indicate that some important changes had taken place for participants in a worksite exercise program, informal indicators of program effects regarding effectiveness or cost-containment may be most persuasive: ". . . *The Physical Fitness Programs are considered to be effective in cost-containment because so many individuals believe they are*" (p. 697). In short, employee feedback to senior executives about the effects that a worksite program has on their health or health habits may serve as the most powerful, immediate reinforcers for continued commitment of resources to health promotion programs. For these programs to become more than passing fads, however, the accruing of more rigorous forms of data becomes a long-term issue. The appropriate mix of different types of evaluation data—valued, required, or used by particular companies—may vary at different program stages, with changes in management, and with multiple other idiosyncratic factors.

Overcoming Manager versus Consultant/Researcher Dilemmas

This section summarizes several manager and consultant/researcher dilemmas. For managers, practical problems emerge because of the absence of adequate "hard data" about efficacy of health promotion programs and activities. Lacking this information, managers must still decide whether (and how) to allocate monies for program implementation and evaluation. Top management, however, may be unprepared or unmotivated to invest the substantial time and energy required for well-planned health promotion efforts in large or small companies. Rather than being sold a "quick fix," executives must understand that health promotion activities and programs require experimentation to customize an approach that reflects organizational goals and budgets, as well as employee needs and expectations (Novelli & Ziska, 1982; Merwin & Northrop, 1982). Conversely, researchers as consultants to industry must become "process oriented" in educating their clients, while not turning executives off with excessive detail. Furthermore, enhanced ability to measure and substantiate the effects of health promotion interventions on employee morale and other indices important to top management represents a continuing challenge (Hollander, Lengermann, & DeMuth, 1985).

METHODS FOR ENHANCING THE PRECISION OF EFFECTIVENESS VARIABLES

A number of problems related to measuring effectiveness of worksite health promotion programs might be cited. This section focuses primarily on researcher dilemmas (Table 20.4). Advances in measuring effectiveness variables must necessarily precede any precise linking of outcomes to costs. Several examples of how particular methods might lead to discrepancies in reported effectiveness outcomes are discussed with regard to (1) target population, (2) participation rates, (3) intervention content, and (4) company context. These four areas are chosen for their representativeness and prevalence as potential problems in method. No attempt is made to be comprehensive regarding all possible method problems. Instead, encouraging researchers to sharpen their procedures for calculating and reporting effectiveness variables in at least four areas is suggested as an important first step.

Target Population

One salient issue related to calculating outcome variables is the match or mismatch between the target population (e.g., by organizational level, sex) and the health promotion intervention. Different techniques are needed, for example, to reach blue-collar and white-collar workers (Fielding, 1983). A well-designed four-week smoking-cessation program may be extremely effective with managers, but totally ineffective at blue-collar levels. Assembly-line workers may dislike doing homework or may perceive a three- to four-week program as too long. By contrast, more extended programs sequenced over several weeks or months tend to be more popular with managerial levels. Furthermore, differential effectiveness of lifestyle change programs for women and men may also affect the calculation of effectiveness variables. The sustained quit rate after smoking-cessation programs typically tends to be lower in women than in men. Simply reporting aggregated statistics without a separate breakdown of the characteristics of the target population may also distort effectiveness measures.

Participation Rates

Variations in how participation rates are measured and counted describe additional reasons for reported differences in effectiveness outcomes. Participation rate measurement depends on answers to a number of questions. Is the participation rate to be based on those people who come to any class? Are participants required to attend a certain number of classes to be considered participants? Is completion of homework assignments mandatory for being considered a participant? How are participants recruited and programs marketed? Responses to all these questions *change* the specific numbers reported as participants and used to calculate program outcomes.

Two examples from weight control and exercise programs, respectively, suggest unfounded and erroneous conclusions, which result when participation rates are not compared and recalculated based on the same criteria. In the first example, average weight loss may be 15 pounds over a 12-week period for Program A, while in Program B the average weight loss is 10 pounds for the same period. At first glance, one would say that Program A was more effective. In Program A, however, there may have been an unpublished stipulation that participants attend at least 50 percent of the classes in order to be counted. In Program B, the denominator may include anyone who attended at least one class. When denominators are recalculated using the same criteria, different results emerge and Program B is thereby judged to be most effective. For Program B, the average weight loss was actually 20 pounds for those who attended 50 percent or more of the sessions. In yet another example, Company X may report 25 percent participation for their exercise facility, defined as employees who used the facilities at least twice a month. For Company Y, there may be 10 percent participation, but these are people who exercise at least three times a week.

Participation rates for Companies X and Y are *not* comparable. In published reports, underlying procedures must be made explicit if outcome comparisons are to be meaningful.

Intervention Content

Published reports are often unclear regarding the intervention techniques employed, number of sessions provided, materials used, timing and content of homework assignments, and presence of any organizational reinforcements to support the intervention. Was a company-wide smoking policy, for example, implemented before, during, or after the initiation of a smoking-cessation program? Descriptions of programs must be explicit in their identification of (1) specific methods employed, (2) presence and content of organizational incentives, and (3) existence of complementary programs such as weight control, which may have related or synergistic effects on the program being analyzed. Without such information, effectiveness comparisons across programs are meaningless because extraneous factors, not stated in scientific reports, may inadvertently bias quantification of outcome variables.

Company Context

Difficulties in deciphering underlying contextual circumstances of programs further complicate evaluation of effectiveness data. Few published studies provide information regarding underlying corporate cultures, historical context, or policies. Is the health promotion program, for example, supported by a letter from the Chief Executive Officer (CEO) or by visible executive participation? Does the current health promotion effort follow on the heels of several poorly attended and unsuccessful programs? Are programs scheduled after work when most people carpool or when flextime results in staggered work schedules? All these extraneous factors may alter participation and yet may be omitted from published reports.

Furthermore, it may be misleading to extrapolate results from one company population (and context) to some other similar population in another organization. Corporations and their employees vary tremendously. A program that works well in one context may not work as well in another because of differing seasonal business cycles, threatened layoffs, or flextime changes. Similarly, successful change programs in a clinical setting may not be equally successful in nonclinical applications. A well-known weight loss program, for example, may consistently produce a 15 percent dropout rate in clinical settings. When the same program is implemented in a company with the same instructors and the same number of sessions, the dropout rate may be substantially more (e.g., 65 percent). Even after three months of adapting the program for the target company, the dropout rate may remain 20 or 30 percent.

Multiple examples in this section underline the importance of making method sections of published reports as specific as possible in their variable definitions and procedures if the emerging worksite health promotion literature is to be useful and replicable.

POLICY IMPLICATIONS AND ETHICS

Various authors (Coates & DeMuth, 1984; Stokes, 1983; Wilbur, 1983) are now indicating that perhaps we have too narrowly conceived worksite health promotion activities as individual behavior change. In a broader context, health promotion may also be viewed as a process involving environmental and economic incentives to help maintain healthy behaviors. If individual behavior change is the sole focus of attention, promising research areas and opportunities for group, organizational, and environmental interventions may be missed. More must be learned about optimal organizational and group strategies to promote wellness and about the cost-effectiveness of these competing approaches. Health promotion and disease prevention, however, have not been priorities and still remain frontier areas for clinical and experimental medicine and phychology. Other lines of inquiry, including motivational research, are also required to understand how to convince middle-age people that it is in their best interest to follow prudent health behavior including exercise, eating properly, and smoking cessation (Gray, 1983).

Environmental and Economic Incentives

The degree to which particular economic and other incentives have an impact on particular health behaviors must be explored and evaluated. Possible long-term effects, for example, of offering reduced insurance premiums for those who practice more healthful behaviors require further analysis (e.g., Brailey, 1980; Stokes, 1983). While community rating procedures give insurees few incentives for altering lifestyles, preferred risk groups—which monetarily reward those who practice healthier behaviors— may show promise (e.g., Stokes, 1983). Unfortunately, interest in economic incentive systems remains primarily ignored and unappreciated in both community and worksite studies (Yates & DeMuth, 1981).

Additional examples of environmental or economic incentives (as possible adjuncts to and reinforcements of worksite health promotion programs) merit attention and include direct cash incentives, cost-sharing programs, free or subsidized services, and employee participation in program development (Thomas, 1981). Thomas's categories are briefly summarized and suggest promising directions for refining evaluation science beyond traditional focus on individual behavioral outcomes.

Direct Cash or Other Incentives.   A number of companies are beginning to experiment with direct cash incentives, such as those offered to employees who quit smoking. Speedcall Corporation of Hayward, California, for example, offered employees quitting smoking an extra $7 per week (Shepard, 1980). Similarly, the Neon Electric Company of Houston, Texas, provided a raise of 50 cents an hour to any employee who gave up the habit. It still needs to be assessed whether such incentives are effective, cost-effective, and merit inclusion as permanent or periodic features of worksite programs. For example, determining ways to optimize periodic use of cash incentives to enhance health behaviors provides a provocative challenge for evaluators. This challenge is analogous to our business colleagues developing formulas for maximizing the effects of television advertisements by using particular timing and sequencing strategies. Additional studies might evaluate the impact of cash bonuses or "wellness days" on absenteeism or other variables important to management. Short- and long-run effects of organizational policies, piggybacked on existing health promotion programs or operating primarily independent from them, also must be analyzed to help ensure better allocations of corporate resources to company-wide versus special-risk group interventions.

Cost-sharing Programs.   Whether employees should help bear the cost burden of worksite health promotion also remains a matter of debate. Some stress that the presence of copayments can improve motivation. This alternative might also be combined with rebates for individuals who might, for instance, attend more than half of the company's health promotion workshops or who improve their fitness levels by some preestablished criteria (Thomas, 1981).

Free or Subsidized Services.   Another possibility for motivating behavior change relates to various forms of subsidized or free services. Scher Brothers Lumber Company in Minneapolis provides its imployees free decaffinated coffee, a free salad bar three days a week, and an assortment of fresh fruit and fruit juices from company vending machines at a subsidized cost of ten cents. The Minnesota Heart Health Program, which operates in several communities, (1) endorses items at local restaurants adhering to more heart healthy menus and cooking styles and (2) tags selected grocery store items to remind consumers of heart-healthy choices while shopping (Blackburn et al., 1984). Such broad-based community projects must ultimately demonstrate cost and outcome accountability of alternative interventions (e.g., worksite programs versus mass media campaigns versus restaurant/grocery store incentives). Reframing the problem in yet another way, Kristein (1982) suggests that "it is likely that beginning health promotion activities early in life, like the American Health Foundations's Know Your Body school health program, will be found to be both more effective and more economical for the nation, in the long run" (p. 731). Ideally, health promotion efforts require attention as early as possible, preferably in the public schools as an integral part of the curriculum.

Employee Participation in Program Development. Yet another proposed direction for motivating or reinforcing employees to change their lifestyles involves participation in health-related decisions. The Johnson & Johnson Company, for instance, encourages interested employees to serve on topic-specific task forces, such as for exercise, stress, or publicity. With staff input, these leadership groups help develop interventions to make the workplace more conducive to healthy lifestyles. For example, the Nutrition and Weight Loss Task Force introduced a salad bar, a breakfast program, and fresh fruit for dessert. Caloric counts are presented on daily menus. Opportunities for greater consumptin of fish and fowl and less beef are also encouraged. The purpose of such participatory groups and programs is to create and sustain a reinforcing environment, rooted in peer encouragement, which favors healthier behaviors. Effects of both activities and related programs are being analyzed to help determine short and long-term behavior changes on a number of health and organizationally relevant indices (Wilbur, 1983).

Worksite and Community Incentives: CHIP

Additional examples of innovative worksite health promotion programs and policies, which combine elements of employee participation in program development and company endorsement of health promotion, derive from the Pennsylvania County Health Improvement Program (CHIP). This program is described briefly here to illustrate many of the evaluative issues raised earlier in the chapter. A comprehensive account of CHIP is available elsewhere (Stunkard, Cohen, & Felix, 1985).

CHIP is a community-based, multiple-risk factor intervention trial designed to decrease mortality and morbidity from cardiovascular disease in a county of 118,000 people in north-central Pennsylvania. It is based on the pioneering programs of the Stanford Three-Community Study, which showed marked reduction in coronary risk factors in two communities compared to a control community, and the North Karelia Project, which showed marked reduction in coronary heart disease and stroke. Five risk factors are targets—smoking, hypertension, elevated levels of low-density lipoprotein cholesterol, overweight, and physical inactivity. Interventions are conducted through the worksites, mass media, the health sector, voluntary organizations, and schools. The program is a cooperative one involving the University of Pennsylvania, Lycoming College, and the community of Lycoming County. It began operation in July 1980 and evaluation of its first three years in comparison to a reference county in another part of the state has been completed.

From the beginning, evaluation has occupied a high priority for CHIP planners and concerns with cost-effectiveness played a major role in program implementation. CHIP planning was guided by the assumption that measurable health benefits would probably be limited and that reasonable cost-effectiveness ratios could be obtained only by the most economical use of resources. Thus, from design as well as from necessity, CHIP has functioned on a modest budget, supported by local resources. Because costs are low, CHIP can be widely replicated.

A key element of CHIP is its worksite health promotion programs. The baseline survey in 1980 revealed few health promotion activities at the work-sites of either Lycoming County or the reference county. In Lycoming County only 35 percent of 47 industries conducted such programs. For the reference county, comparable figures were 41 percent of 27 industries. By 1983 the number of industries conducting health promotion programs in the reference county had fallen by 45 percent, a function of the highly unfavorable economic circumstances afflicting this area of the country. In these same unfavorable circumstances, however, the number of industries implementing programs in Lycoming County increased by 50 percent.

An important conceptual issue described earlier is the desirability of a three-step health promotion model: (1) a needs assessment, (2) program development based on these assessed needs, and (3) feedback to relevant parties. This model was used in CHIP. The first step in this process involved two surveys: an Employee Interest Survey designed to ascertain the types of programs most favored by employees, and a Health Status Appraisal, designed to provide an overview of the health needs of the industry. Not surprisingly, there was a sizable gap between initial employee interests (which initially favored exercise programs) and management desires (which frequently included smoking-cessation programs). These differences were always resolved in favor of the employees' interests, and program development was carefully tailored to their interests. Significantly, as employees gained experience and confidence in their ability to mount health promotion programs, they accorded smoking cessation an increasingly higher priority. Feedback of results to both employees and management maintained their interest and commitment, and these results were used to improve the program.

Every effort was made from the program's inception to help employees not only participate, but also to assume the major role in health promotion activities. A 14-step process to further these efforts has been developed, and much of CHIP's success can be attributed to this process, which has been described more fully elsewhere (Felix, Stunkard, Cohen, & Cooley, 1985). It consists of the following steps:

1. Introduction of the program to all levels of management.
2. Announcement of the new program to employees.
3. Recruitment and organization of a heart health committee that includes both management and labor.

4. In-house promotion and communication strategy planning, either via a company newsletter or by a vehicle developed for this purpose.

5. Development and administration of an Employee Interest Survey and a Health Status Appraisal, carried out by the heart health committee.

6. Formation of risk factor subcommittees to focus on each of the five risk factors.

7. Exploration by the subcommittee of risk factor reduction programs available in the community.

8. Committee review of the findings of the subcommittees and selection of health promotion programs.

9. Development of a program proposal to be presented to management.

10. Presentation of the proposal to management and enlistment of its support in such forms as released time and financial incentives.

11. Scheduling of programs by the heart health Committee and its subcommittees.

12. Promotion of programs and recruitment of participants by the committee and its subcommittees.

13. Program implementation, which has resulted in as many as eight different risk factor reduction programs at one worksite.

14. Evaluation of the programs by the joint efforts of the committee and CHIP, with feedback of information to management and employees to maintain their interest and commitment and to revise the program.

The most successful of the programs developed with this process have been those involving the use of competition to increase the motivation for weight loss, exercise, and smoking cessation. The most intensively studied of these competitions have been those involving weight loss. Twenty-eight weight-loss competitions have been carried out within and between worksites. The results of these competitions appear far superior to those of other worksite weight loss programs, both in terms of lowered attrition and greater weight loss. Table 20.5 shows that the first of these CHIP worksite competitions produced an average weight loss of 13 pounds, with an attrition of no more than 0.5 percent (Brownell, Cohen, Stunkard, Felix, & Cooley, 1984). The contrast with the outcomes of a clinically oriented worksite

Table 20.5.   Comparison of CHIP and Union Program Competition Results

	N	Duration (weeks)	Attrition (%)	Weight Loss (pounds)
CHIP competition	175	12	0.5	13.0
Union program	172	16	42	7.9

program conducted in a union by the same investigators (Brownell, Stunkard, & McKeon, 1985), could not have been more striking. In the union program, the average weight loss was only 7.9 pounds and the attrition rate rose to 42 percent (see Table 20.5).

An important feature of the CHIP worksite competition is the limited use of professional and even of nonprofessional time, with resulting highly favorable cost-effectiveness ratios. Figure 20.1 shows the cost of losing one

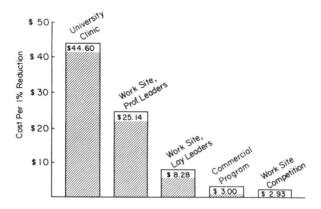

FIGURE 20.1.   The cost per 1-percent reduction in percentage overweight for different weight-loss programs. (From Brownell, Cohen, Stunkard, Felix, & Cooley, 1984. Copyright 1984 by the American Public Health Association. Reprinted by permission.)

percent of body weight in a variety of different weight-loss programs; weight-loss worksite competitions rank with the best of these programs (Brownell, Cohen, Stunkard, Felix, & Cooley, 1984).

Overall, the CHIP experience helps debunk old notions about the most expensive worksite health promotion efforts as being superior. This program persuasively underlines the merits of properly organizing incentive systems and of piggybacking community and worksite programs. Worksite health promotion interventions require organizational consultants who are sensitive to and skilled in simultaneously combining organizational, group, and individual incentives for more positive health behaviors.

Societal and Ethical Context for Health Promotion

While we have identified multiple evaluation opportunities within corporate and worksite settings, nudging forward the state of the science of health promotion is, of course, larger than a worksite problem. Torrens, Breslow, and Fielding (1982), for example, have suggested that universities also need to value, advocate, and pursue policy analysis of health promotion and disease prevention strategies. The authors also note that the Graduate Medical Education National Advisory Committee (GMENAC) report predicted a significant shortage for medical personnel in preventive medicine in 1990 (e.g., an estimated 5,000 preventive medicine specialists available, compared to 6,310 needed). At least 25 percent more personnel are needed than will be available. As a result, Torrens, Breslow, and Fielding (1982) stress that if universities and industries are to play appropriate roles in chronic disease prevention, whether through lifestyle changes or other means, attention must be given to (1) training a greater number of physicians and other health professionals for disease prevention leadership and/or (2) reallocating physicians from overpopulated specialties to the preventive medicine area. These changes would of course necessitate reallocations in how monies are assigned to residency and other training programs in light of national needs for health personnel. Unfortunately, disease prevention concepts still have a relatively weak foothold in our medical system. As a result, the changes advocated by Torrens et al. (1982) are likely to be difficult to achieve in the immediate future.

Policies pertaining to research and demonstration programs related to chronic degenerative disease are still being developed at national, state, community, and worksite levels. While there is popular support for fitness or health promotion concepts, much work remains to be done in developing and maintaining cohesive policies for reinforcing healthful behaviors within and across these levels. This goal may be difficult to achieve due to fragmentation in our health system and underlying ethical questions. The subject of worksite health promotion, for example, raises issues regarding freedom of choice. At worst, companies may be perceived as meddlers intruding in areas of individual choice and privacy. Tension exists regarding the question of whether concern for the social good can overwhelm and submerge the right of individual choice. An issue left unresolved is the degree to which people should be left alone, even if they choose to be unhealthy, in the interest of preventing a health-oriented totalitarianism (Goldschmidt, 1977). A related question is how to balance individual freedom of choice and confidentiality with research progress in the pursuit of health.

CONCLUSION

This chapter explores some of the difficulties in holding worksite health promotion programs accountable for particular outcomes or returns-on-investment. Steps needed for more accurately documenting program costs/outcomes and for identifying the "fittest" programs—worthy of survival and replication—are identified. Conceptual, methodological, and policy issues are discussed with special reference to the researcher and manager perspectives. In the concept section, problems in defining common evaluative terms and in specifying evaluation models appropriate for large-, medium-, and small-size companies are discussed. If many companies consistently implement modest forms of "popular" health promotion and forego substantive evaluation efforts, there will probably be negative long-run consequences to the health promotion field (Chadwick, 1982; Wilbur, 1983). Given the scarcity of resources and demands for optimal data from minimal investments, the resolution of this issue is unclear and many worksites may be poor candidates for consistently "good" evaluation.

In the method section, several examples were given of ways current studies are compromised by methodological flaws or by omissions of important procedures that have direct bearing on the quantification of effectiveness variables. In the policy section, worksite health promotion is discussed in a broader context than individual lifestyle change to include environmental interventions and economic incentives. Consequently, worksite health promotion efforts become part of a comprehensive strategy for reinforcing healthful behaviors at multiple levels (organizational, group, individual) and require data-based evaluations at each level.

Promising directions for enhancing the scientific aspects of evaluation of worksite health promotion activities and programs have been emphasized. The generation of increasingly precise questions and data about worksite health promotion interventions and their short- and long-run impacts are still needed. We need to know more about (1) issues of optimal program mix for particular companies, settings, and employee subgroups, (2) data-based links among health promotion activities, programs, health benefits, and organizational goals, and (3) opportunities for medium- and small-size companies to contribute systematic and comprehensive (not compulsive) evaluation data, targeted to existing management control and information systems. Better evaluation data may in turn expand possibilities for the generation of return-on-investment calculations, derived from mathematical modeling, to better predict which health promotion programs and policies are most cost-effective for individual companies. Advances in these diverse areas will help expand understanding of cost-outcomes and other effects— expected by managers, measured by evaluators, and affecting employee health and health behaviors.

REFERENCES

Banta, H. D., & Luce, B. R. (1983). Assessing the cost-effectiveness of prevention. *Journal of Community Health, 9*(2), 145–146.

Blackburn, H., Luepker, R. V., Kline, F. G., Bracht, N., Carlaw, R., Jacobs, D., Mittlemark, M., Stauffer, L., & Taylor, H. L. (1984). The Minnesota Heart Heath Program: A research and demonstration project in cardiovascular disease prevention. In J. D. Matarazzo, S. M. Weiss, J. A. Herd, N. E. Miller, S. M. Weiss (Eds.), *Behavioral health: A handbook for health enhancement and disease prevention* (pp. 1171–1178). New York; Wiley.

Brailey, A. G. (1980). The promotion of health insurance. *New England Journal of Medicine, 302*, 51–52.

Brownell, K. D., Cohen, R. Y., Stunkard, A. J., Felix, M. R. J., & Cooley, N. B. (1984). Weight loss competitions at the worksite: Impact on weight, morale, and cost effectiveness. *American Journal of Public Health, 74,* 1283–1285.

Brownell, K. D., Stunkard, A. J., & McKeon, (1985). Weight reduction at the worksite: A promise partially fulfilled. American Journal of Psychiatry, *142,* 47–51.

Chadwick, J. H. (1982). Cost effective health promotion at the worksite? In R. S. Parkinson et al. (Eds.), *Managing health promotion in the workplace* (pp. 288–299). Palo Alto, CA: Mayfield.

Coates, T. J., & DeMuth, N. M. (1984). Health psychology: Training, responsibilities, contributions. In J. D. Matarazzo, S. M. Weiss, J. A. Herd, N. E. Miller & S. M. Weiss (Eds.), *Behavioral health: A handbook of health enhancement and disease prevention* (pp. 1196–1200). New York: Wiley.

Cohodes, D. R. (1982). Problems in measuring the cost of illness. *Evaluation and the Health Professions, 5,* 381–392.

DeMuth, N. M., & Yates, B. T. (1981). Improving psychotherapy: Old guilts, new research, and future directions. *Professional Psychology, 1981,* 587–595.

DeMuth, N. M., & Yates, B. T. (1984). Training managerial and technological research skills. *Clinical Psychologist, 37*(1), 33–34.

DeMuth, N. D., Yates, B. T., & Coates, T. C. (1984). Psychologists as managers: Overcoming old guilts and accessing innovative pathways for enhanced skills *Professional Psychology, 15,* 758–768.

Dickerson, O. B., & Mandelblit, C. (1983). A new model for employer-provided health education programs. *Journal of Occupational Medicine, 25,* 471–474.

Felix, M. R. J., Stunkard, A. J., Cohen, R. Y., & Cooley, N. B. (1985). Health promotion at the worksite: A process for establishing programs. *Preventive Medicine, 14,* 99–108.

Fielding, J. E. (1979). Preventive medicine and the bottom line. *Journal of Occupational Medicine, 21,* 79–88.

Fielding, J. E. (1982). Effectiveness of employee health improvement programs. *Journal of Occupational Medicine, 24,* 907–916.

Fielding, J. E. (1983, March). *Effectiveness of four types of employee health improvement programs.* Invited address at the meeting of the Society of Behavioral Medicine, Baltimore, MD.

Fielding, J. E., & Breslow, L. (1983). Health promotion programs sponsored by California employees. *American Journal of Public Health, 73,* 538–542.

Flay, B. R., & Best, J. A. (1982). Overcoming design problems in evaluating health behavior programs. *Evaluation and the Health Professions, 5,* 43–69.

Fuchs, V. R. (1980). What is CBA/CEA, and why are they doing this to us? *New England Journal of Medicine, 303,* 937–938.

Goldschmidt, P. (1977, May). *Future of medical care: Technology.* Paper presented at the meeting of the National Association of Blue Shield Plans National Professional Relations Conference, Chicago, IL.

Gray, H. J. (1983). The role of business in health promotion: A brief overview. *Preventive Medicine, 12,* 654–657.

Green, L. W. (1977). Evaluation and measurement: Some dilemmas for health education. *American Journal of Public Health, 67,* 155–161.

Green, L. W. (1981, June). Needs assessment: Key to effective health education. In Institute for Health Planning, *Planning for health education: A training package.* Videotape prepared at training workshop, Atlanta, GA.

Green, L. W., & Lewis, F. M. (1981). Issues in relating evaluation to theory, policy, and practice in continuing education and health education. *Mobius, 1,* 46–58.

Hollander, R. B., Lengermann, J. J., & DeMuth, N. M. (1985). Cost-effectiveness and cost-benefit analyses of occupational health promotion. In G. S. Everly & R. H. L. Feldman (Eds.), *Occupational health promotion: Health behavior in the workplace.* New York: Wiley.

Kristein, M. M. (1982). The economics of health promotion at the worksite. *Health Education Quarterly, 9* (Suppl. 1-92), 27–36.

Kristein, M. M. (1983). How much can business expect to profit from smoking cessation? *Preventive Medicine, 12,* 358–381.

McDonagh, T. J. (1982). The physician as a manager. *Journal of Occupational Medicine, 24,* 99–103.

Merwin, D. J., & Northrop, B. A. (1982). Health action in the workplace: Complex issues–no simple answers. *Health Education Quarterly, 9,* (Suppl. 1-92), 73–82.

Moser, M., Rafter, J., & Gajewski, J. (1984). Insurance premium reductions: A motivating factor in long-term hypertensive treatment. *Journal of American Medical Association, 251,* 756–757.

Novelli, W. D., & Ziska, D. (1982). Health Promotion in the workplace: An overview. *Health Education Quarterly, 9* (Suppl. 1-92), 20–26.

Parkinson, R., et al. (1982). *Managing health promotion in the workplace.* Palo Alto, CA: Mayfield.

Shepard, D. S. (1980). *Incentives for not smoking: Experience at the corporation.* Paper presented at the Corporate Commitment to Health: First Executive Conference, Washington, DC.

Shepard, D. S., & Thompson, M. S. (1979). First principles of cost-effectiveness analysis in health. *Public Health Reports, 94,* 535–543.

Stoddart, G. L. (1982). Economic evaluation methods and health policy. *Evaluation and the Health Professions, 5,* 393–414.

Stokes, J. (1983). Why not rate health and life insurance premiums by risks? *New England Journal of Medicine, 308,* 393–395.

Stunkard, A. J., Felix, M. R. J., & Cohen, R. Y. (1985). Mobilizing a community for health: The Pennsylvania County Health Improvement Program. In J. C. Rosen & L. J. Solomon (Eds.), *Prevention in health psychology* (pp. 143–190). Hanover, NH: University Press of New England.

Thomas, J. (1981, November). *Promoting health in the work setting.* (Available from Institute for Health Planning, 702 North Blackhawk Avenue, Madison, WI 53705).

Torrens, P. R., Breslow, L., & Fielding, J. E., (1982). The role of universities in personal health improvement. *Preventive Medicine, 11,* 477–484.

Warner, K. E., & Murt, H. A. (1984). Economic incentives for health. *Annual Review of Public Health, 5,* 107–133.

Weinstein, M. C., & Stason, W. B. (1977). Foundations of cost-effectiveness analysis for health and medical practices. *New England Journal of Medicine, 296,* 716–721

Wilbur, C. S. (1983). The Johnson & Johnson Program. *Preventive Medicine, 12,* 672–681.

Wright, C. C. (1982). Cost containment through health promotion programs. *Journal of Occupational Medicine, 24,* 965–968.

Yates, B. T., & DeMuth, N. M. (1981). Alternative funding and incentive mechanisms for health systems. In A. Broskowski, E. Marks, & S. H. Budman (Eds.), *Linking health and mental health,* (pp. 77–99). Beverly Hills, CA: Sage.

Conclusions and Recommendations: A Behavioral Medicine Perspective

MICHAEL F. CATALDO and THOMAS J. COATES

If the major health problems of the American people can be delayed or treated by changes in environments and behaviors, then such approaches are likely to be most appropriate in those settings that impact the most on people's health and behavior. The workplace is one such setting. We have presented descriptions of the conceptual basis, research in specific problem areas, exemplary programs, and critical factors for a behavioral medicine perspective about the relationship between health and industry. Where has the research been conducted and what can we conclude from this research? Where are the most successful applications? And where do we need to go next in terms of both interventions and research?

WHERE HAS THE RESEARCH BEEN CONDUCTED TO DATE, AND WHAT CAN WE CONCLUDE?

These questions can be considered in terms of the *problems, populations, settings,* and *interactions* studied. The majority of the research has been conducted on specific problem areas. However, in general, we can conclude that those problems suspected to be among the greatest contributors to poor health can be changed by altering either aspects of the environment or people's response to the environment. Not known at present is whether these changes can be maintained and if they will result in significant decreases in morbidity and mortality.

453

Good research poses more questions than it answers, and the current situation is no exception. What we still have yet to address successfully with regard to problem-oriented research is, first of all, the methods by which successful modifications of behaviors related to health, once changed, can be maintained. Relapse rates for problems of alcoholism, obesity, dietary modifications for salt and high cholesterol, exercise, and so forth, are notoriously high. We have not yet determined how to maintain behavior changes once they are established. However, we do believe that the solution is obtainable within the current principles of behavioral science. The challenge is to translate those principles into effective workplace programs.

Also still unresolved for most of these problem areas and risk factors is the question of whether changing them will make a difference in morbidity and mortality. Clearly, this question will not be answered until we have been able to change suspected risk factors for a sufficiently long period of time. Nonetheless, as we proceed with our behavioral medicine efforts in industry, we must remember what we can and cannot conclude from the research bases and the degree to which we have a database for implementing interventions in the workplace. Finally, one primary incentive for workplace involvement in health promotion is to decrease escalating health care costs. Whether this objective can be achieved effectively awaits further convincing demonstration.

WHAT ARE THE APPLICATIONS?

Appropriately, the *research* that is most prevalent involves programs for *specific* health-related behaviors. *Program development* has probably most often involved *multifactor* or comprehensive strategies. The efficacy of these differing approaches remains to be determined. Do specific, focused programs that are based on established need affect morbidity and mortality to the same degree as more comprehensive programs? Can the cost of multifactor approaches be reduced if we identify unnecessary components?

While the early subjects of health interventions were most often white-collar workers at the executive level, this has changed over the past few years to include the entire spectrum of workers. Research is clearly needed to identify the efficacy of various approaches with different types of workers in various types of industry. Of substantial importance is the hard-to-reach worker. Most programs can be effective with those who are ready to be reached. Those who fail to respond to programs, the hard-to-reach, may require different strategies to encourage recruitment, participation, and maintenance of change.

Programs with rigorous data-based evaluation components have only been implemented in perhaps a few dozen industry settings. The programs presented here provide an interesting first look at the types of results that can be obtained from industry-based programs. However, it should be noted that the workplace settings have been those of large companies. Eventually, to reach the majority of workers, we must begin to look at companies of 100 employees or less, wherein can be found 75 percent of today's workers. Presumably, programs are being implemented with the financial resources available to large corporations. If these programs are successful in making important changes in workers' health behaviors, and especially if these programs produce clinically relevant changes in biological indices related to morbidity and mortality, then we must begin to consider how to adapt these programs to smaller companies.

Where Do We Need To Go Next?

The chapters presented in this book document the latest research in primary risk factor intervention. This field is quite new and promising and, therefore, needs to continue. In addition to the types of research presented here, we believe that the field must advance through the following types of studies.

1. **Continued Efforts Must Be Made to Bridge Research and Practice.** The field could fail through ineffective programs and practices; industry will fail to continue to support health promotion if risk factors do not change substantially to justify the cost. Research on program components must be factored to isolate and identify effective components. Comprehensive programs must be evaluated carefully, as they are applied in practice. For example, the American Home Association has recently published a comprehensive cardiovascular risk reduction program called *Home at Work*. The program appears to be sound and based on established principles. Its efficacy in practice must be documented.

2. **Conduct Research at All Program Levels.** Industrial programs involve (a) organizational or start-up variables, (b) program strategies, and (c) follow-up and maintenance variables. Effective strategies at all three levels must be conceptualized and investigated.

3. **Components Are Not Limited to Classes or Manuals.** A key aspect of workplace programs is that they offer the advantage of using social and physical environmental variables to encourage and maintain change. Some examples are presented in this book—for example, in the chapters on obesity, smoking, and accident prevention. However, discussions of workplace programs often revolve around health education materials or curriculum content. Research is needed to identify key environmental variables related to health behavior as well as methods for modifying these variables to produce changes in behavior.

4. **Implementation Research and Quality Control Are Essential.** As

research in this area progresses, variable outcomes will be documented that are not totally attributable to specific program components. Personnel responsible for second modes of delivery can influence outcome. Research is needed in both areas—not only so that variance in outcomes can be accounted for but also so that key variables related to quality of programs can be identified. The practice must keep up with the science to ensure credibility.

Author Index

Abrams, D.B., 32, 36, 38, 39, 40, 41, 42, 43, 48, 152, 153, 154, 232, 250
Achber, C., 75
Adair, J.G., 91
Adams, S.K., 309
Adamson, G.J., 363
Agras, W.S., 204, 205
Ahlbom, A., 70, 71, 166
Albino, J.E., 261
Alderman, M.H., 75, 151, 196, 197, 199, 200, 222, 223, 338, 356, 376, 410
Alexander, J.K., 29
Alexious, N.G., 338, 356, 381, 383
Alfredsson, L., 65, 70
Allard, G.B., 104
Allegrante, J.P., 403
Allport, S., 362
Anderson, D.C., 96
Anderson, O.W., 256
Anger, W.K., 303
Anthony, D.F., 380, 381
Anderson, D.W., 309
Andreoli, K.G., 75
Armstrong, D.B., 210
Arndt, R., 428
Arnold, C., 356
Arnold, W.B., 356
Arnoldi, L.B., 76
Aronsson, G., 73
Ashley, F.W., 384
Asma, F.E., 366
Avnet, H.H., 256, 259, 261
Axelrod, C.M., 251

Ayer, W.A., 256, 262
Azar, N., 304

Baer, D.M., 107
Baer, L., 197
Bagley, R.W., 382
Bailey, J.S., 96
Bailit, H.L., 257
Baker, D., 70, 99, 166
Balma, D., 258
Balster, R.L., 24
Bandura, A., 28, 34, 42, 44, 46
Banta, H.D., 434, 435
Baranowski, T., 75
Barefoot, J.C., 71
Barlow, D.H., 89, 102, 124, 131
Barnard, R.J., 380, 381
Barrett, J.E., 24, 59
Barwick, K.D., 101, 128, 303
Baum, A., 56
Baun, W.B., 381
Beaudin, P., 41
Beazoglou, T., 257
Beck, R.N., 389
Beech, A.R., 52
Beehr, T.A., 52, 53, 75
Bellack, A., 304, 385
Benfield, A., 108
Bennett, W., 149
Benson, H., 21, 366
Berkanovic, E., 70
Berkman, L.F., 31, 39
Berkson, D.M., 193

457

Subject Index

ABAB reversal design, 90–91
Absenteeism:
  accident prevention and, 308
  dental health and, 257
  hypertension programs and, 196, 223, 375–376
  Lockheed hypertension study and, 205–206
  physical exercise programs and, 223–224, 381
  psychological problems and, 415
  reversal design, 95
  STAYWELL program and, 325–326
  tobacco smoking and, 380
Academy of Behavioral Medicine Research, 2
Accident prevention, 301–320
  alcohol abuse and, 402
  behavioral intervention and, 302–304
  behavior analysis potential in, 304–309
  evaluation of programs, 389–390
  health promotion programs, 366
  psychological problems and, 415
  statistics on, 301
  step-by-step example in, 316–319
  stress reduction and, 186
  suggested method for, 309–316
  tobacco smoking and, 380
  traditional approach to, 307–309
Active mode, 10, 11–12
Alcohol abuse:
  costs of, 352
  health promotion programs, 366
  LIVE FOR LIFE program, 343, 347

public health perspective on, 402
  see also Substance abuse
American Psychological Association, 1
Appetitive conditions, 20–21
ARIMA, 122–123
Arsenic compounds, 286
Asbestos, 286, 297–300
Assessment, organizational, 37, 38. *See also* Evaluation
Association for the Advancement of Behavior Therapy, 2
Attrition rates, *see* Relapse
Autoregressive integrated moving averages analysis (ARIMA), 122–123
Aversive control, 16
  basic aspects of, 17–18
  biobehavioral applications of, 20–21
  defined, 17
Avoidance condition, 17–18

Baby boom, 400
Baselines, *see* Multiple-baseline design; Reversal design
Before-and-after evaluation, 87–88. *See also* Evaluation
Behavior:
  accident prevention and, 304, 305–306
  biobehavioral principles and, 12
  individual/environmental interface, 10
  multiple-baseline designs and, 108–112
  occupational cancer risk and, 286, 287, 288–300
  reversal design and, 95–97

467